PLANTS
FROM TEST TUBES

FROM TEST TUBES
AN INTRODUCTION TO
MICROPROPAGATION

Fourth Edition

Lydiane Kyte · John Kleyn
Holly Scoggins · Mark Bridgen

TIMBER PRESS
Portland, Oregon

Frontispiece plants, left to right, coastal redwood, pine, bamboo, apple, rhododendron, iris, potato, primrose; page 11: daylily; page 169: moth orchid; page 223: banana.

Copyright © 2013 by Lydiane Kyte, John Kleyn, Holly Scoggins, and Mark Bridgen. All rights reserved.

Photo and illustration credits appear on page 259.

Hachette Book Group supports the right to free expression and the value of copyright. The purpose of copyright is to encourage writers and artists to produce the creative works that enrich our culture. The scanning, uploading, and distribution of this book without permission is a theft of the author's intellectual property. If you would like permission to use material from the book (other than for review purposes), please contact permissions@hbgusa.com. Thank you for your support of the author's rights.

Timber Press
Workman Publishing
Hachette Book Group, Inc.
1290 Avenue of the Americas
New York, New York 10104
timberpress.com

Timber Press is an imprint of Workman Publishing, a division of Hachette Book Group, Inc. The Timber Press name and logo are registered trademarks of Hachette Book Group, Inc.

Printed in China on responsibly sourced paper
Sixth printing 2024

Book design by Susan Applegate

The publisher is not responsible for websites (or their content) that are not owned by the publisher.

Library of Congress Cataloging-in-Publication Data

    Plants from test tubes: an introduction to micropropagation/Lydiane Kyte . . . . [et al.].—4th ed.
        p. cm.
    Includes bibliographical references and index.
    ISBN 978-1-60469-206-8
    1. Plant micropropagation—Laboratory manuals. 2. Plant tissue culture—Laboratory manuals. I. Kyte, Lydiane.
    SB123.6.K98 2013
    631.5'3—dc23   2012050583

A catalog record for this book is also available from the British Library.

Contents

Preface to the Fourth Edition, 8
Preface to the Third Edition, 9
Preface to the First and Second (Revised) Editions, 10

SECTION ONE The Basics of Tissue Culture

1. Overview, 12
2. The Botanical Basis for Tissue Culture, 21
3. History, 28
4. The Laboratory: Design, Equipment, and Supplies, 38
5. Media Ingredients, 60
6. Media Preparation, 79
7. Explants: Selecting, Managing, and Preparing Stock, 89
8. Sterile Technique and Aseptic Culture, 96
9. The Growing Room: Plant Multiplication and Dealing with Problems, 107
10. Rooting and Preparing for Transplant, 115
11. Dealing with Biological Contaminants, 122
12. Business, 138
13. Plant Biotechnology, 149

SECTION TWO Culture Guide to Selected Plants

Actinidia deliciosa, kiwifruit, 172
Alstroemeria hybrids, Inca lily, 173
Amelanchier alnifolia, serviceberry, 174
Anthurium andraeanum, flamingo flower, 174
Asparagus officinalis, asparagus, 175

Begonia rex-cultorum, rex begonia, 176
Brassica oleracea var. *italica* and var. *botrytis,* broccoli and cauliflower, 177

Catharanthus roseus, Madagascar periwinkle, 178
Cattleya spp. and hybrids, cattleya, 179
Chrysanthemum ×*grandiflorum,* garden mum, 181
Citrus ×*limon,* lemon, 181
Corymbia ficifolia, red-flowering gum, 182

Cucumis sativus, garden cucumber, 183
Cymbidium spp. and hybrids, cymbidium, 183

Dianthus caryophyllus, carnation, 184
Dieffenbachia seguine, dumb cane, 185

Epiphyllum chrysocardium, heart of gold, 186

Ficus carica, fig, 187
Fragaria ×*ananassa*, garden strawberry, 188
Freesia ×*hybrida*, freesia, 188

Gerbera jamesonii, Transvaal daisy, 189

Hemerocallis spp. and hybrids, daylily, 190
Heuchera spp. and hybrids, alumroot, 191
Hippeastrum spp., amaryllis, 191
Hosta spp. and hybrids, hosta, 192
Hypoxis hemerocallidea, African potato, 193

Ipomoea batatas, sweet potato vine, 194
Iris germanica (rhizomatous), bearded iris, 195
Iris ×*hollandica* (bulbous), Dutch iris, 196

Juglans regia, English walnut, 196

Kalanchoe blossfeldiana, kalanchoe, 197
Kalmia latifolia, mountain laurel, 198

Lilium spp., lily, 199

Malus domestica, apple, 200
Mammillaria elongata, golden star cactus, 201
Miltonia, *Odontoglossum*, and *Oncidium*, miltonia, odontoglossum, and oncidium, 201
Miscanthus sinensis, Chinese silver grass, 202
Musa acuminata × *M. balbisiana*, banana, 203

Nandina domestica, heavenly bamboo, 204
Nephrolepis exaltata, Boston fern, 204

Panax ginseng and *P. quinquefolius*, Asian ginseng and American ginseng, 205

Pelargonium graveolens, scented geranium, 206
Phalaenopsis spp. and hybrids, moth orchid, 207
Phlox subulata, moss pink, 208
Phyllostachys aurea, golden bamboo, 208
Pinus spp., pine, 209
Platycerium stemaria, staghorn fern, 210
Primula vulgaris, primrose, 211
Prunus spp., plum, peach, apricot, and cherry, 211

Rhododendron spp. and hybrids, rhododendron and azalea, 212
Rosa spp. and hybrids, rose, 214
Rubus fruticosus, blackberry, 214
Rubus occidentalis, American raspberry, 215

Saintpaulia ionantha, African violet, 216
Sequoia sempervirens, coastal redwood, 217
Solanum tuberosum, potato, 217
Syngonium podophyllum, arrowhead vine, 218
Syringa spp. and hybrids, lilac, 219

Vaccinium corymbosum, highbush blueberry, 220
Vitis vinifera, wine grape, 221

SECTION THREE Resources

Formula Comparison Table, 224
Conversion Formulas and Tables, 226
The Microscope: Introduction, Use, and Care, 228
Professional Organizations, 232
Suppliers, 232
Glossary, 236
Bibliography, 242
Photo and Illustration Credits, 259
Index, 260
About the Authors, 269

Preface to the Fourth Edition

Long before tissue culture was widely practiced, Timber Press recognized the need for a tissue culture primer that could serve the general public and commercial horticulture. To fill that need, *Plants from Test Tubes* was published in 1983. Author Lydiane Kyte drew from her practical experience in the industry to create an accessible yet thorough review of micropropagation techniques, along with hands-on projects. For the third edition, she collaborated with microbiologist John Kleyn to enhance sections related to contamination, as she noted, "the bane of all tissue culture."

Plants from Test Tubes now celebrates its 30th year in print and has gone through numerous printings with a worldwide distribution. Mark Bridgen and I have sought to update the science and resources while retaining much of Lydiane and John's original explanations and trademark conversational, non-technical style.

My contribution to the fourth edition comes on behalf of nursery and greenhouse propagators and growers. New plant introductions drive the market, and the importance of micropropagation cannot be overstated in bringing new and improved plant selections to the ornamental plant market. Mark brings years of laboratory and teaching experience. We have also sought to thoroughly update the resources presented, accounting for corporate mergers and other changes, and have included web URLs.

I thank those in academia and industry who graciously contributed to this revision: research technician Suzanne Piovanno, Gary Seckinger of *Phyto*Technology Laboratories, and George Chan of LumiGrow. Thanks also to my patient partner, Joel Shuman. Mark Bridgen thanks his friend and colleague, plant pathologist Margery Daughtrey. To those who contributed illustrations for this new edition we express our appreciation, especially to AG3, Briggs Nursery, Terra Nova Nurseries, and Thermo Fischer Scientific. Award-winning scientific artist Marjorie Leggitt provided the many new drawings. Finally, our thanks to the many devotees of previous editions, and those who continue to utilize and recommend *Plants from Test Tubes* as a gateway to the fascinating world of micropropagation.

HOLLY SCOGGINS

Preface to the Third Edition

The first edition of *Plants from Test Tubes* came about as a result of the realization by Timber Press that there was a need for a tissue culture primer that could serve the general public and commercial horticulture, areas where tissue culture was not yet widely practiced. The response was overwhelming. The book found its way beyond the United States and Canada, appearing in Egypt, in Brazil, in Australia, in India, and in Europe, where it was even translated into Danish.

The basics of tissue culture have changed very little in the past 20 years. A reliance on sterile technique and on variations of Murashige and Skoog's medium formula continue to be focal points of successful micropropagation. But even in the academic world, where sterile technique is commonplace and articles in technical journals are readily available, the book has proven useful as an introductory text.

This third edition was launched due to an outcry of demand when the second edition went out of print. (The demand echoed in a tissue culture network.) It seemed appropriate to widen the scope of the book and to touch on the explosion in the related field of biotechnology, a field that depends on the use of tissue culture in such areas as the multiplication of transformed plants. Cell culture is also introduced in this edition, as further research suggests the numerous applications of that procedure in industry.

Without the encouragement and contribution of John Kleyn, my collaborator, this edition would not even have been attempted. As a microbiologist, professor, commercial tissue culturist, and consultant, he brings a new dimension to the book. Specifically, he suggests to the reader methods of preventing and identifying microbial contaminants, the bane of all tissue culture.

For their encouragement and helpful suggestions, I would like to thank the many devotees of previous editions and my associates in the International Plant Propagators' Society. John Kleyn joins me in gratitude for the patience of our respective spouses, Bob Kyte and Jan Kleyn.

LYDIANE KYTE

Preface to the First and Second (Revised) Editions

This book was born out of a personal need for such a book when I started into tissue culture. It is, at best, a starting point and, hopefully, a partial answer for those who ask, "What is tissue culture? As a grower, is micropropagation something I can and want to do? And, how do I do it?"

I would like to acknowledge with gratitude Hudson Hartmann (University of California at Davis) for so generously reading this work, providing useful suggestions, and writing the Foreword; Dale Kester (University of California at Davis) and Gayle Suttle (Microplant Nurseries), who so kindly read the manuscript and assisted with helpful comments; Bruce Briggs (Briggs Nursery), Sarah Upham (Native Plants), and countless others for their encouragement and willingness to share hard-earned information; Richard Abel, editor, for his patience reinforcement, and untiring effort to help make this book a credible reality; and Bob, my husband, for his support assistance, and encouragement.

LYDIANE KYTE

ONE
THE BASICS OF TISSUE CULTURE

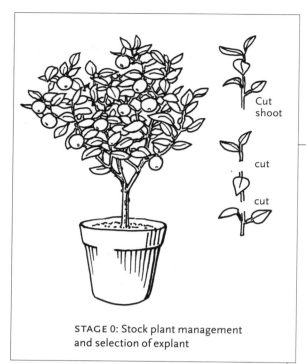
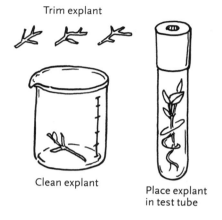

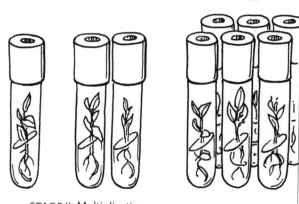
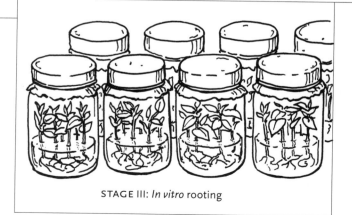

Figure 1-1. The 5 stages of micropropagation.

1 Overview

Plant tissue culture is a field of many facets. Its applications vary from the curious gardener multiplying plants in a modest home kitchen to the renowned scientist working in an elaborate laboratory. It reaches from the orchid hobbyist who has learned to multiply a few personal favorites to a million-dollar industry producing houseplants, ornamentals, and secondary products. Tissue culture is the ever-ready tool for specialists who hybridize plants by either sexual or asexual means. It is a clean and rapid way for genetic engineers to grow material for identifying and manipulating genes or to transfer individual characteristics from one plant to another. It plays a role in a wide array of fields, such as botany, chemistry, physics, genetic engineering, fecular biology, hybrid development, pesticide testing, and food science.

The term *tissue culture* is borrowed from the field of animal studies. It is actually a misnomer because plant micropropagation is concerned with the whole plantlet and not just isolated tissues, although the explant may be a particular tissue. In fact, the term *plant tissue culture* has a much broader definition. First, it is both an art and a science. It is an art because to be successful there are procedures to be learned and practiced; the more that you do it, the better you will be at using these techniques. Plant tissue culture is also a science because it requires training and education to understand what can be accomplished and how to attain the results that you desire. Second, by definition tissue culture includes the aseptic culture of plant cells, tissues, organs, protoplasts, and whole plants. *Aseptic* means that the plants are free from living contaminants such as microorganisms, insects, mites, thrips, and so forth; fungi and bacteria are major pathogens that need to be conquered in tissue culture. The third and final part of tissue culture's definition is that the plants need to be grown under controlled conditions where temperature, humidity, photoperiod, light intensity, and the nutrient medium all supply the ideal growing environment.

Plant tissue culture has many uses: rapid clonal propagation or micropropagation, disease elimination, anther culture for the production of haploid plants, protoplast isolation for hybrid production, embryo culture, production of botanical substances, cell selection and mutation, studies in fundamental plant anatomy, development, and nutrition, plant biotechnology, somatic embryogenesis, synthetic seed production, callus culture, and flower culture. For the purpose of this book, however, tissue culture can be thought of as plantlet culture, or micropropagation. Whether we call it cloning, tissue culture, micropropagation, or growing *in vitro*, the process remains the same: it is a vegetative method for multiplying plants. This method of propagation is commonly accepted by the nursery trade and it has had a significant impact on commercial horticulture.

The number of nurseries and specialists engaged in plant propagation by tissue culture has rapidly escalated from the 1970s onward. The basic scientific principles of plant tissue culture were established—one might say that commercial plant tissue culture had come of age. This exciting technique was a result of active research and was employed by progressive growers, students, hobbyists, and gardeners, while still serving as a valuable tool for the plant scientist. The 2009 Census of Horticultural Specialties administered by the USDA reported 117 tissue culture operations in the United States with more than 94.5 million plantlets sold with a value of $65 million. Hundreds of commercial laboratories are currently involved in micropropagation with some laboratories producing tens of millions of plants per year (Bridgen 1992). Previously there had been doubts about the feasibility of using tissue culture to propagate woody plants (trees and shrubs), but it is estimated that by the mid-1980s "woodies" accounted for about 16 percent of total worldwide tissue culture production (George and Sherrington 1984).

Tissue culture was first used on a large scale by the orchid industry in the 1950s. Some fortuitous early discoveries opened the door to tissue culture for quality orchids, where previously growers had struggled with unpredictable seed or difficult-to-propagate, virus-infected stock. Later, it became clear that nearly any plant would respond to tissue culture as long as the right formula and the right processes were developed for its culture.

With the widespread acceptance of the technique in the nursery business, it is surprising how many people are still not aware of plant tissue culture. Some people ask when they hear or read about tissue culture, What is it? How is it done? Who is doing it and why? You may ask, Can I do this myself? or, Should I even try? What are the costs of building a laboratory? What are the potential financial returns? Some want to know who invented it or where it came from. This book helps answer some of these questions.

Although plant science has become increasingly complex with the explosion of biotechnology, the procedures of tissue culture are not complicated (Figure 1-1). A piece of a plant, which can be anything from a portion of stem, root, leaf, or bud to a single cell, is placed in that tiniest of greenhouses, a test tube. In an environment free from microorganisms and in the presence of a balanced diet of chemicals, that bit of plant, called an explant, can produce plantlets that, in turn, will multiply indefinitely, if given proper care. The medium (plural, media) is the substrate for plant growth, and in the context of plant tissue culture, it refers to the mixture of certain chemical compounds to form a nutrient-rich gel or liquid for growing cultures, whether cells, organs, or plantlets.

The process of shoot culture, from selecting the explant to transplanting to a greenhouse involves 5 basic stages: Stage 0, stock plant selection and preparation; Stage I, establishment of aseptic cultures (disinfection); Stage II, multiplication; Stage III, *in vitro* rooting; and Stage IV, transplanting and acclimatization, or hardening off. These stages can overlap in certain cases, and the requirements of each stage vary widely from plant to plant.

When new material is started in culture, grown *in vitro* (literally, "in glass"), it develops very small juvenile shoots, which are reminiscent of seedlings. A plantlet continues to produce and maintain small stems and leaves throughout its duration in culture. This is fortunate because most mature material would be too unwieldy for micropropagation to succeed in a test tube (Figure 1-2). After multiplication in culture and when transferred to soil outside the laboratory, the plantlets will produce leaves of normal size and assume the mature features of the plants from which they originated.

When plants are multiplied vegetatively (not from seed), whether by tissue culture or by cuttings, all the offspring from a single plant can be classified as a clone. This means that the genetic make-up of each offspring is identical to that of all the other offspring and to that of the single parent. In contrast, plants propagated by seed, resulting from sexual reproduction, are not clones because each seedling and the resulting plant have a unique genetic make-up—a mixture from

2 parents, different from either parent and different from one seed to another. The term *cloning*, with respect to tissue culture, refers to the process of propagating in culture large numbers of selected plants with the same genotype (the same genes or hereditary factors) as the parent plant.

Some problems in plant propagation can originate with the stock plants from which seeds or cuttings are taken. Oftentimes, however, diseases that are normally transmitted from parent to offspring can be eliminated through tissue culture procedures. Cleaning explants removes external contaminants, such as bacteria, fungi, spores, and insects, and using the apical meristem as the explant for the tissue culture eliminates some internal contaminants, such as viruses.

The apical meristem is the new, undifferentiated tissue at the microscopic tip of a shoot. These meristematic cells, which are not yet joined to the plant's vascular system, often are virus free, even in diseased plants, and perhaps they grow faster than the viruses. Thus, removing the few virus-free cells that make up the microscopic dome of the plant's apical meristem and placing them in a culture allows them to grow and produce healthy, disease-free plants. This technique is known as *meristem culture*, a term sometimes wrongly used to denote micropropagation or tissue culture, but which should be limited to indicate cultures started from an apical meristem.

A valid concern arises concerning the genetic stability of tissue cultured plants. Thousands of plants could be cultured at one time, only to reveal some defect when they are planted and growing in the field—a defect resulting from a chemical imbalance or a mutation that was multiplied in culture. As a practical matter, however, cultures have been known to remain healthy and normal for decades, showing no aberrations. Some species are more prone to mutation than others are. Cell and callus cultures are more likely to be genetically unstable than those grown as plantlets. With limited experience in a particular species, a tissue culturist should start with numerous explants, transfer them frequently, limit the number of subcultures, consider the levels of growth regulators that are used in the media, and start with new material annually or as often as experience dictates.

Tissue culture can serve a number of purposes, and growers have started their own commercial production laboratories for a variety of reasons. Growing certain plants from seeds or cuttings can be unacceptable or impractical due to some of the following (and other) factors:

- Seed-grown products lack uniformity
- Seed-grown products are not true-to-type
- Seeds take too long to grow to mature plants
- Seeds are difficult to handle
- Seeds are not available
- Cuttings grow too slowly
- Cuttings have poor survival rate
- Cuttings require too much care

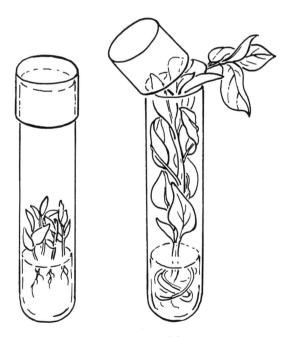

Figure 1-2. If tissue cultured plants did not revert to a juvenile state, most material would never fit into a test tube or jar and plants would be too unwieldy to micropropagate.

- Cuttings are too vulnerable to disease
- There is a shortage of stock plants from which to take cuttings because there is only one hybrid, only one virus-free plant, or only one desirable mutant
- Insufficient space for stock plants
- Not cost effective to maintain the stock plants

Tissue culture is often the only practical way to produce the large numbers of plants required. If an ample supply of seeds is available or if plants from cuttings are acceptable and cutting stock is available and not a problem to maintain, then tissue culture may not be the most practical option because it can be an expensive and labor-intensive process. If plants from seeds are acceptable but there are not enough seeds, then seeds or excised embryos can be used as starting material for tissue culture.

Every year excessive amounts of a grower's time, labor, and space are spent on unproductive seeds, cuttings, and grafts. Significant numbers of young plants are lost to pests, diseases, or other environmental factors. Tissue cultured plantlets are less subject to such attacks and disasters because in the sterile environment of the laboratory they are not exposed to the pathogens or extreme conditions that afflict many plants grown in the field or greenhouse. Material usually comes out of culture as well-started plantlets or microcuttings with a stockpile of nutrients and vigor often superior to that of conventional cuttings. It is no secret that healthy plants are the first line of defense against disease.

Most conventional nursery-grown seeds and cuttings must be grown and propagated during a particular season, and consequently, work schedules in the nursery must revolve around this factor. Controlled-environment greenhouses can alleviate some of these restrictions. Although explant material may be limited to the season in which it can be taken, established tissue cultures can be grown at any time of year, regardless of weather. Propagation conditions in the tissue culture laboratory are close to ideal, and therefore conducive to year-round production, a situation that promotes maximum labor efficiency.

When using conventional propagation methods, one cutting produces one plant, and one seed produces one seedling. In contrast, one explant—one piece of stem, leaf, bud, root, or seed, one meristem, or even one cell—theoretically can produce an infinite number of plants (Figure 1-3). Consequently, fewer stock plants are required to provide the explants needed to produce thousands of new plants. With the ever-increasing value of land and plants, more than ever the grower needs to put a monetary value on every bit of ground and every stock plant. Because tissue culture requires a minimal amount of plant material to start with, substantial savings can be realized not only by allocating less money to the acquisition and maintenance of stock plants, but also in the space and time required to maintain the stock plants.

In the greenhouse, cuttings can take weeks or months to grow and root. The relatively rapid multiplication typical of tissue culture often makes it possible to produce and sell true-to-form plants quicker and in greater numbers than you could by other means. The actual production time for tissue cultured plants varies depending on the particular plant. Whereas houseplants can start multiplying in culture within a matter of days or weeks, a woody plant can take several months before it begins to multiply. To give one conservative example, a single *Rhododendron* explant might take 4–6 months to start multiplying.

Once multiplication gets underway, assuming exponential multiplication (a doubling of material every month), from a single explant you would have 1024 plants after 10 months, 2048 plants after 11 months, and so on. Of course, you would hope to have more than one successful explant, and you should expect the explants to do more than double each time they are transferred to new cultures. In any case, while cultures are multiplying in such great numbers in the laboratory, greenhouse space is available for growing other crops.

Tissue culture avoids an enormous amount of the daily care that is required with cuttings and seedlings. Cultures usually need to be divided and transferred to a fresh medium every 4–6 weeks, but between transfers, there is no need to water or tend to the cultures other than casual surveillance. How different this is from the daily watering and fertilizing requirements accompanying greenhouse growing! Many people who work in tissue culture laboratories enjoy this benefit because it allows for weekends off and vacations away without worrying about day-to-day maintenance.

For anyone who likes plants and wants a hobby, tissue culture can be a satisfying avocation. People can find great pleasure in watching and caring for plants in culture. It is an inviting field for the elderly or the handicapped because there are very few physical demands and the laboratory climate is generally comfortable. Hobbyist breeders can easily use this technique to assist with their hybrid production. Retirees, especially gardeners who want to take life a little easier, can find many hours of interest and satisfaction in discovering how their favorite plant responds to tissue culture, all with a relatively small cash outlay and minimal training.

Most large laboratories involved in tissue culture have their own research and development (R & D) departments to support the production operation. There are refinements of nutrients to be worked out and procedures for plants to be outlined. The research department may be responsible for detecting and identifying contaminants in the production cultures, or it may be responsible for "fingerprinting," the identification of plants by separating genes or proteins in an electric field (electrophoresis).

The spectacular and often proprietary achievements that have been made behind the closed doors of elite laboratories do not prevent the curious minded from indulging in greater exploration, however. Hobbyists, amateurs not bound by the constraints of production, are free to pursue any avenue their aptitude, time, and budget will allow. Each will approach the field with unique background and insight, whether it is from academic or practical experience. Some will have studied chemistry, botany, or microbiology, others will have studied the ways of plants in the field, greenhouse, or garden, but all will have 2 characteristics in common: wonder and curiosity.

Much remains to be learned and studied in the field of plant tissue culture. Culture media are always subject to change and improvement. Techniques need to be improved upon and carefully described. Information needs to be made readily available to potential tissue culturists (except, of course, for priority information and patents, which is another subject). Countless plants have never been tissue cultured, many of which are in danger of extinction. What better legacy than to save some of these plants by means of tissue

Figure 1-3. Tissue cultured fern clones at Casa Flora, Dallas, Texas. One explant has the potential to produce an infinite number of plants.

culture? These and many other areas are increasingly open to the amateur.

The simplest tissue culture hobby is the multiplication of easy, fast-growing plant material, such as *Kalanchoe*, Boston fern (*Nephrolepis*), African violet (*Saintpaulia*), or *Begonia*; next in order of complexity are carnation (*Dianthus*), strawberry (*Fragaria*), or *Syngonium*. The chemist looking for a tissue culture hobby may be challenged to explore the field of plant by-products, such as dyes, flavorings, medicinals, and oils. The more studious hybridizer might be attracted to hybridizing plants by using embryo culture, anther culture, mutation breeding, or even the difficult process of fusing protoplasts (cells without cell walls). These procedures require more knowledge and equipment than simple micropropagation, but again, today they need not be confined to commercial laboratories or academia.

Gardening is one of the most popular hobbies. Conventional gardening is limited by the seasons of the year, but tissue culture knows no season. Gardeners who propagate by tissue culture will delight in year-round micropropagation. If successful, they may find they have even more plants than anticipated. Excess plants can be shared with friends or offered for sale, and many ventures that start out as hobbies may turn into businesses.

The hobbyist or amateur gardener need not feel restricted in tissue culture pursuits for want of a transfer hood and a laboratory or any other fancy equipment, such as is required by commercial operations. Small-scale tissue culture is often carried out without benefit of a laboratory or special equipment. It can be done by almost anyone in almost any house. Material can be transferred on a desk or table in a clean, dust- and draft-free room of a home. Transferring cultures in a homemade chamber, with a glass or clear plastic front and just enough room for gloved hands to enter, is a reasonable method for the hobbyist. Furthermore, commercially available premixed culture media, plus a pressure cooker or a microwave, forceps or tweezers, a paring knife, a few test tubes or jars, household bleach solution, and a lighted shelf, along with a lot of determination, are enough to bring about exciting discoveries for the amateur tissue culturist.

In contrast to the hobbyist, the commercial grower is compelled to make tissue culture a profitable enterprise. Growers with limited resources who must make a living from a small operation are finding that a large number of plants can be propagated by tissue culture with a minimal amount of space and outlay of capital. Oftentimes growers employ tissue culture techniques for 1 or 2 cultivars consistent with their operation, and then build a reputation for these cultured specialties. A few such plants that have made reputations for growers include tropical plants such as cannas, bananas, and elephant ears, carnations, ferns, iris, fruit-tree rootstock, orchids, lilacs, and rhododendrons. In most cases, these growers used tissue culture propagation for plants to grow on to field-ready size; in other words, to develop rooted plantlets that will grow-on in the field to full-sized plants. Purchasing plants that

Figure 1-4. A clear plastic berry cup makes an inexpensive greenhouse for the hobbyist and amateur tissue culturist.

are already started allows for a quicker return on capital for the commercial grower. On the other hand, in-house growing-on of plantlets from one's own laboratory is important to a nursery's normal market and precludes the competition that may be generated by selling the plantlets to other nurseries.

Growers, plant brokers, salespeople, greenhouse suppliers, and others interested in marketing plants have a variety of new products to offer:

1. Plantlets still in culture.
2. Plants directly out of culture, rooted or unrooted (microcuttings).
3. Plants hardened-off, or acclimatized, to greenhouse conditions.

Plants that have been tissue cultured correctly are true-to-type and vigorous; they can also be disease-free and less expensive than conventionally propagated material.

Sometimes a grower welcomes the opportunity to buy rooted tissue cultured plantlets from other sources. In lieu of building their own laboratories, some small-scale growers with growing-on capability (that is, the ability to take rooted plantlets and grow them to fuller size) will buy rooted plantlets from established commercial tissue culture laboratories, especially in the case of hard-to-start or -root or newly introduced plants. Many growers with the proper controlled-environment facilities can buy plantlets just out of culture (Stage III) from a commercial laboratory and then harden them off to finish and sell the plants (Figure 1-5). The wary tissue culturist, however, will guard against opportunists who might buy young tissue cultured plantlets, only to return the material to culture themselves, thus avoiding start-up costs while becoming an active competitor.

Before starting a laboratory, it is very important to make sure there is a market for the proposed tissue cultured products, and then proceed gradually. It can be all too easy to buy quantities of equipment and supplies, only to discover that there is insufficient market demand for the product, that the plant will not respond to culture, or that contamination is too difficult to remedy, or any other unforeseen problems. Many pitfalls can be avoided with a well-thought-out plan.

Pathogen-free plants from culture open the door to a freer exchange of plants between states and between countries. Plant tissue cultures have gained acceptance in world trade because the danger of introducing disease is very limited. Border restrictions on entrance of plants in soil and soil-less media do not apply to plantlets in culture. The foreign exchange of plants will likely increase as new hybrids are developed asexually from feats of genetic engineering and hybridization. The free exchange of tissue cultures will have a significant impact on global food problems by allowing more improved cultivars to be brought more rapidly to growers worldwide.

Although the application of tissue culture to farm crops is not as common, progress has been made, especially in the area of small fruits and

Figure 1-5. A healthy plantlet of *Canna* 'Australia' at the successful conclusion of Stage III culture.

tree fruits. For example, in apple, to maintain clonal characteristics, desirable scions (pieces of stem) are grafted onto special rootstocks, which are usually propagated by layering or cuttings. The rootstocks are cloned to maintain those that are known to influence certain aspects of scion growth, such as hardiness, disease resistance, or dwarfing. Commercial tissue culture laboratories are currently culturing rootstocks with great success. If tissue cultured rootstocks prove economical, they will likely replace layered rootstocks. In fact, fruit trees are already being tissue cultured and grown on their own roots. If these efforts continue to be successful, grafting may come to be a cumbersome process of the past.

Whether the motivation for growing plants by tissue culture is profit, research, or personal satisfaction, the potential is there to produce a significantly greater number of healthier plants in less space, with less labor, and at less cost than by other means of vegetative propagation. The potential of plants is far greater than we know. It is as if the grower were a potter, little knowing what could be molded from the clay.

2 The Botanical Basis for Tissue Culture

The remarkable diversity of naturally occurring vegetative reproduction reflects the amazing capability and potential of plants for multiplication. The same factors that are involved in multiplication and growth initiation in nature are involved in the greenhouse and in tissue culture. The natural capability of plants to multiply by asexual means is the basis for multiplication *in vitro*. No new phenomena have been invented for these processes. There is no mixing of gene traits, as occurs in sexual reproduction. Vegetative reproduction, whether occurring naturally or through human intervention, is initiated in stems, buds, roots, or leaves.

There are both positive and negative aspects to asexual reproduction when compared to sexual reproduction. Valuable characteristics are retained in clonal propagation, but at the cost of genetic diversity. The mixing of gene traits is the driving force behind biodiversity. The end use of the plant may help determine if vegetative propagation is suitable. Environmental restoration or reintroduction programs benefit from genetic diversity, thus seed-grown plants are desirable. Plants bred or selected for specific foliage, flower, or fruit characteristics must be clonally propagated to ensure consistency. Genetic resources (germplasm) for horticultural and agronomic crops are preserved via clonal means.

Throughout history, careful observation of plant behavior and the study of plant anatomy and physiology have revealed a seemingly limitless amount of information that has taught us to manipulate certain natural phenomena to serve our own purposes. We take cuttings, we make divisions, we layer, we make grafts, and we tissue culture—in short, we promote vegetative reproduction mimicking a natural phenomenon (Figure 2-1).

Many plant parts are suitable for some form of tissue culture (Figure 2-2). Stems, leaves, nodes, buds, roots, and reproductive parts have all been utilized in tissue culture protocols, one way or another.

Figure 2-1. Spider plant (*Chlorophytum comosum* 'Variegatum'). Any time a plant reproduces itself vegetatively, it produces a clone.

Plant stems have tremendous potential for regeneration. They grow in many different forms and habits: long or short, slender or stout, round, flat, or square, above ground or underground, trailing or upright. One of the methods of vegetative propagation used most often by growers is that of growing new plants from stem cuttings (Figure 2-3). Stem cuttings are shoots or sections of stems that root when they are inserted into a potting mix, which may be a mixture of peat, soil, or bark with an aggregate such as sand or perlite. Often an auxin (a growth regulator) is applied to the base of the shoot to hasten and enhance rooting. Roots usually grow from the nodes (the part of the stem from which buds or leaves originate) when placed below the surface of the medium, although some plants can generate roots straight from the stem, as with coleus (*Solenostemon*). Woody cuttings are sometimes encouraged to root by wounding. To wound a cutting, a thin slice of epidermis (outer layer of tissue) is peeled away with a knife on 1 or 2 sides of the lower part of the stem. This will often cause the stem to grow callus, which can help root initiation.

Small stem cuttings are frequently used as starting material for microproprogation. Microcuttings, tiny cuttings taken from *in vitro* material, are a product of tissue culture that can be planted and grown-on to mature plants. The size of microcuttings can be critical to the success of growth; 1–2 in. (2.5–5 cm) is usually best (more on microcuttings in chapter 8).

Layering is another form of vegetative reproduction. It occurs frequently in nature and is widely used by growers as well. A branch is said to layer when it comes in contact with the soil, forms roots, and grows on to become a new plant. Wild blackberries (*Rubus*) rapidly expand their territory by layering. Layering is used commercially to propagate filberts (*Corylus*), grapes (*Vitis*), black raspberries and trailing blackberries (*Rubus*), currants (*Ribes*), apple (*Malus*) rootstocks, and some ornamentals such as *Forsythia* and vine-type foliage plants such as Swedish ivy

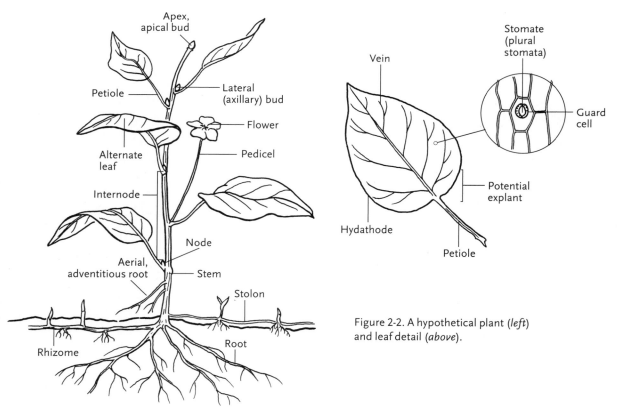

Figure 2-2. A hypothetical plant (*left*) and leaf detail (*above*).

(*Plectranthus*), *Philodendron*, pothos (*Epipremnum*), and so on, although the majority of tropical foliage plants are now products of tissue culture). Air layering is a method whereby a stem is induced to root without it being in contact with the soil. The stem is wounded, auxin is applied, a layer of moss is wrapped around the area, and the entire site is covered in plastic and tied to prevent water loss. When roots are formed, the stem section can be cut away from the main plant and planted. Many tropical ornamentals with upright habit are suitable for layering such as rubber plant (*Ficus*), croton (*Codiaeum*), *Dieffenbachia*, and *Hibiscus*. The process of layering further demonstrates the ability of stems to root.

Nature can also be manipulated by growers in the process of grafting. In grafting, growers attach a scion (a shoot or bud) of one plant onto the understock or rootstock (rooted stem) of another plant to obtain the desired traits of both. The more closely the scion and rootstock are related, the greater the chance of a successful graft. Many woody ornamentals and fruit trees are produced via grafting; the latest frontier is grafting of herbaceous plants like tomatoes and pumpkins. A great deal of work and research concerning the compatibility or incompatibility of various plants has provided better understanding of which scions will graft successfully with which rootstocks. Several different methods can be employed in grafting, but whatever method is used, it is important that the cambial layers match—the cambium is the thin layer of tissue between the wood and the bark of a stem. Micrografting involves the grafting of tissue cultured material in aseptic conditions, a craft that requires great skill.

True bulbs, such as those of lilies (*Lilium*), tulips (*Tulipa*), and daffodils (*Narcissus*), are underground storage organs that function as modified stems. A bulb consists of scales (modified leaves) attached to a basal plate (a flat modified stem at the base of the bulb and the source of roots). Bulbs multiply naturally by growing bulblets in

Figure 2-3. Grower-assisted clones from stem cuttings. Growers have been cloning by cuttings for centuries.

the axils of their scales. Growers can encourage bulblet formation in some bulbs (such as daffodils and squills [*Scilla*]) by cutting the bulbs from top to bottom, then further dividing them laterally. Each segment will produce bulblets if it contains a piece of the basal plate. Growers can promote bulblet formation in lilies by scaling, a process whereby the scales are removed from mature bulbs, dipped in a rooting hormone, and placed under growing conditions in which they will produce bulblets. Scoring and scooping are the means by which bulblet formation can be promoted in hyacinths (*Hyacinthus*) and squills. Scoring is merely making cuts across the basal plate. Scooping is the process of cutting a cone-shaped piece out from the base. These methods induce the adventitious formation of more bulblets than would occur naturally without intervention.

Corms are swollen underground stems, complete with nodes, internodes (the section of stem between 2 nodes), and lateral buds; they do not have scales like true bulbs. Plants that produce corms, such as *Gladiolus* and *Crocus*, multiply naturally by producing cormels (miniature corms).

In addition to bulbs and corms, several other modified stem structures function in vegetative reproduction. Among these are rhizomes (horizontal underground stems) and tubers (swollen or thickened fleshy parts of a rhizome). Potatoes (*Solanum tuberosum*), *Caladium*, and gloxinia (*Sinningia*) produce stem tubers. A piece of tuber or fleshy rhizome will generate a new plant if it contains an "eye," or bud. Other rhizomes are long and slender, such as those of lily-of-the-valley (*Convallaria*), Inca lily (*Alstroemeria*), false Solomon's seal (*Polygonatum*), bearded irises, and most perennial grasses. These rhizomes have long internodes with terminal and lateral buds, which allow the plants to multiply effectively. Stolons, or runners, are long, prostrate, aboveground, modified stems. Examples of stoloniferous plants are creeping dogwood (*Cornus canadensis*), perennials such as bugleweed (*Ajuga reptans*) and spotted dead-nettle (*Lamium maculatum*), strawberry (*Fragaria*), Bermuda grass (*Cynodon dactylon*), and white clover (*Trifolium repens*). The sole purpose of stolons or runners is the production of new plants.

Roots take up water and nutrients from soil and anchor plants firmly in the ground, but they also serve as useful structures for vegetative reproduction, both in nature and in commercial growing. Growers use root cuttings to propagate gooseberries (*Ribes*), raspberries (*Rubus*), horseradish (*Armoracia rusticana*), flowering quince (*Chaenomeles*), bayberries (*Myrica*), and aspen (*Populus*), to name a few. The tuberous roots of sweet potatoes (*Ipomoea batatas*), *Dahlia*, and tuberous *Begonia* are modified roots; buds form on the stem end of these root tubers and take nourishment from the food stored in the tubers as they grow into new plants. Tissue culture of roots is an important research tool for studying root development, mycorrhizae (beneficial fungi that associate with roots), and other soil organism–root interactions.

Leaves too will occasionally produce new plants. Leaf cuttings are made from rex begonias, rhizomatous begonias, *Sedum* species and hybrids, African violets (*Saintpaulia ionantha*), and *Sansevieria*, among others. In a few cases, new plants will grow from leaves without being separated from the parent plant; this happens in the piggyback plant (*Tolmiea menziesii*), American walking fern (*Asplenium rhizophyllum*), and the Bryophyllum group of the genus *Kalanchoe*. Leaves and parts of leaves are a common source of explants in tissue culture.

The foregoing examples from garden, field, and greenhouse illustrate the inherent potential, the power, and the inclination of plants for vegetative multiplication. These examples help to reveal the diverse anatomy of plants and some of their reproductive habits, and they serve as clues to areas and structures from which explants may be taken.

The multiplication of plants *in vitro* does not establish any new processes within the plants.

Tissue culture simply directs and assists the natural potential within the plant to put forth new growth and to multiply in a highly efficient and predictable way. In contrast to plants produced from seeds via sexual reproduction, new plants produced through vegetative reproduction are severed extensions of the original plants. Micropropagation is merely another way to propagate plants vegetatively.

New growth is usually initiated in meristematic tissues, which are undifferentiated cells that have not yet been programmed for their ultimate development (Figure 2-4). These cells are located at the tips of stems and roots, in leaf axils, in stems as cambium, on leaf margins, and in callus tissue. Under the influence of genetic make-up, location, light, temperature, nutrients, hormones, and probably many other factors, meristematic cells can differentiate into leaves, stems, roots, and other organs and tissues in an organized fashion. Meristematic tissue is the basis of plant growth and development.

Parenchyma cells, the most common type of plant cell, are thin-walled cells that have the capacity to regenerate and differentiate, to initiate the growth of new and varied tissues or organs for specialized functions. These and other cells that can dedifferentiate, or revert to an undifferentiated state, account for adventitious growth. Adventitious growth refers to the development of new shoots, buds, roots, flowers, or leaves from atypical locations. Adventitious buds are distinctly different from axillary buds because the latter are preformed whereas adventitious buds arise *de novo* (anew). Examples of adventitious growth are aerial roots (roots that emerge from aboveground parts of the plant), buds from roots, plantlets from leaves, and shoots and roots from callus.

Callus is a mass of undifferentiated cells (Figure 2-5). In propagation by stem cuttings, callus forms at the base of the cutting prior to root initiation in some species. Callus can also form in tissue culture. The callus mass can contain embryoids (embryo-like somatic structures

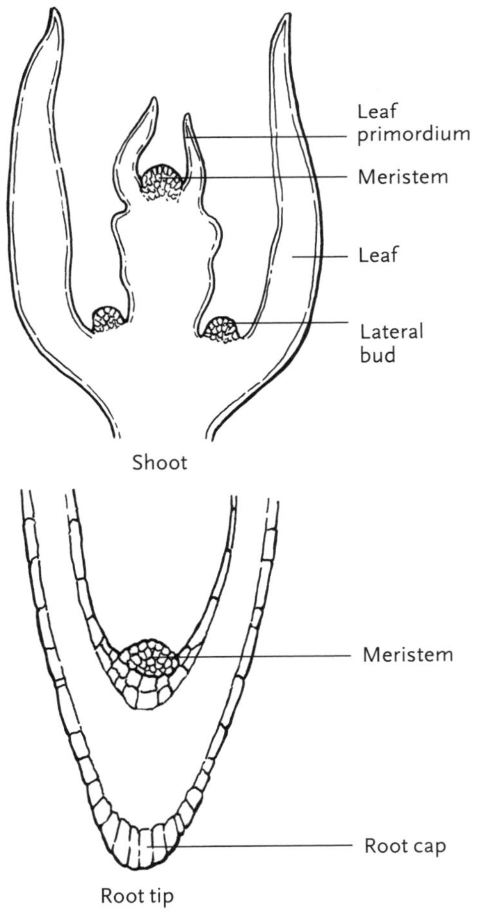

Figure 2-4. Meristematic zones in shoot and root tips.

Figure 2-5. Tobacco (*Nicotiana tabacum*) callus.

THE BOTANICAL BASIS FOR TISSUE CULTURE 25

capable of developing into whole plants), or it can contain shoot or root primordia (the earliest developmental stage of an organ or cell). Callus can also develop cells with an abnormal number of chromosomes; plant breeders often encourage this, but micropropagators who want clones, avoid it. For example, some plants differentiating from a callus culture may be tetraploid (4n, or double the normal number [2n] of chromosomes in vegetative cells), but plants cultured from shoot tips or other organized growing points, without a callus stage, usually do not show this variability. Treating callus with colchicine *in vitro* can increase ploidy; in the case of some ornamentals such as daylily (*Hemerocallis*), polyploidy is desirable because tetraploids are more vigorous, with larger, more substantial flowers (Van Tuyl and Lim 2003).

There are 2 kinds of plant cell division: mitosis and meiosis. Every somatic cell (all cells except reproductive cells in the flowers) is diploid (2n), containing 2 sets of chromosomes. During mitosis, the chromosomes duplicate and then segregate. From this, 2 new cells form, each with chromosomes identical to those of the original cell. (See chapter 13 for further discussion of these processes.)

Meiotic division, or reduction division, is the process of forming sexually reproductive cells. This process occurs in flowers during the formation of seeds. In meiosis, each chromosome of a 2n cell splits in 2. The chromosomes segregate such that one chromosome from each set of 2 goes to each of the 2 new sex cells, or gametes, each of which has, thus, only one set of chromosomes (n). This process allows for the genetic diversity of plants that was discussed in chapter 1.

Whenever cells divide there is the possibility of genetic variability. If a mutation (a change in the genomic sequence) occurs during cell division, and assuming the cell survives the mutation, the mutation is carried in all future divisions. A mutation is a sudden, abnormal change in genetic order that will alter some characteristic, or it can be a change in chromosome number. The effects of mutations are not always noticeable. Most mutations will cause a plant to die or to produce undesirable qualities, such as misshapen fruit or abnormal shoots. However, mutations can be desirable for food and fiber crop improvement or new plant introduction of ornamentals. Mutations can occur naturally as a product of cell division or can be induced by treating a plant with gamma rays or chemicals. Induced mutations are used widely in plant breeding programs. Mutation induction is used not only in plant breeding programs but also in genetic research, including functional genomics and gene discovery. When mutations occur in tissue culture, whether induced or not, the process is called *somaclonal variation*. Inducing mutations *in vitro* will be further discussed in chapter 13, Biotechnology.

One desirable mutation is the chimera, where genetically dissimilar tissue grows adjacent to one another. Some variegated plants, such as rex begonia, have "heritable" patterns that are not produced by mutations, but many popular variegated ornamental plants are the result of chimeras. Variegated *Hosta*, geraniums (*Pelargonium*), *Sansevieria*, and *Croton* are just a few examples of ornamental chimeras (Figure 2-6). Any tissue cultures that are made from sections of variegated plants will exhibit only those characteristics of the section of the plant from which the explant was taken. For example, if cultures arise from the white, variegated part of a plant, they will grow albino. When they are removed from the tissue cultures, they lose their sugar source, cannot photosynthesize, and eventually die. Some thornless blackberries (*Rubus*) are chimeras, in which the aberrant characteristic (that is, the lack of thorns) is limited to the epidermal tissue and can only be reproduced from stem tissues; new stems growing from roots or divisions will have thorns. When utilizing tissue culture to propagate chimeral shoots, care must be taken to use organized growing points like shoot tips and not to encourage callus formation, which leads to adventitious shoots arising from a limited number of cells. Slow rates of propagation and limited use of growth regulators can help maintain clonal integrity (Skirvin and Norton 2008).

Navel oranges (*Citrus*), which are seedless, and seedless grapes (*Vitis*) originated as sports (spontaneous mutations) and can only be reproduced vegetatively if they are to maintain their seedless characteristic. Some abnormal divisions give rise to tetraploids ($4n$), as in the case of some giant apples (*Malus*)—a mutation that proved desirable. Such mutations, however, that are not desirable include the huge, unfruitful grapevines that appear in some vineyards.

The principles of tissue culture are all around us—in nature, in the field, and in the greenhouse. We learn from experience, from other growers, or from reading. We learn about the normal requirements for soil composition, nutrients, light, and temperature for a particular plant species. We study its form, its growth habit, and its method of reproduction. These are only some of the clues that we can utilize in learning how to tissue culture plants. With this background information, we can turn to specific formulas and principles unique to tissue culture. First, we will review how these principles developed—a fascinating history, a story without end.

Figure 2-6. *Hosta* is one of the most popular garden perennials, and tissue culture is the most common method of propagating it. Most of the more than 5000 cultivars registered with the American Hosta Society are variegated (chimeras).

3 History

To some degree, the history of tissue culture involves the entire history of botany, the origin of which is lost in antiquity. A search through the years, however, finds the unique contributions of a certain few individuals especially relevant, and the events with which they were associated come together to provide us with an understanding of and appreciation for the past, as well as for the current state of the art.

Most early botany focused on the therapeutic value of plants, real or imagined. Throughout history, the merchants, missionaries, and crusaders who traveled near and far, returning to Europe with gold and fabrics, perfumes, herbs and spices, and many other goods from foreign cultures, also brought with them amazing stories of plant cures and remedies. With the advent of printing in the 15th century, knowledge and information began to spread as never before. In Europe, the medical traditions of Persia, Egypt, and Greece were studied and taught, focusing on the works of Hippocrates, Aristotle, Theophrastus, Pliny, Dioscorides, Galen, and Avicenna.

Dioscorides was one of the most important figures in this early history of botany. He traveled throughout the Middle East during the first century AD, observing and experimenting with plants. He made salves, extracted poisons, and found antidotes. He collected pistachio (*Pistacia*) resin and heated it for incense. He advocated the study of live plants as opposed to dried ones, as was the custom of most students at the time. He carefully described and recorded in detail the shapes, growth habits, fragrance, color, and beauty of the plants he observed. He also recorded in a lengthy work much of the botany that had preceded him.

The knowledge of Dioscorides and these other ancient scholars inspires awe, respect, and wonder even today. Without benefit of microscopes or modern chemistry, their studies relied on observation, reason, experience, tradition, and occasionally, magic and superstition. In their works can be found elements of today's botany: classification (taxonomy), form (morphology), dissection (anatomy), and function (physiology). There were also growers among these scholars—horticulturists, gardeners, and farmers. In all these individuals and all these fields lie the roots of plant tissue culture.

The microscope, an essential tool for examining and understanding cells and cell structure, is credited as having been invented by Zacharias Jansen, an eyeglass maker in Holland, in 1590. The following century, microscope lenses were significantly improved by Antonie van Leeuwenhoek, a Dutch merchant, linen draper, politician, and inventor, who turned his glass-blowing hobby to the making of lenses. Using a single-lens microscope, which he invented, Leeuwenhoek was the first to observe and describe one-celled microorganisms (Figure 3-1). He observed them in a drop of pond water using a microscope with a magnification of approximately 100×. What Leeuwenhoek observed was one of the most astonishing sights ever seen—the first view of spectacular microscopic life forms, the microbial world. From his careful drawings, we know now that what he observed were protozoans, algae, and bacteria. He examined scrapings from his teeth, in which he observed microscopic forms darting back and forth through the water. In 1695 he recorded, "I then most always saw, with great wonder, that in said matter there were many very little animalcules very prettily a moving" (Robbins et al. 1965). His initial findings were submitted to the Royal Society of London in 1674. Because of his pioneering work, Leeuwenhoek is known as "the father of microbiology."

A contemporary of Leeuwenhoek, Robert Hooke, an English physician, mathematician, and inventor, invented a compound microscope consisting of 2 lenses—an objective lens and an ocular lens (Figure 3-2). His findings also were reported to the Royal Society of London. He observed and illustrated the fruiting structures of molds and mosses, as well as parts from insects and small animals. Hooke was the first to apply the term *cells*, and he identified cells as nature's building blocks of all living tissues, an idea that has come to be a cornerstone of biology. He counted—or perhaps more likely, calculated there to be—1259 cells in 1 in.3 (16 cm3) of cork.

A controversy surrounding how Leeuwenhoek's animalcules and Hooke's cells originate lasted for more than 150 years and had a devastating effect on the progress of biology. Some people argued that these microbes originated from other living cells, but others maintained that they generated spontaneously from nonliving matter (abiogenesis). Several early scientists tried to disprove the theory of spontaneous generation. Among them were Francisco Redi, who in the late 17th century proved that maggots came

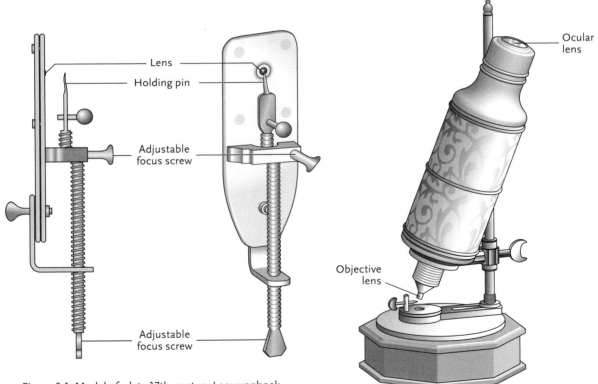

Figure 3-1. Model of a late 17th-century Leeuwenhoek microscope. Early hand-held microscopes were usually smaller than 6 in. (15 cm) long. The object to be viewed was placed on the holding pin and brought into focus with the adjusting screws.

Figure 3-2. Early compound microscope of Hooke's time. Note the absence of a condenser lens.

not from the decaying meat but from flies, and Lazzaro Spallanzani, who, nearly a century later, found that when broth was boiled and sealed in jars, microbes did not develop. Nevertheless, the controversy continued, and it was not until the mid-19th century that the theory of spontaneous generation was disproved.

In 1858 Rudolf Virchow declared that cells arise only from preexisting living cells—a theory that now seems so obvious—but people were not convinced until about 6 years later and the work of Louis Pasteur, a French chemist determined to improve France's wines. Before Pasteur, people generally believed that air alone could introduce microorganisms; Pasteur's swan-necked flask experiment, however, proved to the contrary. Pasteur demonstrated that sterilized sugar or yeast broth could remain sterile indefinitely, even with air present in the flask, so long as no external bacteria or other microorganisms were introduced into the flask. In an effort to improve the flavor and lasting quality of France's wines, Pasteur developed a process of heating wine just enough to kill most of the bacteria—a process we now call pasteurization.

The experiments of John Tyndall, an Irish physicist and friend of Pasteur, reinforced Pasteur's conclusions and helped to silence those who claimed that unheated air, or the infusions themselves, contained a "vital force" that produced microscopic organisms. Demonstrating that different solutions or infusions require different lengths of boiling time for sterilization, Tyndall concluded that some cells were able to exist in 2 forms, one form that was sensitive to boiling and another that was resistant to boiling. These heat-resistant forms are now known as bacterial endospores.

Tyndall showed that the endospores could be destroyed if the hay infusion was boiled at 3 different intervals. The first boiling killed all but the endospores. Following an incubation period in which the endospores could grow, a second boiling killed the germinating endospores. After another incubation period, a third boiling killed any late-germinating endospores. This process of sterilization is called *tyndallization*.

Tyndallization is a tool useful for tissue culture in that it helps to determine the best method for sterilizing media for the growth of certain cultures. Through tyndallization, it can be determined if it is preferable to use a boiled medium or a medium more conventionally sterilized in an autoclave (an enclosed chamber in which to sterilize equipment using high pressure steam to heat substances above their boiling points). Autoclaving can cause problems for some media ingredients because certain chemicals will degrade or change under heat or pressure. For some of his experiments, Tyndall also constructed a chamber, the first recorded forerunner of present-day tissue culture hoods (boxes or chambers in which cultures are transferred aseptically).

During the 1830s Matthias Jacob Schleiden, a botanist, and Theodor Schwann, a zoologist, further studied and speculated on the nature of cells. They observed that among lower plant forms a cell could be detached from the plant and continue to grow on its own. "We must, therefore," wrote Schwann (1839), "ascribe an independent life to the cell as such." Such was the beginning of the theory of cell totipotence, which states that any plant cell has the capability to regenerate the entire plant.

In spite of these inventions, theories, and discoveries, progress was slow, due in part to the use of microscopes that today would be considered primitive. In 1883, Ernst Abbe developed another important advance in the microscope. He greatly improved upon Jansen's binocular microscope by adding a third lens, the condenser lens, located below the microscope stage. By focusing this lens up or down it was possible to concentrate the light on the underside of the specimen, thus increasing image clarity. Abbe further improved image definition by introducing the use of lens immersion oil, which helps to deflect light rays into the lens, essential for magnification with the oil immersion lens (100×).

In any discussion of the history of the field of biology, one must pay tribute to Charles Darwin, the king of all observers and author of *On the Origin of Species*, the famous work on the theory of evolution. In 1880, Darwin and his son Francis first deduced the presence of a hormonal substance in grass coleoptiles (the sheath covering the seedling shoot tip). They observed grass stems bending toward light, yet when the tips were shaded, the stems no longer bent. They proposed that some "influence" controlling the rate of growth flowed from the shoot tip to the growing region located some distance from the tip. Five years later German biochemist Ernst Salkowski was the first to isolate this hormone found in plant shoot tips, but its growth-promoting properties were not fully recognized until years later.

English surgeon Joseph Lister provided the next important development in the field during the latter part of the 19th century. He applied the new theory of the time, which argued that certain microbes (germs) cause disease, to medicine. Using carbolic acid (now better known as phenol) to clean surgical instruments, Lister was the first person to use disinfectants.

Robert Koch, a German physician, further advanced the practice and understanding of bacteriology and sterile technique, the art of working with cultures in an environment free from microorganisms. He proved that specific organisms were responsible for causing specific diseases; anthrax, he demonstrated, is caused by *Bacillus anthracis*, and tuberculosis by *Mycobacterium tuberculosis*.

Justus von Liebig, a German physical chemist, theorized in 1840 that the minerals found in soil were essential to plant growth. His particular contribution was the concept of limiting factors, which stated that if any one essential plant nutrient is missing, all the others are of little benefit. Twenty-five years later, Wilhelm Knop (1865) developed a nutrient solution based on soil analysis. This solution was used by early experimenters in soil-less culture, and it is still used today.

In 1902 German botanist Gottlieb Haberlandt used Knop's medium, supplemented with sucrose, asparagine (an amino acid), and peptone, to grow cells, but the cells lived only a few weeks. Haberlandt also correctly predicted that plant cells in culture would be able to vegetatively divide and form embryos and whole plants. Two years later E. Hanning, another German botanist, anticipating the procedure of embryo rescue, successfully cultured premature, excised crucifer embryos. He observed the embryos produce small, weak plantlets in culture instead of developing into normal embryos. He called this *precocious germination*.

As the 20th century progressed, the field of plant tissue culture embarked on an era of exponential growth (see Table 1). Of early commercial significance was the germination of orchid seeds and seedling growth on a medium composed of agar, a polysaccharide gel derived from certain algae, in aseptic culture. This feat was reported independently but almost simultaneously by Lewis Knudson, Nöel Bernard, and Hans Burgeff in the early 1920s. At about the same time, Walter Kotte, a student of Haberlandt's, and William J. Robbins independently accomplished limited success in the culture of root tips.

Early experiments confirmed that tissue cultured plants while in culture were heterotrophic (unable to manufacture their own food from inorganic substances), as opposed to autotrophic (capable of manufacturing their own food). Unlike plants that are growing in soil, plants in culture cannot manufacture proteins and carbohydrates from inorganic nutrients. Perhaps empirically, it was discovered that sugar and undefined substances such as coconut milk, yeasts, and fruit juices were able to support culture growth, but inorganic chemicals alone could not. Robbins reported that tomato (*Lycopersicon*) root tip cultures were aided by the addition of yeast (*Saccharomyces*) to the medium. Later analysis revealed

Table 1. **Highlights in plant tissue culture from 1900 to 2010**

1900 Theoretical basis for totipotency described

1920 Successful root and shoot tip culture
Ectogenesis/embryo rescue demonstrated
First true tissue culture (from cambium of *Acer pseudoplatanus* and other woody plants)

1940 Callus growth maintained in culture
Coconut water first used for culture of young embryos; later with monocots and other recalcitrant tissues
Manipulation of auxin and kinetin ratios influences root and shoot formation
Single cells successfully cultured from shaken callus cultures
Induction of somatic embryos from callus
Cell suspension in carboys used for large-scale culture
Use of meristem culture to obtain virus-free dahlia

1960 Meristem culture used for propagation of orchids
Creation of the highly effective Murashige and Skoog (MS) salt medium, suitable for many species
In vitro pollination and fertilization achieved with poppy
Haploid *Datura* plants resulting from anther culture (androgenesis)
Totipotency realized as whole tobacco plant produced from a single cell
Investigations into production of secondary metabolites *in vitro*
Creation of interspecific hybrid of tobacco via protoplast fusion
Micropropagation described as three stages: establishment of explant, multiplication, rooting and hardening of plantlets

1980 Somaclonal variation described as a method of plant improvement
Multiplication of DNA sequences *in vitro* via polymerase chain reaction (PCR)
Tumor-inducing (Ti) plasmid (*Agrobacterium tumefaciens*) used as a vector for transformation; kanamycin resistance gene used in selection of transformed cells, resulting in transformed plants
Initial work on transformation by various methods including DNA microinjection, particle bombardment (biolistics), and *Agrobacterium*
Cryopreservation of cell cultures
Planting of the first commercial transgenic crops including "Roundup-ready" cotton

2000 First complete plant genome (*Arabidopsis*)
Transgenic rice containing beta-carotene (golden rice) is developed to combat severe Vitamin A deficiencies prevalent in some developing countries
StarLink Bt corn (includes gene sequence of *Bacillus thurigiensis* as a systemic pesticide) developed for ethanol production, not human consumption, is found in taco shells
World acreage for biotech crops: soybean 57%, maize 25%, cotton 13%, and canola 5%

that yeasts contain certain desirable vitamins, particularly thiamine (vitamin B_1).

Callus culture of carrots (*Daucus carota*), a classical subject of investigation, was reported by 2 physicians, Ferdinand Blumenthal and Paula Meyer, in 1924. They were primarily interested in the pathological implications of such a study, comparing callus to tumor growth, but in 1927, L. Rehwald demonstrated the cultivation of callus from carrot slices, irrespective of the pathological considerations.

Drawing on the Darwins' earlier observations of coleoptiles and on the work of Salkowski, in 1911 Peter Boysen-Jensen demonstrated that the growth-promoting substance found in plant shoot tips would diffuse across a wound covered with gelatin, and he realized that the substance could diffuse out of the tissue and into a collecting gel. In 1928, Frits Went collected this growth substance from coleoptile tips in tiny blocks of agar. When he placed an agar block containing the substance asymmetrically on top of a decapitated shoot, the shoot proceeded to grow more on the side with the block than on the side without the agar block. The greater the number of tips that were used for collecting the substance in a particular agar block, the greater was the curvature in the growth of the shoot. In 1934 Fritz Kögl, Arie Jan Haagen-Smit, and Hanni Erxleben were the first to isolate and chemically analyze the substance. They identified it as a plant hormone, or auxin, which they named indole-3-acetic acid (IAA). Five years later, Roger J. Gautheret and Pierre Nobécourt, both in France, independently reported indefinite growth of callus from carrot cambium when using auxin in the nutrient medium. Gautheret (1942, 1982) had been the first to successfully culture plant tissue. In 1934, he grew cambial (meristematic) tissue of the sycamore maple (*Acer pseudoplatanus*) and, later, of pussy willows (*Salix*) and elders (*Sambucus*), some of which continued in culture for more than a year.

Philip R. White is acknowledged as the father of tissue culture in the United States. He was the first to grow excised root tips of tomatoes (*Lycopersicon*) in continuous culture. The inclusion of glycine (an amino acid), pyridoxine (vitamin B6), and nicotinic acid (niacin, vitamin B_3) in his liquid root tip media were important and effective additions to the experiment. In 1939, he reported successful culture of tobacco (*Nicotiana*) callus. In collaboration with Armin Braun, White demonstrated the similarity of tumor cells in plants and animals by growing tumor tissue from tobacco. White's *A Handbook of Plant Tissue Culture*, originally published in 1943 and revised in 1963, brought together the accumulated knowledge of plant tissue culture as it stood at the time.

Many scientists in the mid-20th century were working on culturing embryos extracted from seeds (embryo rescue) or attempting to stimulate the spontaneous production of embryos from undifferentiated cells (embryogenesis). In light of this current state of the research, a logical next step seemed to be to investigate the use of coconut milk, the liquid endosperm (nutritive liquid or tissue of seeds) from coconuts (*Cocos*)—a ready-made, natural nutrient medium within seeds that nourishes the embryo. Coconut milk was first used in 1941 by Johannes van Overbeek, Marie E. Conklin, and Albert F. Blakeslee, who discovered that it stimulated callus formation in cultures of excised embryos of jimson weed (*Datura stramonium*).

Frederick C. Steward, a renowned plant physiologist at Cornell University in New York, was so impressed by the dramatic effects of coconut milk in carrot culture media that he set aside his other objectives to dedicate himself to the study of growth factors in this and other liquid endosperms. Among the active materials he extracted from the coconut milk were several organic ingredients that are now commonly included in purified form in many tissue culture media. Coconut milk is still used in some orchid culture media. Steward was also able to confirm the concept of Thorpe with plant cells by showing that small pieces of carrot tissue could dedifferentiate (revert back to differentiate) and then grow into specialized plant cell types.

The orchid industry was the first to apply micropropagation on a commercial level. Georges Morel and C. Martin cultured virus-free dahlia shoots and potato (*Solanum tuberosum*) plants by meristem culture, following protocols set forth by Ernest A. Ball in 1946. In 1960, Morel and Martin applied their findings to orchids. Not only did they succeed in freeing the orchids of viruses, but unexpectedly, they also observed multiplication of their cultures, a phenomenon (micropropagation) that in time would revolutionize many features of the entire horticultural industry. Orchids, for example, have become more plentiful and less expensive because of this discovery, and many tested, disease-free cultivars are now readily available because of micropropagation.

In spite of these early accomplishments, the need to define media ingredients and their proper proportions for successful commercial application remained an important issue. Several discoveries during the 1950s addressed such concerns. In 1955, Carlos O. Miller discovered kinetin, a hormone that promotes bud formation, and the first of a group of plant growth regulators known as cytokinins. Frits Went collaborated with Folke K. Skoog to examine the bud-inhibiting effects of auxin and its interaction with kinetin. Went and Kenneth V. Thimann demonstrated the root-initiating properties of the auxin IAA. Later, in 1957, Skoog and Miller published "Chemical Regulation of Growth and Organ Formation in Plant Tissues Cultured *In vitro*," in which they discussed the desired auxin-to-cytokinin ratios for the growth of cultures.

Skoog's name is immediately recognized by anyone involved in tissue culture for his part in the Murashige and Skoog medium formula, commonly referred to as M & S or MS medium. The medium, now used almost universally, was reported in the classic 1962 article "A Revised Medium for Rapid Growth and Bioassays with Tobacco Tissue Cultures." Higher in salts (especially nitrate and ammonium) than previous media, with additional subsequent modifications, the MS medium proved key to culturing many more plants than had previously been possible. Elfriede Linsmaier collaborated with Skoog shortly thereafter with a report of their systematic study of the organic requirements of tobacco (*Nicotiana*) callus (Linsmaier and Skoog 1965). They indicated some appropriate adjustments of the MS formula, and the groundwork was laid for an even wider application of this versatile formula.

Toshio Murashige, a former student of Skoog's at the University of Wisconsin and later a professor at the University of California at Riverside, did much to bridge the gap between research and industrial applications. He is credited for developing 3 (I, II, III) of the now-expanded 5 stages of micropropagation (see Figure 1-1). These stages not only describe the procedural steps of micropropagation, but they also coincide with phases in the process where environmental or media changes are required; they are explained in detail later in this book.

New Technologies

In the broad field of the natural sciences, there is an increasing recognition of the interrelatedness and interdependency of the many particular fields and disciplines; physics, chemistry, botany, zoology, mycology, genetics, and others can no longer be isolated and departmentalized, but must be studied in context with one another. In light of this, new developments in plant research inspire the use of tissue culture in several overlapping areas. Some of these that are of broad interest include anther, callus, cell, and embryo culture, protoplast fusion, mycology, algology, secondary products, and genetic engineering. See chapter 13 for further discussion of some of these applications.

Liquid cell suspension cultures have particular significance for the mass production of cells.

One common source of cells for cell suspension is from friable (crumbly) callus, although specific cells, such as from leaf mesophyll (the thin, soft tissue between the upper and lower epidermis of the leaf), are also grown in suspension. Cells in suspension can form embryoids (somatic embryos) in the process of somatic (non-sexual) embryogenesis. Embryos may multiply and/or be induced to form plantlets in the process of morphogenesis. They may also be artificially coated with a "seed coat" for the production of synthetic seeds. Many hybrid plants produce embryos that do not mature to viable seeds. These embryos can be "rescued," removed from the seed in an immature stage, and then grown in culture.

Suspension cultures have been enhanced by new methods that can continuously introduce fresh medium into the suspension culture, thereby enabling the production of thousands of cells or embryos in a single container with a minimum of manual transfer. This is one way that tissue culture can compete with the plentiful seed production in nature.

Interest in the tissue culturing of anthers or pollen to obtain haploid clones (plants composed of cells with half the normal number of chromosomes of vegetative cells) is spurred by the practical applications of such haploid cultures (Figure 3-3). Haploid (*n*) plants are sterile, but if the chromosomes duplicate, either spontaneously or by induction, the plants will be diploid (2*n*), which is normal for the vegetative state, and their progeny will be true to form. By conventional means, it takes several generations of inbreeding to obtain a pure line; therefore, plant breeders are very interested in anther culture.

Genetic engineering has also brought about several new useful products and processes. The 3 most common ways of altering the genetic make-up plants are protoplast fusion, mutagenesis (the production of cell mutations) using chemicals or radiation, and genetic engineering whereby plants are transformed by the introduction of foreign genes.

When the wall of a cell is dissolved away, the remaining membrane and its contents comprise a protoplast. Because they lack cell walls, protoplasts are more permeable and, as such, are particularly useful for studying the intake of or resistance to toxins, nuclear material, viruses, bacteria, or fungi. Of special interest to plant breeders is the ability that protoplasts have to fuse with other protoplasts not only of the same clone, but also with those of other species or genera. Before 1960, cell walls had to be removed by physical means (microsurgery). Cells were plasmolyzed by placing them in a solution that caused the protoplast to shrink away from the wall. The cells were then randomly cut up with a sharp scalpel, and the undamaged protoplasts were recovered with a pipet. In England in 1960, Edward C. Cocking developed and published a method for removing cell walls by chemical (enzymatic) methods. Complete plants have been regenerated from protoplast cultures in at least 30 plants, including tobacco (*Nicotiana*), wheat (*Triticum*), carrot (*Daucus carota*), *Asparagus*, potato (*Solanum tuberosum*), and tomato (*Lycopersicon*).

After many failed attempts by numerous experimenters to induce fusion between protoplasts from different sources, in 1974 Kuo Nan Kao and

Figure 3-3. Anther culture for use in the potato breeding program at Virginia Tech.

M. R. Michayluk discovered that when a solution of polyethylene glycol is added to a mixed protoplast culture, followed by high calcium dilution, a significant percentage of protoplasts would fuse. A few years later, other researchers reported that electrical impulses are another inducing agent.

With these methods, protoplasts of tomato plants have been fused with protoplasts of potato plants, and the resulting product was able to differentiate into complete hybrid plants. The objective was not to create a new "freak," but rather to convey certain disease-resistant qualities of one plant to another by an intermediary plant. So far, only few fusions between different species have developed into mature plants, but the potential is there to produce plants with greater food value, more resistance to pests, and better ornamental qualities.

Mutations in plants occur naturally at a very low rate (about one cell in a million). They can be induced by artificial means, however, and tissue culture offers a convenient method of handling plant material when it is treated with mutagens (mutant-inducing agents) such as radiation, ultraviolet light, or mutagenic chemicals. Following treatment, cultures are incubated and then tested for certain characteristics, such as resistance to toxins, salts, herbicides, antibiotics, or disease. The cultures can also be tested for their tolerance to heat or cold, their hormone and nutrient requirements, or their ability to synthesize secondary products, or metabolites (products of metabolism, the life processes of all living matter).

The production of secondary products is now a common application of plant tissue culture. Some cell cultures can be induced or genetically engineered under controlled conditions to produce useful compounds in greater quantity and at less cost than can be obtained from whole plants. A sampling of the types of secondary products potentially available from cell cultures includes flavorings, pigments, medicinals, gums, resins, antibiotics, insecticides, fungicides, alkaloids, enzymes, cosmetics, and oils. The products are extracted chemically and physically either from the cells or from the medium, depending on whether or not they are released from the cells.

The origin of genetic engineering in plants can be traced to the observation of the condition of hyperplasia, an abnormal increase in cell number causing irregular swelling or growth, which is often a response to disease. The tumors caused by crown gall (*Agrobacterium tumefaciens*) are a familiar example of cells that have undergone natural genetic transformation.

Because it is possible to determine the specific gene or genes responsible for certain plant characteristics, a genetic engineer can insert into a plant the genes that code for a desired trait or traits. These genes can be obtained from another plant, from a bacterium, or even from animal cells. They may code for cold hardiness, insect resistance, disease resistance, proteins, enzymes, medicinals, flower color, and many other qualities (Figure 3-4). The importance of this research must not be underestimated.

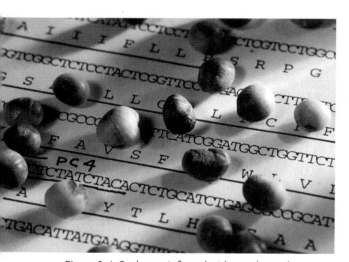

Figure 3-4. Soybeans infected with purple seed stain rest on a printout showing the genetic code for cercospora facilitator protein. This gene from the fungal pathogen *Cercospora kikuchii* is responsible for transporting the disease.

In the 1980s, Kary B. Mullis (1990) invented PCR (polymerase chain reaction), for which he was awarded a Nobel Prize. This outstanding process automates the exponential multiplication of DNA (deoxyribonucleic acid), a carrier of genetic information. The double-stranded fragments are denatured (broken down) by high temperature and are dissociated to single strands. Each new piece serves as a template for another new copy. Polymerase, the enzyme (a specialized protein that acts as an organic catalyst in specific chemical reactions) used in the duplication part of this process, was extracted from bacteria from hot springs. Because the enzyme can withstand the high temperature that is required for the denaturation part of the process, it is a key player in the duplication of single-strand DNA.

Also in the early 1980s, the gene gun or biolistic particle delivery system was developed by John C. Sanford and others at Cornell University for plant transformation and was used to inject genetic information into cells. This state-of-the-art technology often referred to as biolistics can transform organelles.

The 1990s saw an explosion in the number and types of plant species upon which tissue culture approaches were used. Food crops, ornamental plants, forest species, and more have been subjected to a wide variety of *in vitro* techniques for both propagation and genetic modification purposes. For much more on this topic see chapter 13.

For every discipline, there are organizations composed of like-minded practitioners, and plant tissue culture is no exception. The Tissue Culture Association was founded in 1946 and later became the Society for In Vitro Biology (SIVB). The SIVB produces the journal *In vitro Cellular and Developmental Biology—Plant*. Another journal featuring the latest research is *Plant Cell, Tissue and Organ Culture* (PCTOC: Journal of Plant Biotechnology). The International Plant Propagator's Society (IPPS) and the American Society for Horticultural Science (ASHS) also publish articles about plant tissue culture in their proceedings and journals. (See "Professional Organizations" at the back of the book for more information.)

This brief summary of some of the most notable activities involving tissue culture should stimulate the imagination and alert the commercial propagator or student to the broad opportunities offered by the new technologies. The tangible products of these research efforts will come as no surprise to the knowledgeable propagator. They appear in the nursery trade, in the corn and wheat fields, in the kitchen, in the pharmacy, in the flower shop, and in the bowls of the hungry.

4 The Laboratory: Design, Equipment, and Supplies

Any one considering a tissue culture operation will approach the project with a unique background and with different goals, limitations, and resources than another person. Before planning a laboratory, you should determine the magnitude of the operation, the purpose or goals, and to some extent, the plants to be cultured. How elaborate a facility do you want or can you afford?

A hobbyist can reap great pleasure, and a little pocket money, from a small laboratory that serves personal or local interests. At the other extreme, a large laboratory serving the domestic market or shipping to other countries has as its major concerns the potential market and the lab's distance from customers. A preliminary assessment of potential markets is an important early step; find out in advance who will buy your product and at what price. It is often advisable to establish a pilot plant operation to assess the feasibility of producing the specific plants you have selected to grow.

If possible, visit government, university, or commercial plant tissue culture facilities. Observing these large-scale facilities will help you to formulate your own plans and can serve as models. Also, visit nurseries, greenhouses, garden centers, and florists, both wholesale and retail, to discover what plants are in demand or are hard to get. Learn from government agencies or exporters about the current regulations for shipping from state to state or to foreign countries. At one time tissue cultured plantlets moved freely through customs; but now it is understood that even cultures that appear healthy can harbor viruses or other unsuspected pathogenic, systemic, microbial contaminants that may not become evident until much later, when the plants are growing in the field or garden. Consequently, regulations have become stricter and now vary between state and national governments. For the United States, the Plant Protection and Quarantine (PPG) division of the United States Department of Agriculture Animal and Plant Health Inspection Service (USDA APHIS) provides phytosanitary certification for export/import of tissue cultured plantlets. (For more information, see http://www.aphis.usda.gov/import_export/plants/plant_exports/faqs.shtml.)

Location

Individual circumstances will, of course, determine the location of a proposed laboratory, but some general concerns need to be addressed in any location. It may be appropriate to set up a small laboratory first until the proper techniques and markets are developed.

Careful planning is an important first step when considering the size and location of a laboratory. Are there existing buildings or greenhouses that can be converted to a laboratory and growing-on facilities? A convenient location for a small laboratory is a room in the basement of a house, garage, or remodeled office. The minimum area required for media preparation, transfer, and primary growth shelves is about 150 ft.^{2} (14 m^2) (Bridgen and Bartok 1987). Walls may have to be installed to separate different areas. Good locations will be isolated from foot traffic, have thermostatically controlled temperature control, water and drains for a sink, adequate electrical service (a minimum 100 amp service is recommended), provisions for ventilation, and good lighting.

When building a new laboratory, especially if it is large, check with local authorities about zoning and building permits. Locate the building away from sources of contamination such as soil mixing areas, pesticide storage, or dust and chemicals from fields. A clear span building allows for a flexible arrangement of walls. The floor should be concrete or capable of carrying 50 lbs/ft.^{2} (240 kg/m^2). Walls and ceilings should be insulated to at least R-15 and be covered inside with a water-resistant material. Windows, if desired, may be placed wherever convenient in the media preparation and glassware washing rooms, but not in the transfer or growth rooms. The heating system should be capable of maintaining a room temperature at 21°C (70°F) in the coldest part of the year and an air conditioning system may be necessary. A minimum of 0.75 in. (2 cm) water service is needed. Connection to a septic system or sanitary sewer should be provided.

In all aspects of planning, cleanliness should be considered one of the most important factors. Air quality and direction of airflow are of particular concern because unsterilized air carries microbial contaminants. Select an area that is relatively free from dust, smoke, molds, spores, and chemicals. Driveways and parking areas should be close enough for convenience, but far enough away that blowing dust will not affect the facility. If possible, an enclosed entrance should precede the laboratory; sticky mats can be laid at the entrance to help collect dirt from the outside. It is good procedure to remove shoes upon entering the laboratory.

A hobbyist may start tissue culturing in a home kitchen, using a homemade transfer chamber (Figure 4-1) and a bookcase with lighted shelves. This setup will serve quite well for a while, but the serious amateur who masters the techniques and achieves rewarding results in limited facilities will soon want to advance to a more sophisticated laboratory for pleasure and profit.

Design

A plant tissue culture laboratory consists of 3 distinct areas:

1. A space for preparing media.
2. A location for transferring cultures.
3. An area for growing the cultures (primary growth room).

Usually these are located in 3 separate rooms, but when just starting out or in the face of limiting circumstances, these areas may occupy the same room. The traffic pattern and work flow in the laboratory must be considered to maximize cleanliness. The cleanest rooms—the transfer room and the growing room—should not be entered directly from the outside of the building; they should be enclosed with doors leading to each. Ideally, the media preparation area will lead to the aseptic transfer room, which would lead to the primary growth room. In addition to the 3 main areas, other facilities to consider include bathrooms, plant preparation area, office space, and break rooms.

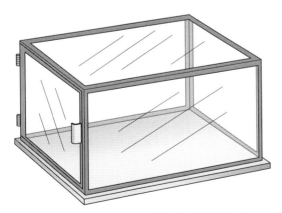

Figure 4-1. A still-air box (transfer chamber) for the hobbyist.

The media preparation area resembles a kitchen. It should have a sink with hot and cold running water, a refrigerator, a dishwasher, and a heat source such as a stove or hot plate (with stirrer). You will need a pressure cooker or an autoclave for high-pressure heat sterilization. Counter space should be made available for a balance, a pH meter, a microwave oven, and a hot plate/stirrer, as well as room to prepare the media. There should be a place for water treatment and storage. Counters should be a height that is comfortable to work while standing (34–36 in./85–90 cm) and deep enough (24 in./60 cm) to hold equipment. Cabinet space is required for storing glassware, chemicals, and other supplies. The preparation of media requires several steps and space must be available to accommodate these activities. Chemicals must be weighed and added to water. A medium containing agar must be heated so that the agar will melt. After the agar has melted and the chemicals are dissolved, the medium must be dispensed, sterilized, and stored, ready for use.

The transfer room contains the transfer chamber or hood, which is usually a bench with a partially enclosed area that is provided with sterile air. It is here that the technician starts, divides, and trims the cultures and transfers them from one container to other containers. Between transfers, the cultures stay in the warm, clean growing area, where well-lighted shelves hold the vessels (test tubes, bottles, flasks, or petri dishes) in which the cultures are growing.

The transfer room and the culture growing room should be isolated as much as possible from outside doors and from significant foot traffic. Interior pass-through windows between rooms can reduce foot traffic, help prevent contamination, and minimize the opening of doors, but they are not practical for larger operations, in which case large numbers of containers should be moved on carts. The use of tables or carts is an effective way of increasing counter space in any of the 3 areas. The size of cart can vary from that of a tea cart to a bank of shelves on casters. The carts should be sturdy, easy to maneuver, and easy to clean. All doorways should be wide enough to accommodate carts.

Windows may be placed conveniently in the media preparation room and, sometimes, in the transfer room, but are not desirable in the culture growing room because of the opportunity to introduce contaminants and because outdoor light and temperatures may reduce the ability to control the growing environment. An objection to windows in the transfer room is that the irregular surfaces on window frames can be difficult to clean properly; on the other hand, many workers find it a more comfortable environment if they can see out through a window. Light-colored walls and music also can help create a more comfortable working environment. Studies have shown that work efficiency drops off after 6 hours, so the best possible environment should be built into the transfer room. In any case, if there are windows in the transfer room, they must be tightly sealed.

Routine cleaning of the laboratory is important to keep down losses from contamination. In all rooms, the walls, floors, and ceilings should be light colored and easily washed with detergent and standard disinfectants. Walls can be painted with acrylic or urethane epoxy paint, and cement floors can be painted with epoxy or urethane floor enamel or have inlaid linoleum installed. Banks of shelves should be set high enough off the floor to allow for convenient cleaning of the floor underneath and to help avoid contamination from the floor.

Any outside air that is admitted into the 3 major rooms should be filtered, preferably by HEPA (high efficiency particulate air) filters, to reduce the particle count to an acceptable level. The particle count for the transfer room and growth rooms should be 10,000 particles or less per cubic meter of air. Regular furnace filters can be installed over air intakes to the media preparation room where 100,000 particles per cubic meter of air are adequate. Rooms used for office and shipping do not require filtration of room air.

Power and wiring

Planning for electrical equipment and fire safety always deserves the best available professional advice. This is especially true for a tissue culture facility because of its high demand for electrical equipment and, in some instances, the use of open flame. Temperature and fire alarms should be connected directly to telephone lines for fast warnings of problems. Most of the wiring can be for 110 volts, but heating, cooling, water treatment equipment, and autoclaves often require 220 volts.

Heating and cooling can be supplied by electric wall heaters or furnaces and air conditioners, but a heat-pump system is far better and well worth the initial expense. A heat pump should contain appropriate filters, such as an in-line electrostatic air filter and, possibly, HEPA filters for transfer and growing rooms. Information regarding developments in heating, lighting, and energy conservation can be obtained from your power company and from heating and lighting equipment manufacturers. It is wise to confer with them before making final decisions about equipment choices, which need to satisfy the plants as well as the workers who must occupy the same environments; for example, cool-white light is desirable for plants, but daylight-quality fluorescent fixtures may be preferable for people.

Electrical power failure and safety are other major concerns. It may be necessary to obtain a portable generator to keep the growing room at the correct temperature in the event of a power failure. Safety plans should include built-in escape routes, an emergency plan, and the location of firefighting equipment.

Growing room lighting and temperatures

The primary growth room is the location where the plant cultures will grow during Stages I through III; a secondary growth room is less clean and is used for Stage IV. Temperature control, relative humidity, lighting units, and shelves need to be included in the culture room. All of these environmental considerations will vary depending on the size of the room.

There are several options for lighting the grow room (Figure 4-2). Ideally, select a light source with spectra in the photosynthetically active region (PAR), which comprises light energy between 400 and 700 nanometers (nm). Fluorescent is the most common, but incandescent will work in a pinch. Light-emitting diode (LED) bulbs and systems have also been adapted specifically for greenhouses, grow rooms, and grow carts. Strips or bars of LED lights work with multi-level shelf systems. Panel systems with ballasts can be mounted over benches and shelves.

Providing the optimum amount of light requires

A foot-candle is a unit of measure of light intensity, based on the illumination on one square foot of surface that is one foot away from a standard candle. One foot-candle is roughly equivalent to 10 lux (0.093 f.c. = 1 lux). Neither foot-candle nor lux are specific to the light available for plants to use in photosynthesis because plants operate in a much narrower band of light known as PAR (photosynthetically active radiation). Light in the PAR is measured in units such as micromoles per square meter per second (μmol m^{-2} sec^{-1}), microeinsteins per square meter per second (μE m^{-2} sec^{-1}), or watts per square meter (wm^{-2}). Although some light meters express light intensity in these newer units of measure, foot-candle measurements are retained in this work for their simplicity and because foot-candle–reading meters are inexpensive, readily available, and still widely used.

some calculations. For example, 2 cool-white fluorescent fixtures, each 8 ft. (2.4 m) long and spaced 20 in. (50 cm) apart in a growing room, will provide 100–300 foot-candles (1000–3000 lux) of light.

Another consideration in the lab is the distance from the light to the cultures. With 2 or even 4 fluorescent tubes over a 4-ft. (1.2-m) wide shelf, it is not possible to have uniform lighting reach all of the cultures. The amount of light varies with the distance from the light source, the type and size of culture container, and the density and size of the cultures. On shelves separated by 16–18 in. (40–45 cm), the fluorescent tube will be about 10 in. (25 cm) above the shelf, taking into account the depth of the fixture. The light intensity on the cultures directly below the lamp might be 3 times that reaching the cultures at the edge of the shelf.

Cool-white fluorescent lamps are still primarily used in commercial tissue culture laboratories in the United States. You will see labs using warm-white tubes and sometimes a mixture of both. There has been a transition to use the more economical T-8 tubes (cool-white). In China, commercial labs are beginning to use cold cathode fluorescent lamps (CCFL). The problem with the standard cool-white is that they emit little to nothing in the far-red wavelength. Some research claims about a 15% increase in production with the inclusion of lighting emitting more in the far-red range. Wide spectrum fluorescent tubes provide more light in the read and yellow range, but are much more expensive than standard fluorescent lamps.

Special fluorescent fixtures can modify or eliminate the problem of heat given off by ballasts. The temperatures of hot spots in the shelves

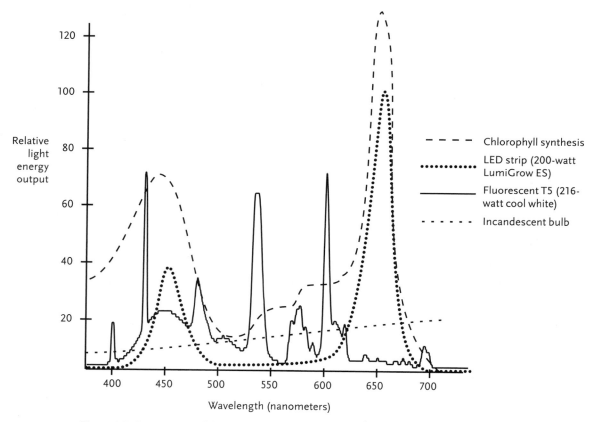

Figure 4-2. Comparison of the energy requirement of chlorophyll synthesis and the spectral energy output of 3 different light sources.

directly over the ballasts can be so elevated that they will cook the cultures; therefore, the ballasts are often removed and placed at the ends of the shelves in a protective metal box or installed outside of the room. Air spaces, 2–4 in. (5–10 cm) between the lights and shelves, will decrease bottom heat on upper shelves and condensation in culture vessels. Some solid-state ballasts offer the advantages of less heat, no flicker, and no noise. Consult your local lighting or electrical authorities for the latest technical information.

Culture growing room temperatures generally should be kept between 75°F and 85°F (24°C–29°C), with 16 hours of light and 8 hours of darkness, although this photoperiod cycle will vary with the particular plants being grown. A centrally located thermostat controls heating and cooling equipment. Such equipment usually has a blower that can be left on continuously to provide good air circulation, thus minimizing hot areas over lights and ballasts or cold areas near the floor. Sometimes shelves are equipped with miniature inflatable air ducts to help circulate the air.

A 24-hour timer can be used to automatically control the lighting and help balance the heating and cooling demands of the laboratory. For the purpose of energy efficiency, the timer can be set so that the 8-hour period of darkness occurs during the daytime and the 16 hours of light fall mostly during the night. By this arrangement, the heat given off by the lights at night helps keep the growing room warm when heat is most needed and allows the room to be cooler in the daytime, to save cooling costs. Usually a culture's light-cycle requirements will not be noticeably affected if you turn on growing room lights periodically during the day for doing chores. A light meter should also be purchased to measure light distribution and intensity; it can be purchased for less than $100.

Examples of laboratory designs

A simple plan for an operation incorporating some of these design features is shown in Figure 4-3. This was the original lab of Lydiane Kyte. The plan lends itself to a variety of options, and the total area is adequate for culturing a million plants. The laboratory was built inside an existing metal building. The media preparation room is 16 × 12 ft. (4.8 × 3.6 m), the transfer room is 8 × 12 ft. (2.4 × 3.6 m), and the culture growing room is 18 × 12 ft. (5.4 × 3.6 m). The transfer chamber can be widened to accommodate 2 workstations, allowing 2 technicians to transfer at the same time. If adjacent counter space is limited, carts are a good supplement, especially for finished test tubes or jars ready to be moved to the growing room. The transfer room can also be expanded into the adjacent entry area. Either entry area can be developed further for storage, office, autoclave, water treatment, or rest rooms. Cupboards above the bench or counter in the media preparation room are desirable, but they may not be required initially.

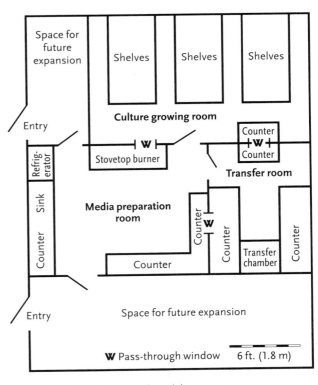

Figure 4-3. A small tissue culture laboratory.

A large laboratory has the same design concepts as a small lab, but is targeted to produce millions of plantlets a year and house 10 or more employees. Figure 4-4, which shows the original laboratory for Briggs Nursery in Olympia, Washington, meets these criteria. The lab was built inside a metal building, which was 80 × 60 ft. (24 × 18 m). Most laboratories are not of this magnitude, of course, but the plan can be scaled down and the concepts and product flow are worthy of consideration.

The transfer room, growing room, and research areas in this large laboratory are under positive air pressure. The ballasts for the fluorescent lights in the growing room are located in the attic area, reducing the excess heat in the growing area and limiting any direct hazards from the ballasts, which have been known to explode. The clean-up room is where contaminated cultures are autoclaved, dishes are washed, plantlets are sent to the greenhouse, and containers are returned for cleaning and re-use. Four counter-top autoclaves are located in the clean-up room and the media preparation room.

The office also serves as a conference room and library. The hallway along 3 sides of the laboratory provides escape, access, and insulation. Sliding glass doors provide access and visibility between the transfer room and media preparation room, between the transfer room and growing room, and between the office and research room.

Visitors to the laboratory are ushered from the main entrance, past the locker area, to windows through which they can view the transfer room and the growing room. In the middle of the transfer room are 8 transfer hoods placed back-to-back, each hood 6-ft. (1.8-m) long.

The laboratory of Micro Plantae in Puna, India (completed in 1993), has some innovations worthy of note. It has a production capability of 10 million plantlets per year, using 2 shifts a day with a work force of 20 people per shift in the transfer room. Two autoclaves, each 4 × 6 ft. (1.2 × 1.8 m), are used for media sterilization. The autoclaves can be opened at both ends, thereby enabling removal of sterilized media directly into the transfer room storage area. This arrangement avoids diminishing the air quality and cleanliness in the transfer room. The glass bead sterilizers for transfer tools are mounted below the level of the surface of the transfer hood, again helping to maintain a sterile working surface. Each hood contains an ultraviolet light unit that is turned on for 30 minutes before each shift to sterilize the bench surface. (Ultraviolet lights are rarely recommended due to the health hazard they pose to eyes and skin.) Carts are used to move trays containing small culture jars to various production areas. When leaving either the transfer or growing rooms, the carts are moved through a pressurized interlock (double door) area.

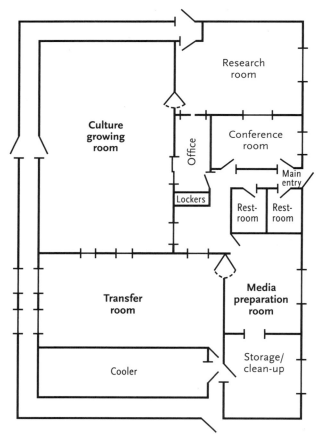

Figure 4-4. A large tissue culture laboratory.

Attached to the transfer room at the Micro Plantae laboratory is an office that has a computer for entering production data, such as the location of each batch of culture jars in the growing room. In an operation of this magnitude, such record keeping is essential.

The growing rooms measure approximately 30 × 30 × 20 ft. (9 × 9 × 6 m). The shelves extend to the ceiling, necessitating the use of ladders to reach the higher shelves. By using wire-mesh trays and shelves, plus an excellent air-circulation system, the laboratory is able to regulate air temperatures throughout the growing rooms to within 1°F or 2°F (1°C).

The above factors, together with the maintenance of positive air pressure in both the transfer room and the growing room, allow for an air cleanliness standard in the Micro Plantae laboratory of 10,000 particles per cubic meter of air (as detected by an active quality-control program). No doubt, an in-depth operator-training program also helps Micro Plantae keep the number of contaminated containers at a level of 1% or less.

If you have a new laboratory, with perhaps only a laminar airflow hood (transfer chamber) and a small kitchen laboratory, do not be discouraged. The designs that are mentioned at first may seem complicated and overwhelming. Rest assured that you will feel more comfortable as you become more familiar with the laboratory.

Equipment

A variety of equipment is available for micropropagation laboratories. The budget, experience, and amount of space will determine the type and amount of equipment purchased. Many of the pieces of equipment that are listed are not essential when first building a laboratory. Alternatives to the expensive equipment are listed.

Water purification equipment

Water is the largest component of tissue culture media, so high quality water is of critical importance in establishing a successful operation. Tap water often contains dissolved minerals, particulates, and organic matter. These substances must be removed before using the water in tissue culture media because they can upset the precise balance of nutrients in media formulas, or they can be toxic. Any imbalance can affect the pH of the medium and promote precipitates (insoluble chemicals), which make the nutrients in media unavailable to the plantlets.

A convenient and economical way to obtain pure water is to buy bottled distilled water. Both deionized and distilled water are available at local supermarkets in gallon jugs, usually for less than US $1.50 per gallon (3.8 liters). It is advisable to buy bottled purified water when starting on a small scale and limited budget, but for a permanent operation a still or deionizer should be purchased.

Rainwater can also be used, but it must be carefully tested, particularly with regard to the means of collection. Rainwater can pick up undesirable elements from a metal roof, or it can grow undetected organisms that give off undesirable organic compounds, either of which might not be eliminated even with subsequent treatment.

One common method of ascertaining the purity of water is to measure its conductivity. Conductivity is a measure of the ability of water to conduct electricity. An electrical current passes through water by the movement of ions, so the conductivity level indicates the concentration of ions dissolved in the water. Conductivity is often measured in units called siemens (S). Its reciprocal, ohm-centimeters (ohm-cm), which measures resistivity, is an optional form of measurement. The lower the conductivity (that is, the lower the number of siemens), the greater the purity of the water. Because siemens are such relatively large units, the values are more readily expressed in microsiemens (µS). Distilled water has a conductance of 0.5 to 4.0 µS; a conductivity level as high as 4.0 µS is acceptable for most tissue culturing. Levels higher than this are often tolerated in cultures, but the best measure of acceptable water quality is plant performance.

A conductivity meter is an essential tool for any tissue culture operation. A conductivity meter costs about US $400; sometimes a meter is built into the water purification system. Most companies dealing in purified water and purification equipment will test water samples without charge. Some government laboratories will also provide this service.

The 3 most common methods used to remove dissolved chemicals from water are distillation, deionization, and reverse osmosis. Sometimes a combination of 2 or 3 methods is required.

Water purified by distillation has been the standard laboratory-grade water for many years. Figure 4-5 shows a Barnstead classic still. Water is boiled in a still, leaving nonvolatile salts in the boiler. The steam passes through a condensing coil, and the condensed distilled water is collected in a container, ready to use. One Barnstead still (product number A440266, MP1), which costs about US $1200, has been referred to as the "tissue culture still." It produces 1.4 liters (a little less than half a gallon) of distilled water per hour. A less-expensive (US $700) compact glass still capable of similar production is available from Carolina Biological Supply Company (70 1699). Sometimes it is necessary to run water through a glass still twice to achieve the desired purity.

Deionization is another effective way to remove dissolved chemicals from water. A single mixed resin bed tank, or cartridge, can provide adequate purity for normal tissue culture production. The system is connected to existing water lines and provides a continuous flow of deionized water. Water passes through a mixture of positive and negative ion-exchange resins. The undesirable ions from the incoming water are exchanged for H^+ and OH^- ions on the resins. These H^+ and OH^- ions combine to form water (H_2O), while the undesirable ions are left on the resins. If the feed water is relatively free of impurities, the resin beds (or cartridges) should work efficiently for several months. Resin bed exchangers are usually equipped with a warning light to indicate when the resins are spent. The light is illuminated when the conductivity level of the product water is becoming too high. If the conductivity of the product water is too high that indicates the resins are saturated with undesirable ions. Companies that sell pure water and water purification equipment rent and sell mixed resin tanks, which they will regenerate. They also sell disposable cartridges, but these can be expensive.

Various deionizing systems that use disposable cartridges are available from scientific supply companies (Figure 4-6). One such system, Barnstead's Bantam deionizer, offers a demineralizer, which includes a stand to hold the cartridges, a conductivity meter with indicator, and a connecting hose, all for about US $1000. Disposable cartridges run about US $125.

Reverse osmosis, a third method of water puri-

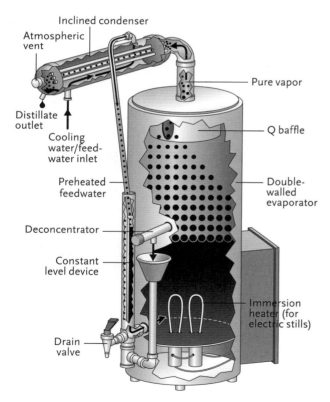

Figure 4-5. Barnstead classic still.

fication, is usually used in combination with a deionizer or still. It does not have the refinement of the other 2 methods, but it provides excellent pretreatment. Osmosis is the diffusion of a fluid through a membrane into a solution of higher ion concentration to equalize concentrations on both sides of the membrane. Reverse osmosis forces the solution through a membrane in the opposite direction, filtering out the impurities. Raw feed water is run through the reverse osmosis system to remove most of the impurities prior to flowing through a deionizer or entering a still. The efficiency of the subsequent still or deionizer is increased significantly when the water first goes through reverse osmosis. Reverse osmosis systems are initially expensive, costing over US $1000, but combined with a still or deionizer, and given the right feed water (nonacidic, low in calcium and iron), extra-high-grade water (0.06 µS) can be obtained. Manufacturers claim that the cost of replacing deionizing cartridges is reduced by a factor of 8 when using reverse osmosis pretreatment. This piece of equipment is a high investment and may not be necessary.

pH meter

The acidity or alkalinity (pH) of media is crucial in tissue culture and is specific to the requirements of specific plants, just as it is in soils and potting mixes. For example, azaleas and rhododendrons (*Rhododendron*), mountain laurel (*Kalmia*), and blueberries (*Vaccinium*) are tissue cultured in a relatively acidic medium (pH 4.5–5.0), whereas strawberries (*Fragaria*) and most other plants require a less-acidic medium (pH 5.7–5.8).

Any laboratory should invest in a pH meter. A pH meter costs from US $50 and up, while a high-quality meter costs about US $750. A beginner, however, can get by simply with pH indicator paper. Recommended pH ranges for test papers are 2.9–5.2, 4.9–6.9, and 5.5–8.0. These small ranges will measure adjustments of media to approximate pH requirements.

Balance

Unless premixed chemicals are to be used, a precision balance will be required to measure accurately some of the small amounts of the chemicals required for tissue culture media, although the accuracy required of analytical chemists is not usually required in plant tissue culture. Most analytical balances cost over US $1000. These expensive, electronic one-pan balances are fast and precise and are a worthwhile investment.

Portable balances are a less-expensive option. For example, a portable balance with 0.01-gram accuracy, such as Ohaus Scout Pro Balance (H-2709), is available for about US $400. Though less precise than an analytical balance, it is adequate for weighing quantities of 0.01 to 10 g (10 to 10,000 mg).

The dilution method is one alternative for weighing smaller amounts. By this method, if, for example, 10 mg of a compound are required per liter of medium, place 100 mg (0.1 g) of the compound in 100 milliliters (ml) of water. Stir it well to be sure the compound is properly dissolved in the solution, and then use 10 ml of this

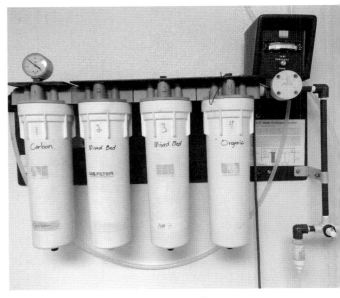

Figure 4-6. A series of cartridges in a Milli-Q water station removes contaminants from tap water, yielding pure water for cell culture.

solution in a liter of medium. The remaining solution can be refrigerated for future use.

When weighing amounts of more than 10 g, use a triple-beam balance, saving the more sensitive balance for the smaller quantities. A good triple-beam balance such as the Ohaus Dial-O-Gram, costs about US $200.

Hot plate/stirrer

Agar, a gelatinous polysaccharide extracted from certain red algae, is required as a solidifying agent in many media formulas. In media preparation, agar in powder form is added to the water together with the other chemicals. This mixture must be boiled briefly to melt the agar and dissolve the chemicals to form a homogeneous mixture. Agar melts at boiling temperature 212°F (100°C), so unless the mixture is stirred constantly and effectively until boiling, the agar will settle and stick to the bottom of a flask and burn. For that reason, a combination hot plate and automatic stirrer is among the most useful tools for preparing media. A rotating magnet built into the hot plate/stirrer causes a magnetic stir bar, placed inside the flask, to rotate and so prevent sticking (Figure 4-7). The stirrer and hot plate features can be used at the same time or independently. The stir bar should be used in heat-resistant glass containers on the hot plate; metal containers generally will not work on a hot plate/stirrer because the stir bar is magnetic. An additional and very important use for the automatic stirrer (without the hot plate feature) is to provide agitation during the cleaning of explant material. Small hot plate/stirrers start around US $300.

A gas burner or an electric stove will heat media much faster than a small hot plate, so it is more efficient for larger quantities of media. If you use a gas burner, you will need a large cooking vessel, possibly a canner, and a means of stirring. An electric kitchen beater is an inexpensive option. It can be mounted on a board so that it does not have to be hand held. A preferred but more costly option is to buy a top-mounted motor and stirring rod, which costs about US $500. Larger hot plate/stirrers work well and cost about US $1200.

A hand-held stirring rod, beater, or spoon will also work for stirring media, but there is a greater likelihood of the agar sticking, unless the solution is in a double boiler (the pan of agar set within a pan of water). There is also a greater hazard of being burned should a hot medium splatter. The time saved by using a stirring device is worth the money spent!

There is another option for melting agar that is especially useful when making larger amounts of medium. To make 3 or more liters of medium, place a flask containing agar and 1 liter of water in a microwave oven, stirring occasionally. When the agar is melted, add the microwaved solution to the medium, which has already been mixed and heated except for the agar. Be sure to allow room for the agar solution in the 3 liters of medium.

Media dispenser

A small laboratory can easily forego high-priced equipment for dispensing media. A 10-ml polypropylene pipet can be purchased for about US $16. A simple heat-resistant Pyrex pitcher

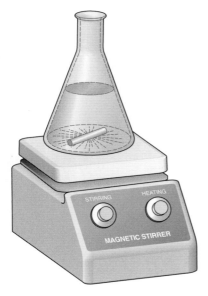

Figure 4-7. A hot plate/stirrer, showing the rotating magnetic bar in a flask of medium.

will serve very well to pour a few liters of a hot medium into test tubes or jars, and it costs only US $12. A coffee urn can also sometimes be used, or use a reservoir for media that is gravity fed through tubing, with a pinch clamp to control the flow.

An automatic pipetter to dispense media is a laborsaving device for large-scale production. Whereas a good technician with a quick eye and a steady hand can pour 100 test tubes in 10 minutes from a 1-liter pitcher, the automatic dispenser will fill 500 tubes in the same amount of time. An automatic pipetter costs about US $1100.

Explant cleaning equipment

A hot plate/stirrer (with the heat turned off) usually provides sufficient agitation for cleaning explants; however, a tightly closed jar containing explants and cleaning solution, shaken by hand, is a very effective method as well. Alternatively, a mechanical shaker or rotator can be built or purchased and will provide effective agitation over a long time.

A few laboratories use vacuum pumps to help disinfect explants. The suction of the vacuum (25 millimeters of mercury) improves contact between the cleaning solution and the explant and helps to disengage contaminants. A vacuum pump or aspirator is also useful for ultra-filtration of chemicals that are unstable when heated. The need for a vacuum pump, or aspirator pump, should be established before investing in one because they are cumbersome to use, and too high a vacuum can injure explants.

An ultrasonic cleaner is another option to help remove contaminants. The action of the ultrasonic cleaner causes different materials to vibrate at different rates, thus dissociating them. The contaminants are literally shaken loose from the plant material. A small ultrasonic cleaner costs about US $100. Here again, caution is advised or the tissues will be injured.

Sterilizing equipment for media

An autoclave or pressure cooker is a vital part of a tissue culture laboratory because spores from fungi and bacteria will only be killed at a temperature of 121°F (49°C) and 15 pounds of pressure per square inch (psi) (103 kilopascals; kPA). A modest laboratory with no more than 3 people can readily use a household pressure cooker (Figure 4-8) for sterilizing test tubes or jars containing media. A 21-liter (5.5-gallon) pressure cooker costs less than US $100. A burner or stovetop unit is necessary for heating the cooker. A wire basket (Figure 4-8) to hold the test tubes must be built or purchased when using such a cooker. The basket consists of a wire frame with handles, lined with 0.5-in. (13-mm) hardware cloth (wire mesh) to prevent the test tubes from falling out.

Autoclaves, which are common devices for sterilizing in tissue culture, come in many sizes (Figure 4-9). Most are automated, more efficient, and more accurate than pressure cookers. A larger operation should have an autoclave, but unless the size and automation of the autoclave appear to provide significant savings in labor, you may as

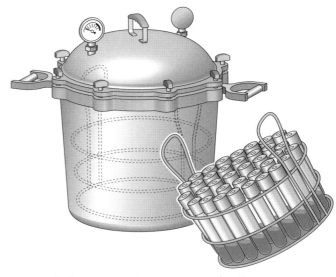

Figure 4-8. A pressure cooker or canner and a wire basket containing test tubes. The basket and test tubes are slanted after sterilization so that the surface of the medium will be slanted when it cools and solidifies.

well use pressure cookers or a stovetop autoclave (US $260). Tabletop autoclaves start around US $3500 (8 liter); large volume autoclaves start at US $10,000. The size of the autoclave needed can be calculated by estimating the number of containers to be processed in a day, how many containers will fit in a particular autoclave, and the length of the total processing time (heating, holding, and shutting down) for that autoclave. Self-generating steam autoclaves are very dependable and faster to operate than autoclaves that require an external steam source. Most autoclaves have options for fast or slow exhaust. Anytime a liquid is being autoclaved, the autoclave should be set at slow exhaust to prevent the liquids from boiling out of their vessels. Nonliquid items like forceps, scalpels, and aluminum foil can be autoclaved at fast exhaust.

Some organic chemicals are heat labile (unstable) and will break down when exposed to the high temperatures of certain sterilization procedures. Therefore, some media ingredients must be sterilized by cold sterilization using a special syringe and filter. The problem of heat lability varies with the chemical, the temperature, and the duration for which the temperature is applied. Cold sterilization is accomplished in the sterile environment of the transfer hood. A solution of the chemical is loaded into a syringe, and a sterile membrane filter unit is then attached to the syringe. The solution is forced through the membrane filter into a sterile container, ready to be added to a sterilized medium, whether liquid or agar, which has been cooled but not solidified so that the additive can still be mixed in.

Some people cold sterilize hormones and vitamins before adding them to sterilized media, but this is usually not necessary. The major application of cold sterilization is in the use of antibiotics, most of which are very heat labile. Whatever sterilization method is used, whether hot or cold, it is important not to over-sterilize the medium.

Sterilizing equipment for transfer tools

In the past, tools used in the transfer hood primarily were sterilized by dipping them in alcohol, followed by flaming (burning off the alcohol) using a Bunsen burner or alcohol lamp. Alcohol used for disinfecting instruments in this manner should be 95 percent ethyl or isopropyl alcohol, because at that strength it will burn off more easily than will a more dilute form. Do not use methyl alcohol because it is extremely poisonous, even just to breathe the fumes.

The Touch-O-Matic burner is a modern modification of the old familiar Bunsen burner and costs about US $150. It saves gas and minimizes the danger of an open flame. A pinpoint pilot flame burns continuously and the higher flame is lit by resting one's hand on a disc while

Figure 4-9. A large-volume, freestanding steam autoclave.

flaming an instrument. A primitive alcohol lamp can also be improvised by using a covered container and a wick, or it can be purchased for about US $9. These burners all involve open flames, which are hazardous especially where alcohol is used for disinfection. They are less safe, but also less expensive, than some modern sterilizing equipment.

For sterilizing implements in the transfer hood, the preference today is the glass bead sterilizer, an insulated pot with a component that heats glass beads within it (Figure 4-10). A glass bead sterilizer costs about US $415.

Another less desirable choice than a glass bead sterilizer is an infrared sterilizer such as the Bacti-Cinerator (Figure 4-11). Serious finger burns can happen if forceps and scalpels are left inside the machine for too long. The Bacti-Cinerator costs about US $400, and replacement heater elements, which may be needed once or twice a year, cost about US $70. The Bacti-Cinerator sterilizes instruments in 5 seconds at 1600°F (870°C) as the instruments are inserted into the red-hot, hollow cone of the cylinder. The glass bead sterilizers are similarly heated.

Bleach solutions are satisfactory for sterilizing tools in some situations using standard household bleach consisting of 5.4% sodium hypochlorite (check the label because some solutions are 6%). Two concentrations of bleach solutions should be used with this method: a 10% solution (1 part bleach in 10 parts solution, as in 10 ml of bleach mixed with 90 ml of water; also referred to as 1:10 bleach) to soak the instruments and a 1% solution (1 part bleach in 100 parts solution, or 1:100 bleach) to rinse them. This is the method used in the procedures described in this book because it is the least expensive option and serves as a good teaching tool. Remember, however, that bleach will burn the skin and the fumes are hazardous and undesirable.

Rotator or shaker

Some cultures grow better in liquid media than on solid agar media. Such cultures often need to be gently agitated, which aerates the medium and thereby prevents the culture from "drowning." Agitation also disperses the waste products of the culture. Cultures can be gently agitated by means of a rotator (Figure 4-12) at one rotation per minute (rpm), or by a shaker. Shakers are motorized to provide a back-and-forth motion.

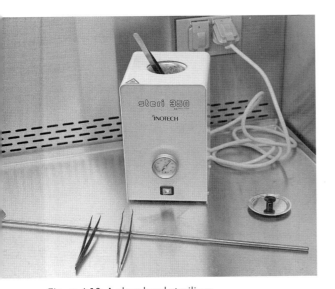

Figure 4-10. A glass bead sterilizer.

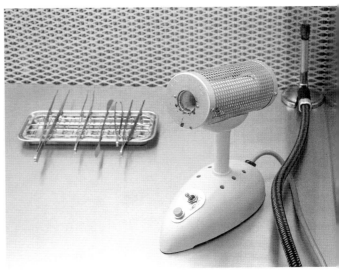

Figure 4-11. A Bacti-Cinerator sterilizer and Bunsen burner.

Agitation is particularly useful if you are attempting to grow callus tissue. The mixing disorients the tissue and, in so doing, inhibits plantlet formation while allowing for increased callus production. A rotator or shaker is also useful for cleaning explants, especially for any overnight treatment. Tabletop orbital shakers with platforms, variable speed, and timers start around US $1200, depending on the size and whether racks for holding test tubes or flasks are purchased as well (Figure 4-13).

Dishwasher

A conventional built-in household dishwasher, costing about US $500, will handle the dishwashing requirements of a small laboratory (fewer than 2000 test tubes per day). If the prongs on the bottom rack of the dishwasher are bent down, 4 test tube racks each holding 40 test tubes—160 test tubes total—can be washed at one time. Tubes in racks must be covered with wire mesh and turned upside-down in the dishwasher. A commercial laboratory dishwasher of the same capacity costs at least 4 times as much as a kitchen dishwasher and yields few advantages. However, a large commercial washer should be considered when the volume of glassware used warrants such a purchase.

Figure 4-12. A CEL-GRO tissue culture rotator.

Refrigerator

A household refrigerator or small tabletop refrigerator is necessary to store perishable chemicals and stock solutions. A larger refrigerator is required if media or cultures are to be refrigerated as well. Food and beverages should not be stored inside laboratory refrigerators; not only is this not safe, but this is not allowed by government mandate.

A walk-in refrigerator is a valuable means of treating or delaying the growth of cold-resistant cultures. Chilling is often called for if plantlets are ready too soon to be transferred to the greenhouse, or if stock cultures need to be slowed down or stored. Cultures of some hardwood species benefit from cold treatment before acclimatization.

An expansion coil with fan, compressor, and temperature controller are basic off-the-shelf components that can be sized to your needs. A 24×24 in. (60×60 cm) exchanger (expansion coil) and a 0.75-horsepower compressor will refrigerate an insulated room $10 \times 10 \times 7$ ft. ($3 \times 3 \times 2.1$ m).

Figure 4-13. An orbital shaker with adjustable flask holders.

Labeler

Different varieties of plants are not easily identifiable in culture, and it is much more difficult to distinguish one culture from another than it is to tell one plant from another in the greenhouse. It is extremely important, therefore, to identify each container at the time it is processed. A practical labeling procedure is essential, but most systems do have their drawbacks. Although marking pencils will work, the labeling process is slow and any moisture on the glass or the pencil makes writing difficult. If using a marking pen with permanent waterproof ink, the ink must be scrubbed off with a pot scrubber before the container can be reused. If using a marking pen with ink that washes off (as in an overhead-projector marker), the ink may rub off prematurely. Hand stamps are also a problem due to permanent ink. Both markers and hand stamps are too slow for large-scale commercial production. Grocery-store price labelers prove satisfactory in most cases. The least expensive kinds usually cost between US $100 and $200. If labels come off in the dishwasher, however, the dishwasher may clog. There is no one solution to fit every situation. To save time, some laboratories will label entire trays of similar plants with the plant name, ID code (if appropriate), date, and technician's initials.

Microscope

The decision to buy a microscope depends on the type of tissue culture work to be done, as well as on the interest and curiosity of the technician. A stereo dissecting microscope (provides a 3-dimensional image), also called a low-power microscope, is usually required for excising meristems (see chapter 8). A microscope with 20× or 40× magnification costs about US $400. Less expensive magnifiers are available and may meet your particular needs. Some people with exceptionally good eyesight may not need a magnifier at all for excising meristems. A compound microscope (provides a 2-dimensional image and is the classic form of microscope used in high school biology classes) is useless for excising meristems because there is not enough working room between the objective and the specimen.

A compound microscope is not necessary for routine commercial production, but one is essential if you plan to identify contaminants (see chapter 11). Student monocular compound microscopes are available for about US $500. Keep in mind that the less expensive the microscope, the more likely it will have poorer resolution or sharpness. Resolution can be defined as the ability of a microscope lens system to give individuality to 2 objects (for example, 2 bacterial cells) that lie in close proximity to one another. For more information on microscopes, see Appendix B.

Transfer chamber or hood

Laminar airflow transfer hoods are essential for commercial operations (Figure 4-14). They provide a sterile atmosphere in which to work with cultures. Air is forced through a HEPA filter, located at the back of the hood, which strains out particles as small as 0.3 micrometers (μm). The gentle, scarcely detectable air stream flows through the filter, across the work area toward the technician at 100 ft. (30 m) per minute, providing a sterile atmosphere in which the technician works. This piece of equipment also helps to filter the air in the transfer room.

A popular benchtop laminar airflow hood kit for the hobbyist, consisting of all the necessary components with which to assemble a hood, is available through Fungi Perfecti for US $525, and one that is wide enough for 2 technicians is available for US $1500. Other laboratory supply companies sell assorted designs of laminar flow hoods from $3000 up. Some larger units that set on the floor have the option of air intake either at floor level or at the top of the unit. The unit that has air intake on top rather than at floor level is preferred because the air at floor level is likely to have more contaminants than the air nearer the ceiling.

HEPA filters can be purchased separately from

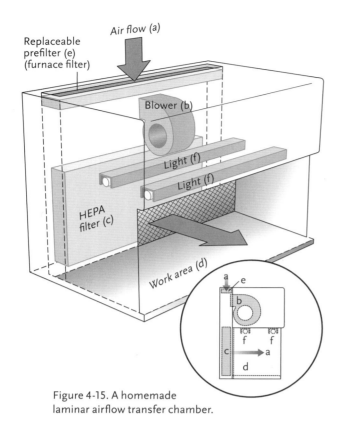

Figure 4-15. A homemade laminar airflow transfer chamber.

the hood. A 24 × 36 in. (60 × 90 cm) HEPA filter costs less than US $300; a 12 × 12 in. (30 × 30 cm) filter costs about US $90.

It is possible to construct a satisfactory laminar flow hood, or have a local cabinet shop make one (Figure 4-15). Additional components necessary for building a transfer hood include prefilters (inexpensive furnace filters), a blower, plywood or acrylic plastic sides and top, a smooth bench top, and a fluorescent light.

A still-air transfer chamber with a slanted glass front and partially enclosed hand access is an alternative for the hobbyist, especially if home built. Some hobbyists have worked successfully with large clear plastic bags. Specially designed, commercially available bags into which sterile air can be blown come with built-in sleeves. Atmos-Bags from Sigma-Aldrich cost about US $45 for a medium size, 39 in. wide by 48 in. long with a 24-in. opening diameter (98 × 120 cm with a

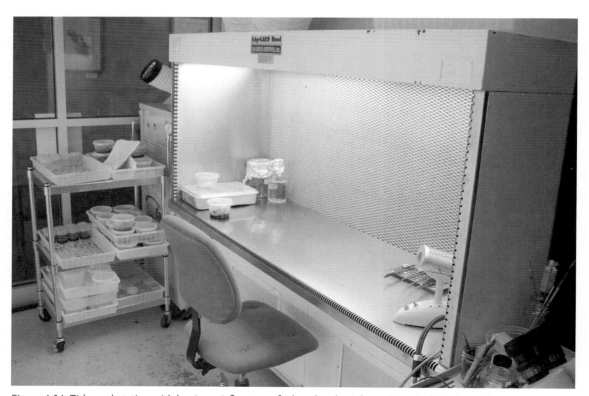

Figure 4-14. Tidy workstation with laminar airflow transfer hood at the Atlanta Botanical Garden in Georgia.

60-cm opening diameter) before inflating, or about 10 ft.³ (280 liters) when inflated.

Growing room shelves

Lighted shelves are needed in the growing room to hold the test tubes or jars of cultures while they are growing. The shelving for light units varies for laboratories depending on the individual needs, budget, and the plants that are being grown. Wood shelving units are inexpensive and easy to build; they should be painted white with an easy-to-clean paint to reflect the room's light. Slotted steel angle supports can be used to build 4 × 8 ft. (1.2 × 2.4 m) shelves of particleboard, plywood, or wire mesh of expanded metal, or 0.25- to 0.5-in. (6–13 mm) hardware cloth. Wire mesh is more expensive than solid shelving but allows better air circulation. Expanded metal is more expensive than wood but provides better air circulation. Tempered glass shelves increase light penetration, but are expensive, decrease air circulation, and are breakable. A room 8 ft. (2.4 m) high will accommodate 5 shelves placed 18 in. (45 cm) apart, with the bottom shelf 4 in. (10 cm) from the floor (this distance underneath provides room for cleaning). One rack, consisting of particleboard shelves, metal framing, and lights, can be purchased for about US $700.

Additional supplies

Basic laboratory equipment includes a variety of glassware and stainless steel tools (Figure 4-16), though today some of these items are made of plastic. Household items often can substitute for traditional glassware and tools used in the laboratory. Such items are appropriate if they are heat and pressure resistant, to allow for sterilization. Heat-resistant glass beakers are good vessels for cleaning explants. Aluminum utensils are undesirable because they can give off aluminum ions into the medium. Remember that there are different kinds of plastic that are available; some can be autoclaved and some cannot. Plastics made of polyethylene cannot be autoclaved,

but polypropylene and polycarbonate can be subjected to high temperatures. None of these plastics can be placed on a hot plate or stove without melting.

Readily available glass jars such as pint-size canning jars and empty baby food jars are perfect for growing plant tissue cultures (Figure 4-17). They are inexpensive, convenient, autoclavable,

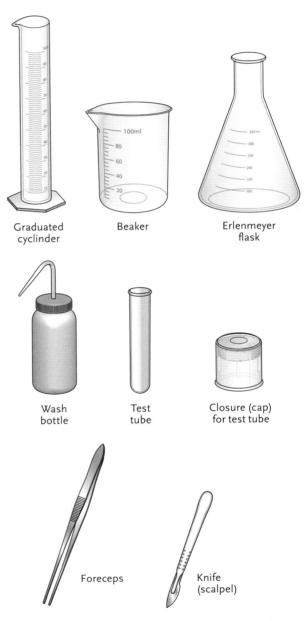

Figure 4-16. Typical laboratory glassware and tools.

and easy to handle. The typical metal lids do not allow sufficient light for the cultures, but there is usually sufficient light entering the sides of the jars for adequate plant growth. If more light is desired, the metal lids can be replaced with rounds of plate glass (generally too expensive and cumbersome for this purpose) or polycarbonate, an autoclavable plastic that is available in 20-mil (0.05-mm) sheets and suitable for cutting rounds to fit the tops of pint jars. For media preparation, the metal rings that come with canning jars can be used to hold the rounds in place. When transferring cultures, remove the ring and the polycarbonate round and place the polycarbonate lid on a sterile paper towel. After positioning the plantlets in the jar, replace the polycarbonate lid and place a square of plastic wrap or a small sandwich bag over the top, and then screw on the metal ring over the wrap. Plastic closures for pint-size jars and baby food jars, made especially for tissue culture purposes, are also available commercially, and can be a good investment, depending on sanitation levels and frequency of transfer.

Autoclavable polycarbonate culture vessels and dishes come in a host of shapes and sizes including space-saving square containers and round, squat Petri dishes. Culture ("test") tubes are made of heat-proof borosilicate glass. Closures for test tubes are affordable, autoclavable polypropylene lids that minimize evaporation while allowing gas exchange. They are available in different colors for easy identification and in a variety of sizes. The closures do not offer maximum protection against contaminants entering the culture tube so they are often wrapped with plastic paraffin film or other film barriers. For the ultimate protection, culture tube closures can be lined on the top end with a 0.25-in. (6-mm) thick

Figure 4-17. A variety of easy-to-find jars that are useful for tissue culture.

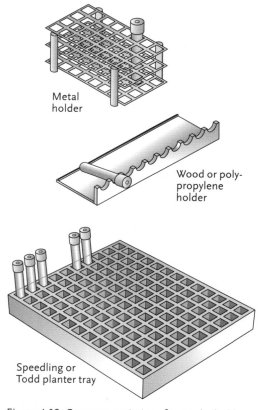

Figure 4-18. Common varieties of test tube holders.

piece of nonabsorbent cotton. Absorbent cotton should not be used because it will take up moisture, thus making it more conducive to contamination. Rather than wrapping individual culture vessels, entire trays can be enclosed in a sterile plastic bag.

Aluminum vertical or horizontal instrument holders or rests can be purchased for less than US $20. They allow tools to cool after sterilization and also hold them at a convenient angle, within easy reach. A metal test tube holder standing on end also works. Do not buy a plastic-coated holder because it will melt with the heat from the instruments. Another option is to make an instrument holder out of glass tubing by heating it and bending it to shape.

Test tube racks are available in a variety of materials (Figure 4-18). A zinc-plated, steel wire rack measuring about 10.5 × 4.5 × 3 in. (26 × 11 × 7.5 cm) and holding 40 test tubes costs about US $30. This holder is almost essential for convenience in washing, storing, and other handling of test tubes. A smaller 10-tube holder is convenient in the transfer chamber or when handling small numbers of tubes. This type of holder can be built with plywood. A plastic Speedling planter tray, with 128 cells for starting plants from seed, can be used as a holder in the growing room, where it allows the test tubes to be slanted, thus permitting exposure to more light than if the test tubes are upright, as they would be in the typical wire-rack test tube holder.

Use and Care of Equipment

Remember that laboratory precision instruments are delicate and demand very careful use—not like tractors or refrigerators, which can stand a certain amount of abuse. Consequently, it is of utmost importance that you read and carefully follow the instructions provided by the equipment manufacturers before you touch a new instrument. If questions arise, direct them to the vendor or the manufacturer. It is in their best interest to have informed, satisfied customers, and it is in your best interest and to your advantage to know how to operate and care for your investment.

Supply List

Most of the suggested supplies listed in Table 2 (see next page) are for a modest 3-person operation culturing about 200,000 plants per year. Your individual selections will depend on your budget and your goals—some supplies you may choose to forego altogether and for others you may prefer to use homemade alternatives. Some of the items listed are available from supermarkets. Most of the items can be purchased from scientific supply companies); a business license may be required to purchase from some suppliers.

Table 2. **Suggested equipment and supplies for a small commercial laboratory**

ITEM	SIZE	NUMBER	TOTAL COST (IN $US)*	USE
Alcohol, isopropyl or ethyl, 70%	1 pint	2	2.50	Clean explants, work surfaces
Aluminum foil, heavy	25-foot roll	1	3.00	Cover items for sterilizing
Autoclave	10 × 18 inches	1	5000.00	Sterilize media
Balance, electronic (Ohaus ScoutPro)	n/a	1	387.75	Weigh small amounts
Balance, triple beam	n/a	1	150.00	Weigh amounts over 10 grams
Beaker (Pyrex)	150 milliliters	6	22.92	Clean explants
Beaker	250 milliliters	6	21.72	Mix chemicals
Beaker	600 milliliters	6	34.20	Miscellaneous dispensing
Bleach (5% sodium hypochlorite)	1 gallon	1	2.00	Clean explants, instruments, and so forth
Bottle, spray	8 ounces	2	5.00	Mist for Stage IV culture
Bottle, wash	125 milliliters	6	58.50	Rinse pH probe
Brush, flask	16 inches	1	8.65	Clean flasks
Brush, test tube	10 inches	12	27.98	Clean test tubes
Conductivity meter	n/a	1	400.00	Check ion content of water
Cotton, nonabsorbent	1 pound	1	14.50	Line top of lids
Cylinder, graduated	10 milliliters	1	5.99	Measure liquids
Cylinder, graduated	100 milliliters	1	5.92	Measure liquids
Detergent, liquid	1 quart	1	1.50	Miscellaneous cleaning
Detergent, powder	1 pound	4	3.00	Dishwasher
Dish, petri, glass	100 millimeters	12	39.84	Microdissection; research
Dish, plastic, glass, or ceramic	2 × 6 × 9 inches	2	6.00	Hold bleach and tools in hood
Dish, sterile, disposable	100 millimeters	500	84.75	Microdissection; callus, fungus growth
Dishwasher	n/a	1	500.00	Wash glassware
Dropper, medicine	1 ounce	12	3.25	Dispense small liquid amounts; adjust pH
Filter, furnace	3 feet × 20 feet × 1 inch	1	14.00/roll	Prefilters for transfer hood
Flask, Erlenmeyer	1000 milliliters	6	53.70	Mix, store stock solutions
Flask, Erlenmeyer	2000 milliliters	1	20.65	Mix media
Flask, Erlenmeyer	4000 milliliters	1	126.73	Mix media
Forceps, stainless	10 inches	2	28.00	Transfer cultures
Forceps, stainless steel (tweezers)	4.75 inches	1	5.25	Microdissection
Gloves, surgical	any	100	12.00	Protect hands
Hot plate/stirrer	n/a	1	260.00	Heat, mix media; clean explants
Jar and lids, baby-food	6 ounces	100	120.00	Grow cultures
Jar, canning (Mason)	1 pint	120	50.00	Grow cultures
Jar holder	n/a	1	3.00	Move hot jars
Knife, plastic handle	6 inches	4	4.00	Divide cultures
Labeler, grocery-store type	n/a	1	100.00	Label cultures
Light meter	n/a	1	90.00	Measure light distribution, intensity
Magenta containers	77 millimeters	100	179.00	Grow cultures

ITEM	SIZE	NUMBER	TOTAL COST (IN $US)*	USE
Microscope, stereo	n/a	1	400.00	Observe meristems for excision
Mop, sponge	n/a	1	11.00	Clean floors
Pens, marking, waterproof	n/a	1	1.00	Permanent label
Pens, marking, not waterproof	n/a	2	3.00	Temporary label
pH meter, handheld, portable	n/a	1	153.75	Measure pH
Pipet, volumetric, polystyrene	1 milliliter	25	11.75	Miscellaneous dispensing
Pipet, volumetric, polystyrene	10 milliliters	25	15.75	Miscellaneous dispensing
Pipet filler	n/a	2	9.00	Pipet safety
Pitcher, Pyrex	1 quart	1	6.00	Dispense media
Polycarbonate sheets	20 mil, 2 × 4 feet	1	7.80	Lids for jars
Pot holder, kitchen	6 × 6 inches	4	8.00	Hold hot items
Pressure cooker	n/a	1	100.00	Sterilize media
Refrigerator	n/a	1	500.00	Store organics, cultures, and other supplies
Scalpel, stainless	8.5 inches	4	74.00	Dissection and transfer
Scoop or spoon	2 ounces	2	1.00	Scoop compounds for weighing
Shaker	n/a	1	425.00	Agitate liquid media; clean explants
Spatula, stainless	200 millimeters	3	11.00	Scoop compounds for weighing
Sponge, household	medium	4	2.00	General cleaning
Sterilizer, Bacti-Cinerator	n/a	1	231.00	Sterilize tools in hood
Sterilizer, glass bead	n/a	1	525.75	Sterilize tools in hood
Sterilizer, Touch-O-Matic	n/a	1	281.75	Sterilize tools in hood
Still, water	n/a	1	500.00	Purify water
Syringe, B-D sterile	10 milliliters	100	20.45	Cold sterilization
Syringe filter, 50-unit pkg.	0.22 microgram	1	154.00	Cold sterilization
Test tubes (culture tubes)	25 × 150 millimeters	125	63.75	Grow explant starts
Test tube closures	25 millimeters	125	34.75	Lids for test tubes
Test tube rack, plywood	holds 10 tubes	2	8.00	Hold tubes in hood
Test tube rack, wire	holds 40 tubes	2	59.50	Hold tubes for washing, storing
Thermometer, digital	−20°C to 110°C	2	21.50	General purpose
Timer, household	60 minutes	2	26.00	Time cookers
Timer/controller	24 hours	1	75.00	Control lights in growing room
Towels, paper	roll	6	3.00	Dry hands, spills
Towels, paper, single-fold	9.5 × 10.25 inches	4000	30.00	Cut cultures on
Tray, household	10 × 16 inches	8	24.00	Hold jars
Tray, seedling, 128 cell	11 × 21.25 × 2 inches	100	70.00	Hold test tubes; plantlets in Stage IV
Tray, Speedling, 128 cell	n/a	6	30.00	Hold test tubes in growing room
Wastebasket, large	n/a	1	6.00	Media room
Wastebasket, medium	n/a	1	4.00	Transfer room
Weighing papers (500/pkg)	3 × 3 inches	1 package	13.75	Hold chemicals for weighing

*Provided by Gary Seckinger of *Phyto*Technology Laboratories. Prices are in 2012 U.S. dollars and are subject to change.

5 Media Ingredients

The tissue culture medium (plural media) is to the micropropagated plantlets as the soil is to the trees and shrubs that grow in it. The medium supplies vital nutrients and water, just as the soil does, but in addition, the medium supplies sugars to substitute for photosynthesis and plant growth regulators to control the type of growth. The medium, along with ideal temperatures, optimum light levels, and favorable photoperiods, provides the perfect milieu for plant growth.

Tissue culture media combine a long list of inorganic minerals, sugars, vitamins, plant growth regulators, and gelling agents into a substrate designed to support and encourage plant growth (Figure 5-1). The standard formulas for tissue culture media were originally derived from hydroponic systems and have been determined by research scientists to provide optimum nutrients and growth regulators for specific plants. The formulas developed by Toshio Murashige and his associates, particularly the one known as Murashige and Skoog medium (MS), are probably the best-known standard formulas and are used primarily for herbaceous foliage plants. The woody plant medium (WPM) developed by Brent McCown and Greg Lloyd is designed, as the name implies, to optimize tissue culture of certain woody plants.

Combining the various chemicals for the media in which plant cultures grow is an art, similar to that of cooking a meal—the ingredients and language are different, and some of the equipment may not be familiar, but recipes (formulas) usually need to be followed in a precise, orderly manner. The nutrients in media affect the health, vigor, and growth of the specialized cells, tissues, and organs as the plantlets differentiate and develop in culture.

When plant tissue culture first began, laboratories were required to mix complex stock solutions and then prepare growing media from these stock solutions. Stock solutions are concentrated chemicals or groups of chemicals from which portions will be added to the final medium. Now, as an alternative to mixing the numerous chemicals, ready-mixed media in either dry powder or liquid concentrate forms are available in various stages of completeness. All that is necessary is the ability to follow a "recipe." Several supply companies now offer various media and media

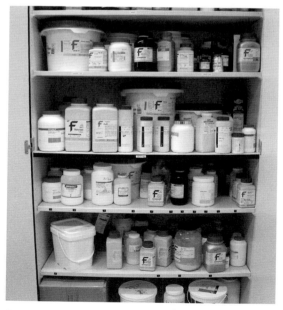

Figure 5-1. The well-stocked media cabinet of a tissue culture laboratory.

components such as salts and vitamins, including all the necessary components, to prepare a medium. A wide array of species-specific formulations is available for plants from African violet to Venus flytrap. One supplier lists more than 25 different media for orchid propagation. The benefits of premixed are obvious: there is no need to keep a large inventory of components, no component will accidentally be left out of a medium, no stock solution is sitting around, and the possibility of measurement errors is greatly reduced. Nutrient-deficient formulations are also available, allowing for experimentation with one or more macro- or micronutrients. If working with a wide variety of species, it also possible to mix formulas "from scratch" because it is less expensive and it is easier to adjust the formula in response to culture performance.

Many people who were afraid or bored in high school chemistry worry about the complexity of preparing tissue culture media. In the field of micropropagation, however, the elements take on a completely new dimension as they relate and translate directly to plant response—a dramatic and exciting process. Read or skim this chapter for what it can mean to you. If a cabinet full of unfamiliar ingredients is too overwhelming, then begin by purchasing premixed media. If the chemistry review is meaningful to you, accept the challenge and put your curiosity to work because the whole world of micropropagation is open to you. As a gardener or plant propagator, you will be in somewhat familiar territory because you have grown plants, used fertilizers, and probably used hormones to root cuttings. With a can-do attitude, you will learn to enjoy the successes and to avoid the failures.

Chemistry Review

Without at least a beginning course in chemistry, it may take a while to become familiarized with the chemical language, symbols, and working principles of mixing media. Initially, the number of chemicals may seem excessive. If you are a novice, however, by starting with a commonplace plant for which a medium formula has been well researched and applied, the task becomes relatively simple. A cookbook approach fortified with determination will help the unsure beginner gain confidence.

Whether mixing the media yourself or using premixed media, it is useful to know something about chemicals and their role in plant growth and development. The information presented here will give beginners an appreciation of the art and science of media formulation. The expert may choose to scan this section for review. Many of the terms and concepts used by chemists are household words; however, a brief review of definitions is desirable to strengthen the foundation for formula mixing.

The metric system

The metric system is the system of measurements used in science, and consequently, it is used in tissue culture. Measurement calculations are simpler than in the English system because the metric system is based on multiples of 10. Committing to memory certain relationships and terms will make the metric system easier for the individual trained in the English system. Here are a few crucial ones: there are 1000 milligrams (mg) in 1 gram (g), 1000 milliliters (ml) in 1 liter, 10 millimeters (mm) in 1 centimeter (cm), and 1000 micrometers (µm; formerly called microns) in 1 mm.

Another important equivalency is 1 cubic centimeter (cm^3) of water has a volume of 1 milliliter (ml) and weighs 1 gram (g). Another easy conversion to remember is that 1% is equal to 10 grams per liter. For comparison with some English system measurements, 1 gram is equivalent to about 0.035 ounces (oz.), 1 liter is about 1.1 quart (qt.), and 1 centimeter is about 0.4 inches (in.). For other metric conversions, see Appendix A. There are a number of online metric converters, such as the one at sciencemadesimple.com.

Significant figures in weights and measures

For weighing chemicals, it is useful to have an understanding of the concept of significant figures. Simply stated, when weighing chemicals, the larger the total amount required, the less important precision and accuracy become. One milligram is an insignificant amount when considering 100 grams; but when 10 milligrams are required, 1 milligram becomes a significant factor, amounting to 10% of the whole. This concept of degree of importance is referred to as significant figures, and it applies to both weights and volumes. Three significant figures in the final measure is adequate for production work; in other words, if, for example, a formula calls for 184.4 g of a compound, 184 g (3 figures) is sufficiently accurate, and 15.6 mg is satisfactory if 15.57 mg are required. This point is being made because often more than 3 figures are given in the scientific literature. People involved in research need to be extremely precise, but people in production usually need not be quite so accurate (Figure 5-2).

Figure 5-2. Measurements to 3 significant figures are adequate for production work. The measurement here could be rounded off to 0.123.

Remember this when adjusting formulas and the calculator spits out 8 decimals.

Some laboratory glassware, particularly beakers and Erlenmeyer flasks, carry the caution that the calibrations on the side are accurate only to within plus or minus 5%. Accuracy within 5% is usually sufficient for our purposes. For the most accurate liquid measurements, volumetric flasks are used, but these are also the most expensive pieces of glassware. After the volumetric flask, the graduated cylinder is the next most accurate. Beakers can be very inaccurate for precise measurements that are needed for stock solutions, but are sufficient for day-to-day use.

Basic units of chemistry

An *atom* is the smallest particle of an element that retains the chemical characteristics of that element. The number of atoms in a chemical formula is always written as a subscript, below and to the right of the letter symbol, as in H_2 or O_2.

An *element* is a substance composed of one type of atom that cannot be separated (decomposed) into simpler substances by any usual chemical means. Oxygen (O_2), copper (Cu), and mercury (Hg) are examples of elements.

A *compound* consists of 2 or more different elements chemically combined in fixed proportions. Some examples of compounds include water (H_2O), composed of hydrogen and oxygen; hydrochloric acid (HCl), which is hydrogen and chlorine; and sucrose ($C_{12}H_{22}O_{11}$), composed of carbon, hydrogen, and oxygen.

A *molecule* is the smallest quantity into which a chemical compound can be divided and still keep its characteristic properties. The simplest molecules have only one atom, as in the case of nitrogen (N), phosphorus (P), or potassium (K). One molecule of water (H_2O) has 3 atoms: 2 hydrogen and one oxygen.

An *ion* is an electrically charged atom or group of atoms. For example, table salt, which is sodium chloride (NaCl), ionizes (chemically separates) in water to form positively charged sodium ions (Na$^+$) and negatively charged chloride ions (Cl$^-$). When water containing calcium carbonate (CaCO$_3$) is deionized for purification, the positive calcium ions (Ca^{2+}) are removed by being attracted to a negatively charged resin bed and the negative carbonate ions (CO^{2-}) are attracted by a positively charged resin bed.

Atomic and molecular weights

The occasional need to use atomic and molecular weights in tissue culture warrants an introduction to the topic. Atoms of different elements have different weights. An atomic weight (relative atomic mass) of an element from a specified source is the ratio of the average mass per atom of the element to one-twelfth of the mass of an atom of ^{12}C (carbon-12; see Table 3). The sum of the atomic weights of all the molecules in a substance is the molecular weight (MW) of a compound.

Sometimes media formulas found in the literature give quantities in molar units rather than in grams per liter, and so an understanding of atomic and molecular weights will help to determine the grams per liter equivalent and ensure correct measurements. A one molar (1 M) solution is 1 gram molecular weight (the molecular weight of a substance expressed in grams; also called a mole) in 1 liter of water. Take, for example, a 1 M solution of sodium hydroxide (NaOH), a compound often used for raising the pH of media. In Table 3 we see that the atomic weight of sodium (Na) is 23, that of oxygen (O) is 16, and that of hydrogen (H) is one; the molecular weight of NaOH is thus 23 + 16 + 1 = 40. Therefore, 40 g of NaOH in 1 liter of water yields a 1 M solution of NaOH. To be precise when measuring, the water plus the NaOH must equal a liter. Some organic chemicals are very complex, and it is difficult to determine their molecular weight. Fortunately, molecular weights of compounds can be found in catalogs of some chemical companies (such as Sigma-Aldrich); molecular calculators can be found online (see webqc.org/mmcalc.php) or in books (see Donnelly and Vidaver 1988).

Another instance in which an understanding of atomic weights is useful is when calculating the water of hydration. Water of hydration is the variable amount of water that is chemically attached to some compounds. It can be anhydrous (no H$_2$O), monohydrate (H$_2$O), dihydrate (2H$_2$O), and so on. The compound manganese sulfate (MnSO$_4$, MW = 151), for example, is manufactured with either 4 molecules of water (MnSO$_4$ · 4H$_2$O, MW = 223) or as the monohydrate (MnSO$_4$ · H$_2$O, MW = 169). If a medium formula calls for 2.23 grams of MnSO$_4$ · H$_2$O, then 1.69 grams of MnSO$_4$ · H$_2$O would provide an equivalent amount of MnSO$_4$ when using the monohydrate form of the compound (169 ÷ 223 × 2.23 = 1.69). Such ratios can be used to calculate other equivalents.

Table 3. **Atomic weights of elements commonly used in tissue culture**

ELEMENT	SYMBOL	ATOMIC WEIGHT
Boron	B	10.80
Calcium	Ca	40.08
Carbon	C	12.00
Chlorine	Cl	35.44
Cobalt	Co	58.93
Copper	Cu	63.55
Hydrogen	H	1.007
Iodine	I	126.90
Iron	Fe	55.85
Magnesium	Mg	24.31
Manganese	Mn	54.94
Molybdenum	Mo	95.96
Nitrogen	N	14.00
Oxygen	O	15.99
Phosphorus	P	30.97
Potassium	K	39.10
Sodium	Na	22.99
Sulfur	S	32.05
Zinc	Zn	65.38

Source: Data from IUPAC Periodic Table of the Elements.

Media formulas published in technical journals are sometimes specified in millimoles (mmol; 0.001 gram molecular weight) per liter or micromoles (µmol) per liter instead of grams per liter. In this case, it is necessary to find the molecular weight of the compound and multiply the moles required by the molecular weight. For example, if 3 mmol of calcium chloride ($CaCl_2$) are required per liter of medium, multiplying the molecular weight of $CaCl_2$ (110) by the number of millimoles required will give the required amount of $CaCl_2$ per liter in milligram (mg) units: 110 × 3 mmol = 330 mg. The advantage of expressing media ingredients in moles is that molecules and ions interact as entities; thus, comparing moles instead of weights provides a more valid basis of comparison of one formula with another.

Principles of acids and bases: Solutions for pH adjustment

The symbol pH designates the degree of acidity or alkalinity of a solution as indicated by the concentration of hydrogen ions; the higher the hydrogen ion concentration, the greater the acidity. Different plant species require different pH levels in media for optimum growth.

A solution is neutral at pH 7, alkaline above pH 7, and acid below pH 7. The numbers are exponential: a solution of pH 6 has 10 times the hydrogen ion (H^+) content as a solution registering pH 7, a pH 5 solution has 10 times as many H^+ ions as a pH 6 solution, and so on. In the opposite direction, a solution at pH 8 has 10 times as many hydroxyl (OH^-) ions as a solution at pH 7, etc. The typical pH range for tissue culture media is between pH 4.5 and pH 5.7 (Figure 5-3).

Medium pH can affect the uptake of components, as well as enzyme-catalyzed chemical reactions. Some genera, such as blueberry (*Vaccinium*) and *Rhododendron*, which perform best in relatively acidic soil, require a media pH in the lower range (4.5 to 5.0). In general, plant response is not too sensitive to changes in pH over a small range. Occasionally, however, a medium that has grown plant material for a period may develop a change in pH detrimental to the culture. In such cases, the culture should be transferred to a fresh medium, and the old medium should be checked to see if the pH might have been a cause of the culture's decline. If a significant change in pH has occurred, it is a sign that the culture should be transferred more often.

When making tissue culture media, use either potassium hydroxide (KOH) or sodium hydroxide (NaOH) solution to raise the pH of media, thus making the media more alkaline when they are too acidic. The KOH solution contains potassium (K^+) ions and hydroxyl (OH^-) ions. The NaOH solution contains sodium and OH^- ions. The OH^- ions of either solution combine with the excess H^+ ions in the medium to form water (H_2O), thus neutralizing the H^+ ions and causing the medium to be more basic (alkaline). Both KOH and NaOH can be purchased in powder form or pre-mixed as a 1 M (sold as a 1 Normal, abbreviated 1N) solution. For KOH, NaOH, and HCl, their normality and molarity are the same. Potassium hydroxide is suitable in some situations, as it does not add additional sodium to the media.

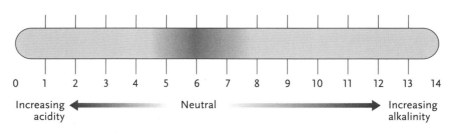

Figure 5-3. pH scale.

When mixing a stock solution of KOH or NaOH, do not handle either of these chemicals without using a forceps, spoon, or spatula as both are corrosive and will burn your skin. To make approximately a 1 M solution of NaOH, add 4 g of NaOH to 100 ml of water (40g/liter). Use great care when making this solution because when NaOH and water mix, the reaction is violent and it may spatter. Never add water to NaOH; always add the NaOH to the water to minimize the caustic spatter that can injure eyes, skin, or anything it touches. You can also purchase a ready-made 1 M solution of NaOH. For NaOH and HCl, the normality and the molarity are the same.

Hydrochloric acid (HCl) solutions are used for lowering the pH of media, making them more acid when they are too alkaline. The HCl solution contains hydrogen (H^+) ions and chloride (Cl^-) ions. The H^+ ions of the HCl solution combine with the excess OH^- ions in the medium to form water, thus neutralizing the OH^- ions and causing the medium to be more acidic. To lower the pH of media, use a 1 M solution of HCl. Rather than risking the dangers of mixing HCl, it is easiest to buy a 1N solution of HCl already prepared. More dilute solutions of HCl are nice to have when adjusting the media pH and can easily be made. A 0.1N (10%) solution is made by mixing 10 ml of the 1N HCl with 90 ml water and a 0.01N (1%) solution is made by placing 1 ml of the 1N HCl with 99 ml water. Then, if the pH of the medium only needs to be lowered slightly, the 1% solution can be used. This is much less aggravating than using a 1N solution.

The amount of potassium hydroxide (KOH), sodium hydroxide (NaOH), or hydrochloric acid (HCl) that is required to change the pH one unit varies considerably, depending on the nature of the solution and at what point on the scale the change is made. The use of these solutions will be further considered in the discussion on making media (see chapter 6). More terms and concepts will be defined as we proceed, but the foregoing will be useful in the following discussion of various important chemicals. Not all the chemicals listed are used regularly in all media. Plants that are growing in culture manufacture many of the chemicals that they need to grow, and different plants have different requirements.

Inorganic Chemicals

Inorganic chemicals and organic chemicals have different structural elements. Most inorganic chemicals have high melting points; few will burn; many are soluble in water; they conduct electricity in water; most are insoluble in organic solvents; and reactions involving inorganic chemicals tend to be very fast.

Growers will recognize immediately those essential elements that all plants require and so will not be surprised to find them in tissue culture media. Six of these elements are major constituents in common fertilizers: nitrogen (N), phosphorus (P), potassium (K), calcium (Ca), magnesium (Mg), and sulfur (S). The 3 remaining major elements—carbon (C), hydrogen (H), and oxygen (O)—are discussed under organic chemicals, in which they play a major role.

It is important here to recall the theory of limiting factors, discussed previously in chapter 3. German physical chemist Justus von Liebig theorized that if one essential element is missing, then all others present are of little or no value. The mobility of an element within a plant can offer clues when attempting to identify deficiencies. Mobile elements are translocated within the plant from the older tissue to new growing points; thus, the older leaves will show deficiency symptoms first. An immobile element cannot be translocated from old tissue to new; when such an element is deficient in the soil or media, the effects will be seen in new shoots and leaves or root tips.

Macronutrients

Nitrogen (N) influences the rate of plant growth and is present in the largest percentage (dry weight basis) of all the essential elements—1%–4%. It is an essential element in the molecular make-up of nucleic acids, proteins, chlorophyll, amino acids, alkaloids, and some plant hormones. As nitrogen is highly mobile within the plant, a deficiency is characterized first by uniformly chlorotic leaves followed by stunted growth. Excess nitrogen promotes vigorous growth but can suppress flower and fruit development. Common sources of nitrogen for tissue culture media are ammonium (NH_4) and nitrate (NO_3) compounds.

Phosphorus (P) is abundant in meristematic and other fast-growing tissues. As part of DNA and ATP (adenosine triphosphate) molecules, phosphorus is an essential element in photosynthesis and respiration, and it affects plant maturation and root growth. Though phosphorus is mobile within the plant, deficiency symptoms can be difficult to detect. Stunted growth and reddish to purple coloring (anthocyanin) can be symptoms of phosphorus deficiency. Potassium phosphate (KH_2PO_4) and sodium phosphate ($NaH_2PO_4 \cdot H_2O$) are routinely included in tissue culture media.

Potassium (K) is necessary for normal cell division and promotes meristematic growth. It plays a role in many reactions within plants, although it is not an actual component of plant protoplasm, fats, or carbohydrates. It helps in synthesizing carbohydrates and proteins, manufacturing chlorophyll, and reducing nitrates. Insufficient potassium results in weak and abnormal plants, sometimes with mottled, curled, or necrotic margins on older leaves. Potassium nitrate (KNO_3) and potassium phosphate (KH_2PO_4) are common sources of potassium in culture media; potassium chloride (KCl) is used occasionally.

Calcium (Ca), in the form of calcium pectate, is an integral component of plant cell walls, where it plays a role in the formation of pectin, a substance that bonds cell walls together. It helps to control permeability and facilitates the movement of carbohydrates and amino acids throughout the plant; it also promotes root development.

It assists in growth and development, as well as in nitrogen assimilation. As calcium oxalate, it ties up oxalic acid, which is a toxic by-product of protein metabolism. Calcium is immobile within the plant and can only be taken up as the plant transpires. Distorted or dead shoot or root tips can be a sign of a lack of calcium. Calcium is usually included in tissue culture media as calcium chloride ($CaCl_2 \cdot 2H_2O$) or as calcium nitrate ($CaNO_3 \cdot 4H_2O$); calcium phosphate (tribasic; $Ca_3(PO_4)_2$) is also sometimes included. Because calcium easily precipitates out of stock solutions, the source used most often is $CaCl_2 \cdot 2H_2O$.

Magnesium (Mg) is the central element in chlorophyll molecules. It is also important as an enzyme activator. Magnesium is a mobile element within the plant and can be translocated to new growth. The older leaves will exhibit interveinal chlorosis. Most tissue culture formulas call for magnesium sulfate ($MgSO_4 \cdot 7H_2O$), commonly known as Epsom salts.

Sulfur (S) is present in some proteins. It promotes root development and deep green foliage. Sulfur is somewhat mobile in plants, and a deficiency results in overall yellowing of the plant. Deficiencies are rare, however, as sulfur is supplied in tissue culture media in conjunction with other elements as various sulfate (SO_4) compounds.

Micronutrients

In addition to the major elements that plants require, a number of other elements are essential to good growth but are needed only in extremely small quantities. They are called minor elements, trace elements, or micronutrients. As the purity of the water and the chemicals used in tissue culture media became more refined with the advances in technology, deficiency symptoms appeared in the media. This occurred because certain trace elements that had previously been supplied as undetected impurities in presumably pure water and chemicals were in fact necessary.

Trace elements are present in soil, water, and

even dust particles in adequate amounts to affect plant growth. Some "chemically pure" compounds that are used in media may contain traces of these elements. If so, they are usually listed as impurities on the label. Several micronutrients are toxic to plants in excess amounts. Diagnosis of deficiency symptoms for these elements is difficult. Metals such as copper, zinc, and iron, can compete with each other for uptake; for others there is a fine line between deficiency and toxicity. The mobility of these elements within the plant varies; most are immobile, with deficiency symptoms first expressed in new tissue, but symptoms are highly variable among species.

Boron (B) is an important micronutrient presumed to play a role in the movement of sugar, water, and hormones. It is also involved with nitrogen metabolism, fruiting, and cell division. Lack of boron produces interesting deficiency symptoms—often a deterioration of internal tissues, as in heart rot of sugar beets (*Beta vulgaris*), cracked stem of celery (*Apium graveolens*), or monkey face in olives (*Olea europaea*). Other symptoms include tip dieback, or thick, curled, brittle, chlorotic leaves. Excessive amounts of boron can cause plant injury or death; thus, some herbicides are borates. Boron is added to tissue culture media in small amounts as boric acid (H_3BO_3).

Chlorine (Cl) helps stimulate photosynthesis and seems to be necessary for growth. Deficiency symptoms are wilted leaves that become yellowed or bronze and die. Plants require chlorine in only minute quantities, but it is included in some tissue culture media in large amounts as calcium chloride ($CaCl_2 \cdot 2H_2O$). These amounts appear to be tolerated by most plants, but some scientists advise caution when including chlorine compounds in some formulas, choosing to get the calcium from sources other than calcium chloride.

Copper (Cu) deficiency results in stunted growth, malformations, twisted and blotched leaves, or dieback of young twigs. Copper is believed to be necessary in energy conversion as it alternates between the cuprous (monovalent) and the cupric (divalent) state. It is involved with chlorophyll synthesis and is found in some enzymes. Only 0.025 mg of cupric sulfate ($CuSO_4 \cdot 5H_2O$) per liter of medium is required to supply the necessary copper in tissue culture of most plants.

Iron (Fe) is involved in chlorophyll synthesis. It also participates in energy conversion in photosynthesis (the process by which carbon dioxide and water, with the help of chlorophyll and light, are converted into carbohydrates and oxygen is released) and respiration as it is reduced from the ferric (trivalent) to the ferrous (divalent) state. As iron is immobile and cannot be translocated from older leaves, plants deficient in iron have yellowed (chlorotic) young leaves, especially between veins (interveinal chlorosis), giving a striped appearance in monocots. In tissue culture media, ferrous sulfate ($FeSO_4 \cdot 7H_2O$) is often mixed with the sodium salt of ethylenediaminetetraacetic acid (Na_2EDTA; $C_{10}H_{16}N_2O_8Na_2 \cdot H_2O$) to sequester the iron (chelation), thereby making it more readily available to the plants. When stock solutions are made for tissue culture media, iron is usually made as a separate stock because of its tendency to precipitate out of solution. Care and attention are necessary when making stock solutions of iron.

Manganese (Mn) deficiency is characterized by mottled yellowing of leaves. This mineral is an essential element in chloroplast membrane. Manganese sulfate ($MnSO_4 \cdot H_2O$) supplies the necessary manganese for tissue culture media.

Molybdenum (Mo) is believed to help convert nitrogen to ammonia and aids in nitrogen fixation. It is required for normal growth and protein synthesis. Its absence is suspect when leaves yellow between veins. Lack of molybdenum will cause whiptail (narrow leaf) condition in cauliflower (*Brassica oleracea* var. *botrytis*), or it can stunt the growth of legumes (family Fabaceae). It is more available to plants in alkaline soils than in acid soils. Molybdenum is added to tissue culture media as sodium molybdate ($Na_2MoO_4 \cdot 2H_2O$), also known as molybdic acid sodium salt. Quantities exceeding 10 parts per million can be injurious to plants.

Zinc (Zn) is an enzyme activator involved in chlorophyll formation, as well as in the production of the auxin indole-3-acetic acid (IAA). Without zinc, roots may be abnormal and leaves may turn mottled bronze or yellow and misshapen. A trace of zinc is included in tissue culture media as zinc sulfate ($ZnSO_4 \cdot 7H_2O$). Large quantities of zinc are toxic to plants.

Other trace elements

Iodine (I) is often added to media as potassium iodide (KI). It is not usually considered an essential element, even though it is a component of some amino acids. Iodine appears to have an adverse affect on *Rhododendron* cultures, so it should be omitted from that medium.

Cobalt (Co) is an element in the complex vitamin B_{12} molecule and is essential to nitrogen fixation (the conversion of atmospheric nitrogen to nitrates by means of nitrogen-fixing bacteria). Cobalt chloride ($CoCl_2 \cdot 6H_2O$) is added to most media in amounts of 0.025 mg per liter.

Organic Chemicals

In biology, the most important molecules are organic molecules. Organic molecules contain at least one carbon atom, which is bonded to another carbon atom or to a hydrogen atom, the 2 atoms each providing half of the electrons. Examples of organic compounds include carbohydrates, hormones, proteins, and enzymes. They can be gases, liquids, or solids with low melting points of less than 680°F (360°C). They are mostly insoluble in water, and they do not conduct electricity in water. Most organic compounds can burn and they usually react slowly.

Plants normally manufacture their own organic compounds, so they rarely need to be fed to outdoor or greenhouse plants. However, plants in culture cannot synthesize all the organic chemicals they need, so organic substances must be added to tissue culture media to augment the plantlets' self-generated (autotrophic) supply.

Carbohydrates

Carbohydrates include such organic chemicals as sugars, starches, and cellulose. In varying amounts and configurations, carbon (C), hydrogen (H), and oxygen (O) are the primary elements that make up the molecules of carbohydrate compounds. These elements are generously supplied as carbon dioxide (CO_2) in air and as water (H_2O).

Sucrose ($C_{12}H_{22}O_{11}$), a disaccharide, is a common carbohydrate found in abundance in plant tissue and the carbohydrate of choice for media preparation. A polysaccharide is a chain of one or more simple sugars linked together in its chemical structure. Sucrose consists of 2 chemically bound monosaccharides, fructose ($C_6H_{12}O_6$) and glucose ($C_6H_{12}O_6$), and is an indirect product of photosynthesis. Plants growing in culture cannot manufacture all the sugar they require to grow, so a high concentration of sucrose, generally 30 g per liter or 3%, is suggested for most media formulas and 20 g per liter (2%) is the second most recommended.

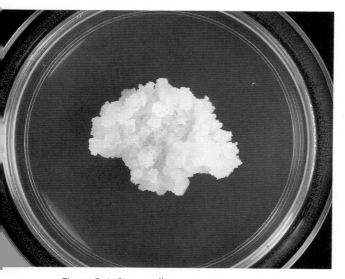

Figure 5-4. Citrus cells growing in a standard MS medium.

Sugar from sugar cane (*Saccharum officinarum*) or from sugar beet (*Beta vulgaris*), each virtually 100% pure sucrose as purchased from the grocery store, is good for most tissue culture media. Occasionally fructose or glucose is substituted for sugar in media.

D-Mannitol is a sugar alcohol used as a nutrient and osmoticum (osmosis control), particularly when inducing plant protoplast formation and fusion.

D-Sorbitol is a sugar alcohol that is the primary translocatable (able to move through phloem and xylem tissue) carbohydrate in some plants. It is occasionally added to media.

Vitamins

The vitamin B complex contains essential compounds for plant metabolism and growth. The growth substances found in yeast extract, which was commonly used in culture media in the past, are now identified as thiamine (vitamin B_1), nicotinic acid (niacin, or vitamin B_3), and pyridoxine (vitamin B_6), all of which are members of the vitamin B complex.

Adenine (vitamin B_4 or 6-aminopurine; $C_5H_5N_5$) is important to cells as part of the nuclear substances (DNA and RNA). It has a weak cytokinin effect. Adenine is used in culture media as adenine sulfate ($AdSO_4$; $[C_5H_5N_5]_2 \cdot H_2SO_4 \cdot 2H_2O$) to promote shoot formation.

D-Biotin (a B vitamin or vitamin H; $C_{10}H_{16}N_2O_3S$) is important in fat, protein, and carbohydrate metabolism. It is commonly used in media.

Choline ($C_5H_{15}NO_2$) is an alkaloid (a compound with alkaline properties) within the vitamin B complex. It occurs naturally in lecithin, which is chemically related to fats but also contains phosphorus and nitrogen. Choline chloride ($C_5H_{14}NOCl$) is occasionally specified in media formulas.

Cyanocobalamin (vitamin B_{12}; $C_{63}H_{88}CoN_{14}O_{14}P$) is sometimes added to culture media as a possible growth promoter.

Folic acid (vitamin Bc or vitamin M; $C_{19}H_{19}N_7O_6$) is found in leaves and other plant tissues. It functions as a B vitamin and demonstrates coenzyme (an organic molecule associated with enzymes) activity.

Myo-inositol ($C_6H_{12}O_6$), a sugar alcohol in the B complex, is required in many media. In its phosphate form, it is part of various membranes, particularly those of organelles such as chloroplast. Myo-inositol, though not essential to growth, is often beneficial in stimulating cell growth when added to tissue culture media at 50–5000 mg per liter (100 mg per liter is average).

Nicotinic acid (niacin or vitamin B_3; $C_6H_5NO_2$) is a component of coenzymes active in light-energy reactions. Various media require nicotinic acid, usually in amounts ranging from 0.1 to 10 mg per liter.

PABA (para-aminobenzoic acid, vitamin Bx; $C_7H_7NO_2$), occasionally used in culture media, serves as an antiseptic and a preservative. It also plays a role in folic acid metabolism.

D-Pantothenic acid (vitamin B_5; $C_9H_{16}NO_5$) is active as a coenzyme in fat metabolism. It is added as the calcium salt ($C_9H_{16}NO_5 \cdot 2Ca$) to some plant tissue culture media. It should be cold sterilized.

Pyridoxine (vitamin B_6; $C_8H_{11}NO_3$) also serves as a coenzyme in some metabolic pathways (chemical reactions of metabolism). It is usually included in culture media as the hydrochloride form ($C_8H_{11}NO_3 \cdot HCl$).

Riboflavin (vitamin B_2 or vitamin G; $C_{17}H_{20}N_4O_6$) is active in carbohydrate metabolism and is essential to cellular respiration.

Thiamine (vitamin B_1; $C_{12}H_{17}ClN_4OS$) is essential for most culture media because it functions as a coenzyme to assist the organic acid cycle of respiration (known as citric acid cycle or Krebs cycle). Concentrations ranging from 0.1 to 10.0 mg of thiamine hydrochloride per liter are specified for most tissue culture media.

L-Ascorbic acid (vitamin C; $C_6H_8O_6$) has some

disinfectant qualities, but its chief use in tissue culture is as an antioxidant to prevent phenolic oxidation (browning of plants that contain phenolics). It should not be used for extended periods because it can become an oxidant itself.

(+)-α-Tocopherol (vitamin E; $C_{29}H_{50}O_2$) is occasionally used in culture media. It is known to promote dispersion in suspension cultures of ladino clover (*Trifolium repens* 'Royal Ladino') and soybean (*Glycine max*) (Oswald et al. 1977). Later studies showed that cell aggregation was increased when vitamin E was removed from the medium.

Plant growth regulators

Plant growth regulators (PGRs), or phytohormones, are not nutrients; they are chemicals that are used to alter the growth of a plant or plant part. Many are endogenous hormones that are produced naturally in plants. Plants that are growing *in vitro*, however, usually do not manufacture sufficient quantities of growth regulators, so they must be added selectively to culture media. Auxins, cytokinins, gibberellins, abscisic acid (ABA), and ethylene are among the substances used as hormones or growth regulators. The auxins and cytokinins are used most frequently in media. In general, auxins promote cell enlargement and root initiation, whereas cytokinins promote cell division and shoot initiation. This is a simplistic summary in view of the diversity of growth regulators and their effects.

There is a wide range of interactions between auxins and cytokinins. The ratio of auxin to cytokinin is manipulated *in vitro* to influence shoot and root growth. The 2 also interact with other chemicals and are affected by environmental factors, such as light and temperature. Under some conditions, an auxin might even react as a cytokinin and a cytokinin as an auxin. It is important that tissue culture media contain the right kinds of hormones and in the correct ratio for each variety of plant being cultured. Occasionally, cultures become habituated, a state in which they lose the need for the hormones in the medium or simply do not respond to them.

Auxins are phytohormones that influence cell enlargement, root initiation, and adventitious bud formation. They induce apical dominance and suppress the initiation of lateral buds. Auxins are commonly used in tissue culture media, either combined with cytokinins during the multiplication stage (Stage II) or without cytokinins for the rooting stage (Stage III). Auxins are also utilized in somatic embryogenesis. Dissolve auxins in a few drops of a 1 M solution of either sodium hydroxide (NaOH) or potassium hydroxide (KOH) before adding it to media.

IAA (indole-3-acetic acid; $C_{10}H_9NO_2$) occurs naturally in plants. It is best known to plant propagators as a dip to promote rooting of cuttings, and it is often used for rooting in tissue culture media. It is unstable in light, so it should be stored in darkness.

IBA (indole-3-butyric acid; $C_{12}H_{13}NO_2$) is another rooting hormone commonly used by propagators. It is produced naturally in plants and also made synthetically. More stable than IAA, IBA is the preferred hormone for root induction in tissue culture.

NAA (1-naphthaleneacetic acid; $C_{12}H_{10}O_2$) is a synthetic auxin-type root-inducing compound that is sometimes used in media, especially to promote callus growth.

2,4-D (2,4-dichlorophenoxyacetic acid; $C_8H_6Cl_2O_3$) is a synthetic auxin known primarily as a weed killer. It has been widely used in plant tissue culture media to induce callus growth.

Cytokinins are growth regulators that are required in tissue culture media for cell division, shoot multiplication, and axillary bud proliferation. They help delay senescence (aging), and they influence auxin transport. If cultures are too spindly, increased cytokinin will help foster shorter, stouter stems. Cytokinins dissolve in about 1 M hydrochloric acid (HCl). They are usually omitted from media for the rooting stage (Stage III), but carryover effects from Stage II can occur, including reduced rooting and survival (Kane et al. 2008) and increased branching when plants are grown *ex vitro* after Stage IV. Although not a cytokinin, adenine sulfate (see discussion under "Vitamins,") has been shown to have cytokinin-like stimulatory effects on plant tissue cultures and is used in some media for enhanced Stage II growth.

BA (6-benzyladenine; $C_{12}H_9N_5O$) or BAP (6-benzylaminopurine; $C_{12}H_{11}N_5$) is used in culture media to promote axillary bud growth. It is a synthetic plant growth regulator and is probably the most common cytokinin used. 6-Benzylaminopurine riboside ($C_{17}H_{19}N_5O_4$) can be used as an alternative to BA.

2iP (IPA; 6-[g, g-dimethylallylamino]purine or N6-[2-isopentenyl]adenine; $C_{10}H_{13}N_5$) is commonly used in plant tissue culture media. With some plants, like azaleas and rhododendrons (*Rhododendron*) and mountain laurel (*Kalmia latifolia*), 2iP is much better to use than BA. It is made synthetically, but it has been found in RNA and in a pathogenic bacterium (*Corynebacterium fasscians*). It causes rapid cell division and consequent irregular growth in some higher plants.

Kinetin (KIN; 6-furfurylaminopurine; $C_{10}H_9N_5O$) induces cell division when auxin is also present. It is frequently used in plant tissue culture media to promote callus formation and shoot proliferation.

Zeatin (6-[4-hydroxy-3-methylbut-2-enlyamino]purine; $C_{10}H_{13}N_5O$) was first discovered in corn seed endosperm. It is a common alternative cytokinin to 2iP or BA for use in media.

Thidiazuron (TDZ; 1-Phenyl-3-(1,2,3-thidiazol-5-yl)urea) is among the most active cytokinin-like substances for woody plant tissue culture. It facilitates efficient micropropagation of many recalcitrant woody species. Very low concentrations (<1 μm) can induce greater axillary proliferation than many other cytokinins; however, TDZ may inhibit shoot elongation.

BPA or PBA (N-benzyl-9-[2-tetrahydropyranyl] adenine or 6-benzylamino-9-[2-tetrahydropyranyl]-9H-purine; $C_{17}H_{19}N_5O$) is a synthetic cytokinin occasionally used in cultures for axillary bud and callus proliferation.

Gibberellins are a group of naturally occurring substances that influence cell enlargement and stem elongation. Kurosawa noted in 1926 that secretions from a fungus (*Gibberella fujikuroi*) resulted in abnormally rapid growth in rice (*Oryza*) seedlings. The substance was gibberellic acid (GA_3; $C_{19}H_{22}O_6$), which was later isolated in crystalline state from both fungi and higher plants. Thirty-four gibberellins have been chemically identified. Some of them appear in embryos, where they initiate production of the enzyme alpha amylase, which converts starches to sugars and stimulates other enzymes. GA_3 is sometimes used as a growth regulator to supplement auxins and cytokinins especially when elongation of plantlets is desired.

Other growth-regulating substances. ABA (abscisic acid; $C_{15}H_{20}O_4$), a naturally occurring compound in plants, plays both inhibitory and stimulatory roles *in vitro*, depending on the concentration used and/or the species. With whole plants, ABA controls stomatal closure, bud dormancy, and seed dormancy. *In vitro*, it is sometimes used to assist with somatic embryogenesis.

Ancymidol ($C_{15}H_{16}N_2O_2$), available commercially as A-Rest, has a growth-retardant, cytokinin effect in tissue culture. In micropropagation of *Hosta*, ancymidol added in Stage II culture reduced leaf size, shifting growth to more desirable shoot buds (Maki et al. 2005). Ancymidol is a restricted use pesticide in some states and may require a pesticide applicator's license to use.

Brassinosteroids (BRs) are a fairly recently identified group of plant hormones. Both plant-extracted and synthetic analogues have been used *in vitro* to enhance shoot regeneration and increase cell division in callus.

CPA (para-chlorophenoxyacetic acid; $C_8H_7ClO_3$) is an auxin-type growth regulator.

Ethylene (C_2H_4) is a gas produced naturally by plants. It has growth regulatory properties and is involved with fruit ripening, flowering, and leaf abscission. The understanding and manipulation of ethylene's effect on morphogenesis and its interaction with other phytohormones in plant tissue culture is expanding (Kumar, Lakshmanan et al. 1998). Its build-up in culture vessels can be detrimental, however. Ethylene production is greater than normal in vitrified cultures (Kevers and Gaspar 1985). (Vitrification, also known as hyperhydration, is a phenomenon in which tissues develop a glassy, swollen appearance.)

Jasmonates, especially jasmonic acid (JA; $C_{12}H_{18}O_3$), have been identified as signaling hormones in plant response to wounding. The understanding and use of jasmonates to regulate plant growth is an evolving field (Avanci et al. 2010).

Paclobutrazol (PBZ; DL-1,2,4-triazole-3-alanine) is useful as an addition to Stage III media to help minimize the need for acclimatization (Oliphant 1990, E. F. Smith et al. 1992). This growth-retardant substance has a cytokinin effect in many tissue cultures.

Phloroglucinol (1,3,5-trihydroxybenzene) is sometimes used as an antioxidant (R. H. Smith 1992) and in the treatment of vitrification (Phan and Hagadus 1986). It is also somewhat bactericidal and sometimes promotes culture growth (Donnelly and Vidaver 1988).

Salicylic acid (2-hydroxybenzoic acid), which is produced within the plant upon injury or contact with a pathogen, regulates systemic acquired resistance (SAR). Researchers have reported various effects *in vitro*, including enhanced salt/osmotic stress tolerance.

TIBA (2,3,5-triiodobenzoic acid), an antiauxin, inhibits auxin movement and may be growth promoting in culture. It also may inhibit unwanted callus.

Amino acids

Amino acids are building blocks of proteins, some of which combine with nucleic acids to form nucleoproteins. Although many more amino acids can be synthesized in laboratories, only 20 essential amino acids are found in nature. These are all in the L chemical form. Several are particularly pertinent to cell and protoplast culture. Others are only occasionally called for in certain media.

L-Alanine ($C_3H_7NO_2$) has helped to increase the number of embryos in some cell cultures.

L-Arginine ($C_6H_{14}N_4O_2$) reportedly has assisted root initiation.

L-Asparagine ($C_4H_8N_2O_3$) is sometimes used in cell culture of soybean (*Glycine max*). When combined with proline, l-asparagine and certain other amino acids assist in embryogenesis in alfalfa (*Medicago sativa*).

L-Cysteine ($C_3H_7NO_2S$), a sulfur-containing amino acid, is occasionally called for in media formulas. It is added to media as the hydrochloride ($C_3H_7NO_2S \cdot HCl$).

L-Glutamine (L-2-aminoglutaramic acid; $C_5H_{10}N_2O_3$) can contribute to somatic embryogenesis.

Glycine (aminoacetic acid; $C_2H_5NO_2$) is another amino acid specified in some media. It is sometimes used in begonia media and has been used in a medium for poinsettia cell culture.

L-Lysine (L-2,6-diaminohexanoic acid; $C_6H_{14}N_2O_2$) combined with proline has enhanced the quality and number of embryos produced.

L-Proline ($C_5H_9NO_2$) is often used in combination with other amino acids for embryo culture.

L-Serine ($C_3H_7NO_3$) has been used in microspore (pollen) culture to grow haploid embryos.

L-Tyrosine (L-3-[4-hydroxyphenyl]alanine; $C_9H_{11}NO_3$) is effective in shoot initiation and is useful as a nitrogen source.

Antibiotics

An antibiotic is a substance produced by plants or microorganisms (or made synthetically) that has a toxic effect on other microorganisms by retarding or preventing the growth of such microorganisms, or by killing them. Antibiotics are not used routinely against contaminants in micropropagation because often they are ineffective, kill the culture, or induce chromosomal instability. Furthermore, it is recommended that antibiotics not be used on a routine basis because of the dangers that exist from the build-up of resistance occurring in the plants and in the people who handle these chemicals. However, some antibiotics have been useful on occasion. More commonly, they are used in gene-transfer experiments to control or eliminate *Agrobacterium* (R. H. Smith 1992).

Ampicillin has been used in amounts of 100 to 400 mg per liter in culture of *Petunia* protoplasts (Dixon and Gonzales 1994).

Carbenicillin at 500 mg per liter and augmentin at 250 mg per liter are most useful against gram-negative bacteria. Both antibiotics were used advantageously in some genetic-engineering experiments for controlling *Agrobacterium* (R. H. Smith 1992).

Cefotaxime has been used to eliminate certain gram-negative bacteria in woody plants. It has been used at 25 mg per liter in combination with tetracycline (at 6 mg per liter), rifampicin (at 6 mg per liter) dissolved in dimethyl sulfoxide (DMSO; C_2H_6OS), and polymyxin B (at 25 mg per liter). These should be cold sterilized because they degrade when heat treated.

Gentamicin sulfate is described as an autoclavable antibacterial when used at 50 mg per liter. It has been used in enzyme solutions for *Petunia* protoplasts at 10 mg per liter, combined with ampicillin (400 mg per liter) and tetracycline (10 mg per liter) (Dixon and Gonzales 1994).

G418 (brand name Geneticin), hygromycin B, kanamycin, and neomycin are used as selection agents in plant transformation systems (see chapter 13).

Methylolurea is useful against certain yeasts. It is available commercially as methylolurea solution or UF-85. It is a fertilizer with nitrogen derived from products of urea-formaldehyde reactions. Most antibiotics do not withstand heat sterilizing, but this one is autoclavable.

Polymyxin B is occasionally used in cell culture.

Ribavirin (brand name Virazole; 1-b-D-ribofuranosyl-1,2,4-triazole-3-carboxamide) is a broad-spectrum antiviral agent that is particularly effective against some potato viruses.

Streptomycin, used at 20 to 100 mg per liter, may be useful even if autoclaved.

Gelling agents

The source of solidifying agent can be important and can influence the type of growth response. Solidifying agents vary in source (algae, bacteria, plants), amount to be used per liter, cost, physical characteristics (melting point, remelting, clarity, and so on), convenience, and effects on the culture's growth and development. Sometimes media are prepared without a gelling agent, resulting in a liquid medium which is better for the growth of some micropropagated plants, Inca lily (*Alstroemeria*) being one example. Many liquid media cultures need to be placed on a rotary table or shaker to prevent the plantlets from sinking and suffocating; however, plantlets that are not completely submerged can grow well without worrying about agitating the medium.

Agar, also known as agar-agar or gum agar, is a mixture of polysaccharides derived from extracts of several species of red algae. As a gelling agent, the agar in tissue culture media is strong enough to support the culture, yet liquid enough to allow air and nutrients to diffuse through the media to the plantlets. Agar does not react with media components and is not digested by plant enzymes. Although agar is supposedly an inert material, in tissue culture media it frequently contains traces of other elements. Because it is not completely defined chemically, any medium containing true agar (as opposed to synthetic agar) must be considered only a partially defined medium.

The concentration of impurities in agar varies

with the source of the algae and the method of manufacture. Other impurities discovered in some agars include sulfate, calcium, magnesium, and iron ions (Rechcigl 1978). Purified agars sold by tissue culture suppliers are usually sufficiently pure for plant tissue culture.

Typically, 6–8 g of agar per liter of medium is usually satisfactory, but gel strength will vary with both the medium formula being used and the source and grade of agar. Gel strength was found to vary by 35% between the lowest and the highest gels in a group of agar samples from different sources. The gel should be firm enough that test tubes will hold a good slant without being sloppy, yet soft enough that plant material can be pressed gently and easily for good contact with the agar. Agar media with low salts or hormones tend to be firmer than those with high salt and hormone contents. Media with a low pH (around pH 4.5) will tend to be softer than media with a higher pH (pH 5.7). To some extent, depending on the plant, gel strength can affect plant performance. Hard agar may slow nutrient movement through the medium and soft agar may cause vitrification (hyperhydration).

Gellan gum (brand names Gelrite, Gelzan, AppliedGel, Phytagel) is a highly refined, anionic heteropolysaccharide (having more than one type of monosaccharide) substitute for agar. A medium made with gellan gum is a very clear gel when it solidifies after autoclaving. Another difference from agar is that less than half of the amount of gellan gum is needed to obtain an equivalent gel strength (2.5 g per liter). A practical difference from agar is that when mixing the gel, the water solution needs to be cold for the gum to dissolve and divalent cations such as calcium (Ca) and magnesium (Mg) need to be present for the medium to solidify. The exact texture and quality of the final medium is also affected by the divalent cations.

Agargel, a blend of agar and Gelrite, is one of several blends on the market; it should be used at 3.5–5 g per liter. This blend can be made in the laboratory by mixing Gelrite (or an equivalent gellan gum) with agar in ratios of 3:1 to 5:1 of Gelrite to agar, depending on the desired strength. The cost is less than it is for general-purpose agar alone because less Gelrite is required to provide the same gel strength.

Cornstarch has been reported as an agar substitute (Stanley 1995). Ingrid Fordham of the U.S. Department of Agriculture's Agricultural Research Service Fruit Laboratory has found that it is as effective or better than agar as a substrate for certain apples (*Malus*), berries, and pears (*Pyrus*). Because cornstarch will melt at 77° F (25°C), she recommends 50 g of cornstarch plus 0.5 g of Gelrite for 1 liter of medium (Pasqualetto et al. 1986).

Chemically undefined constituents of tissue culture media

When making plant tissue culture media, we attempt to define its constituents. However, is the exact composition of a medium ever known? Several undefined and complex constituents can be added to media and then it gets confusing as to what exactly is responsible for the growth response.

Activated charcoal (AC) is frequently added to rooting media to adsorb root-inhibiting agents and to provide the absence of light to promote rooting. Sometimes activated charcoal is added to Stage I media to adsorb toxic phenolics and to Stage II media intermittently for curative purposes. It is usually added at 0.6–1.0 g per liter and may inhibit the gelling of agar. Activated charcoal not only absorbs toxic compounds, but it can also bind plant growth regulators in media and negate their effect.

Casein hydrolysate (brand name Edamin) is an undefined protein mixture that is occasionally used in media as a non-specific source of organic nitrogen. Casein, a phosphoprotein in milk, is hydrolyzed (treated with water) to form this weak acid.

Coconut milk is the liquid endosperm of coconut (*Cocos nucifera*). This undefined medium was used with success in early plant tissue culture and is still used today especially for orchid seed germination *in vitro*. Coconut milk was first reported in 1941 by Johannes van Overbeek in tissue culture of jimson weed (*Datura*) embryos. In 1948, S. M. Caplin and Frederick C. Steward grew carrot phloem explants using coconut milk and casein hydrolysate as basal salt supplements. Steward achieved 80-fold multiplication of carrot roots in 3 weeks using coconut milk. Coconut milk has also been used widely in orchid culture. In the early 1960s, Carl Withner first cultured *Phalaenopsis* orchids on Vacin and Went's medium, which contains 15% coconut milk.

Coconut milk is available as coconut water from many commercial suppliers or it can be extracted from a coconut. To do this, obtain a green coconut that has reached full size (as young or fresh as you can find in the store), using the heaviest one you can find. Shake the coconut before you buy it to make sure there is liquid in it—you will hear it sloshing. Pound 2 holes in the eyes with a clean nail. Pour out the liquid. Filter it through filter paper and store in the freezer.

Corn milk (corn endosperm) is a rare additive to tissue culture media. It is suggested that the presence of the cytokinin zeatin in corn milk may be the source of benefit. Potato extract is commonly added to media for monocotyledon plants and anther culture. Yeast extract is a natural source of vitamins. It is purchased as a powder and is occasionally used in media. Fruit juices, like orange juice and tomato juice, have also been added to tissue culture media, as well as banana homogenate for orchid micropropagation. These juices have not been as beneficial as those previously listed have.

Purchasing Chemicals

Because the number of chemicals required for tissue culture media can be somewhat overwhelming, the list in Table 4 is divided into convenient reference sections for those who wish to create media from "scratch" or supplement incomplete formulations. The prices are in 2012 U.S. dollars and may vary. The list is comprehensive in that it includes some chemicals that are seldom required; these are given for those who wish to experiment or for the few occasions in which they are prescribed.

Only those chemicals commonly used in culture media should be stocked initially; for example, those used in MS medium, in McCown and Lloyd's woody plant medium (WPM; see *Kalmia*), or in Anderson's medium (see *Rhododendron*). Other important compounds that you will encounter in the various media formulas in Section Two are also listed in Table 4, and some others that come up in Section Two are not listed. At first, do not buy more chemicals than you know you will need for particular media.

If purchasing pre-mixed formulations, note the terminology varies among products, suppliers, and manufacturers. For example, *Phyto*Technology Laboratories, a plant tissue culture–specific manufacturer/supplier, uses the word *medium* to indicate a formulation with all components present except for the gelling agent. *Basal salt* refers to a mixture that contains mineral nutrients but lacks organic compounds; a *basal medium* is a basal salt plus some but not all organic compounds, and a *modified basal salt* or medium varies in one or more ways from the originally published formula.

Some of the organic compounds in Table 4 are considered contaminants, the quantities of which are regulated in drinking water by the United States Environmental Protection Agency and other agencies around the world.

Quick and Easy

For the person who thinks tissue culture is all too complicated, here is a the ultimate do-it-yourself

Table 4. Chemicals and compounds used in tissue culture media

Inorganic chemicals used frequently

COMPOUND	AMOUNT TO BUY	COST (IN $US)*
Ammonium nitrate (NH_4NO_3)†	500 grams	30.75
Ammonium sulfate ($[NH_4]_2SO_4$)	500 grams	36.75
Boric acid (H_3BO_3)	500 grams	18.50
Calcium chloride ($CaCl_2 \cdot 2H_2O$)	500 grams	29.75
Calcium nitrate ($Ca[NO_3]_2 \cdot 4H_2O$)†	500 grams	30.79
Cobalt chloride ($CoCl_2 \cdot 6H_2O$)	25 grams	15.75
Cupric sulfate ($CuSO_4 \cdot 5H_2O$)	250 grams	23.85
Ethylenediaminetetraacetic acid, disodium salt (Na_2EDTA)	100 grams	23.80
Ferrous sulfate ($FeSO_4 \cdot 7H_2O$)	100 grams	11.75
Magnesium sulfate ($MgSO_4 \cdot 7H_2O$)	500 grams	22.95
Manganese sulfate ($MnSO_4 \cdot H_2O$)	100 grams	9.75
Potassium iodide (KI)	100 grams	29.85
Potassium nitrate (KNO_3)†	500 grams	35.75
Potassium phosphate (monobasic, anhydrous; KH_2PO_4)	100 grams	13.80
Sodium molybdate (dihydrous; $Na_2MoO_4 \cdot 2H_2O$)	100 grams	23.85
Sodium phosphate (monobasic; NaH_2PO_4)	100 grams	17.75
Zinc sulfate ($ZnSO_4 \cdot 7H_2O$)	100 grams	16.75

Organic chemicals used frequently

COMPOUND	AMOUNT TO BUY	COST (IN $US)*
Adenine hemisulfate ($C_5H_5O_5 \cdot H_2SO_4 \cdot 2H_2O$)	25 grams	44.50
Agar (Agar-Agar or Gum Agar)	1 kilogram	130.00
Agar–Gellan gum blend (Agargel or equivalent)	500 grams	90.00
6-Benzylaminopurine (BA, BAP)‡	5 grams	9.00
Carrageenan	1 kilogram	80.00
Charcoal, activated	500 grams	26.00
6-(γ,γ-Dimethylallyamino) purine (2iP)‡	1 gram	64.00
Gellan gum (Gelzan, Phytagel, CultureGel, or equivalent)	500 grams	70.00
Indole-3-acetic acid (IAA)‡	5 grams	13.75
Indole-3-butyric acid (IBA)‡	5 grams	15.75
myo-Inositol	100 grams	26.75
Kinetin (6-furfurylaminopurine)‡	1 gram	19.75
1-Naphthaleneacetic acid (NAA)‡	25 grams	13.75
Nicotinic acid (niacin, vitamin B_3)	100 grams	13.75
Pyridoxine HCl (vitamin B_6)	25 grams	24.75
Sucrose (sugar)	4.5 kilograms	24.00
Thiamine HCl (vitamin B_1)	25 grams	16.75
Thidiazuron (1-phenyl-3-[1,2,3-thiadiazol-5-yl]urea)	100 milligrams	31.75

Compounds used infrequently

COMPOUND	AMOUNT TO BUY	COST (IN $US)*
L-Alanine	100 milligrams	11.60
para-Aminobenzoic acid (PABA)	5 grams	15.90
Ancymidol (A-Rest)	25 milligrams	40.00
L-Arginine HCl	25 grams	15.70
L-Ascorbic acid (vitamin C)	100 grams	8.75
6-Benzylaminopurine riboside (alternative to BA)	1 gram	128.00
N-Benzyl-9-(2-tetrahydropyranyl)adenine (BPA, PBA)	1 gram	12.75
D-Biotin (vitamin H)	100 milligrams	16.30
Calcium phosphate (tribasic; $Ca_{10}[PO_4]_6[OH]_2$)	100 grams	60.75
Carbenicillin	5 grams	55.75
Casein hydrolysate	100 grams	13.75
Cefotaxime	1 gram	17.75
para-Chlorophenoxyacetic acid (CPA)	25 grams	17.75
Choline chloride ($C_5H_{14}NOCl$)	5 grams	12.75
Citric acid	25 grams	4.75
Coconut water	500 milliliters	12.75
L-Cysteine HCl	10 grams	25.50
Cyanocobalamin (vitamin B_{12})	500 milligrams	31.75
2,4-Dichlorophenoxyacetic acid (2,4-D)	25 grams	12.50
Ferric tartrate ($Fe_2[C_4H_4O_6]_3$)	100 grams	33.70
Folic acid (vitamin Bc)	5 grams	13.75
Gentamicin sulfate	250 milligrams	31.80
Gibberellic acid (GA_3)	500 milligrams	28.75

COMPOUND	AMOUNT TO BUY	COST (IN $US)*
L-Glutamine	100 grams	19.50
Glycine (aminoacetic acid)	100 grams	16.75
L-Lysine	100 grams	15.75
D-Pantothenic acid (vitamin B_5)	5 grams	0.50
Phloroglucinol	25 grams	29.20
Potassium chloride (KCl)	250 grams	22.30
L-Proline	100 grams	37.50
Riboflavin (vitamin B_2)	25 grams	17.75
L-Serine	25 grams	22.75
Streptomycin	25 grams	13.50
2-Thiouracil (4-hydroxy-2-mercaptopyrimidine; an anti-viral agent)	25 grams	34.70
(+)-α-Tocopherol (vitamin E)	5 grams	11.30
L-Tyrosine	5 grams	10.75
Yeast extract	100 grams	48.00
Zeatin	10.0 milligrams	36.75

Nonmedia chemicals

COMPOUND	AMOUNT TO BUY	COST (IN $US)*
Alcohol, ethyl	1 pint	30.00
Alcohol, isopropyl (70%)	1 pint	1.25
Hydrochloric acid (HCl; 1.0 N for pH adjustment)†	500 milliliters	16.75
Hydrogen peroxide (3%; H_2O_2)	1 pint	0.75
Sodium hydroxide (NaOH; 1.0 N for pH adjustment)†	500 milliliters	16.75
Sodium hypochlorite (bleach)	1 gallon	1.00
Tween 20 (polyoxyethylenesorbitan, monolaurate; wetting agent)	100 milliliters	17.75

* Provided by Gary Seckinger of *Phyto*Technology Laboratories. Prices are in 2012 U.S. dollars and are subject to change.
† Hazardous Materials (HAZMAT). Many suppliers restrict delivery to non-residential addresses.
‡ 1 mg/1 ml solutions of these plant growth regulators are available from many tissue culture media suppliers.

strategy, in which the grocery store, pharmacy, and health food store provide the sources for all the ingredients (Bridgen 1986). The following off-the-shelf recipe is suggested:

⅛ cup table sugar
1 cup tap water
½ cup nutrient solution (¼ teaspoon all-purpose 10–10–10 fertilizer dissolved in 1 gallon water)
125 mg tablet of inositol (half of a 250-mg tablet)
¼ vitamin tablet with thiamine
2 tablespoons agar flakes

Combine the ingredients in a flask or beaker. Boil, while stirring until the agar has melted. Dispense medium into pint canning jars or baby-food jars. Cover and process in a pressure cooker, according to the directions of the cooker manufacturer, for 15 minutes at 15 pounds pressure. Tweezers and razor blades can be sterilized in the cooker at the same time; wrap them in aluminum foil before placing in the cooker. At the same time, sterilize pint jars of water to use for explant cleaning.

Using this recipe, Mark Bridgen has successfully propagated Boston fern (*Nephrolepis exaltata*) rhizome tips, African violet (*Saintpaulia*) leaf and petiole sections, and wandering Jew (*Tradescantia fluminensis*) shoot tips. Before proceeding, see chapters 7 and 8 for information on explant preparation and sterile technique, and Section Two for guidelines on particular plants. For the fun of it, experiment with other explants or ingredients. If coconut milk were substituted for some of the tap water, different kinds of growth would be expected.

6 Media Preparation

This chapter presents, in detail, the procedures to follow for mixing media. They are written primarily for the novice, but those familiar with general laboratory practices may also find these descriptions useful.

It is important to use caution, care, and common sense when preparing tissue culture media. Work slowly, especially at first, to avoid costly mistakes, such as spilling or wasting chemicals, using the wrong ingredients, or burning yourself. With practice, you will become familiar with laboratory language and concepts, you will learn what to expect of your equipment and how to use it, and you will come to develop a respect for and knowledge of the various chemicals. The concentration of salts, for example, particularly the level of ammonium (NH_4), can have a profound affect on culture performance. Reducing ammonium nitrate (NH_4NO_3) and potassium nitrate (KNO_3) levels in Murashige and Skoog's (MS) formula to quarter strength has been found to be beneficial in some cultures of Douglas fir (*Pseudotsuga menziesii*), *Rhododendron*, tomatoes (*Lycopersicon*), *Arabidopsis*, and *Torenia*, to mention a few.

The media formulas for *Begonia rex* in this chapter are examples of the formulas you will find in the literature, or later in Section Two of this book. As you work through the procedures, the concepts will become more meaningful. *Begonia rex*, a common houseplant, is the example used for the micropropagation procedures described in this and subsequent chapters because it is easy to tissue culture and is commonly available where houseplants are sold. *Begonia* plants are easy to propagate by cuttings. However, for efficient mass production or for reproduction of an exceptional individual plant, micropropagation of *Begonia* is an effective procedure. In general, if a plant is easy to propagate by conventional cuttings, it usually will respond well and easily to tissue culture.

Media Preparation from Stock Solutions

Stock solutions are concentrated chemical solutions prepared ahead of time and used to make several batches of media. They may be made in liter quantities of 10, 100, or 1000 times the concentration required in the final formula (below we are making concentrations of 100 times). Stock solutions for macronutrients, micronutrients, and vitamins are also commercially available pre-mixed, at concentrations ranging from 10 times to 1000 times. Having stock solutions eliminates the need to weigh so many different chemicals every time you want to make a batch of medium. In addition, the quantities will be more accurate because they are on a larger scale than would be required for a single batch of medium, and thus minor inaccuracies have less impact. Some of the media ingredients will precipitate (form insoluble compounds and go out of solution) if mixed together in concentrated form, so each group is made up of chemicals that usually will not precipitate at the concentration of the stocks. To discourage precipitation when making stock solutions, always put some water in the flask before adding any chemicals.

Experts do not agree on the combinations of

stock ingredients, or on how long the solutions can be stored. Most stocks can be stored for a limited time without adverse reactions. If they have a short shelf life, as do organics, then the chemicals will be stable for a longer time if they are stored in a refrigerator; hormones tend to have a particularly short shelf life, which is why they are made here in small amounts of 25 mg per 250 ml of water. If the stock ingredients have a longer shelf life, as do the inorganic salts, then they can be stored in a cabinet, but they run a greater risk of growing contaminants because of the warmer temperature there. On the other hand, if the salt solutions are bordering on forming precipitates, they will do so in the cold of the refrigerator because the solubility decreases with decreases in temperature. If solutions form precipitates, they can be brought up to room temperature or heated further and then used, providing the precipitates dissolve. If the stock does not go into solution easily or if it precipitates, heating it on a hot plate/stirrer will often solve the problem.

If you spill some chemicals or remove more than you need, it is bad practice to return the excess to the bottle from which it came because it may have mixed with other substances. Therefore, it is good protocol never to return anything back into a stock solution. It is also important to clean the spoon or spatula between weighing different chemicals. A fine paintbrush is handy for brushing the last bits of chemical from the weighing paper into the flask.

MS (Murashige and Skoog) stock solutions for *Begonia rex* media

EQUIPMENT AND SUPPLIES

1 1-liter Erlenmeyer flask
2 400-ml beakers
Distilled water
Hot plate/stirrer and stir bar
2 balances (one for small measurements and one for measurements of more than 1 g)
Weighing papers
1 spatula
Forceps
Wash bottle containing distilled water
Medicine dropper
1 M sodium hydroxide (NaOH) (to dissolve auxins; see chapter 5)
1 M HCl (to dissolve cytokinins; see chapter 5)
Containers, with lids (for storing stock solutions)

MS SALTS STOCKS

These include 5 stocks: nitrate, sulfate, halide, phosphage, and iron.

NITRATE STOCK
1 liter at 100× concentration

Potassium nitrate (KNO_3)	190 g
Ammonium nitrate (NH_4NO_3)	165 g

PROTOCOL

1. Pour 500–700 ml of distilled water into a 1-liter flask.
2. Slide the magnetic stir bar into the flask.
3. Place the flask on the magnetic stirrer and turn on the stirrer.
4. Place a weighing paper (or piece of aluminum foil) on the balance and tare (weigh) it, then set the balance for 190 g more than the weighing paper. Some of the more expensive electronic balances will allow you to automatically zero out the weight of the weighing paper. In this situation, the weight of the empty paper is the tare weight, the weight of the chemical is the net weight, and when

added together they give you the total or gross weight. If your balance has this capability, then zero out the weighing paper and go to the next step.

5. Using a spatula (or spoon or knife) weigh out 190 g of potassium nitrate (KNO_3).
6. Pour the KNO_3 into the flask on the stir plate.
7. Place a weighing paper on the balance and tare it, then set the balance for 165 g more than the weighing paper. Alternatively, zero out the weight of the weighing paper if your balance has that capability.
8. Weigh and add 165 g of ammonium nitrate (NH_4NO_3). At this point, the total weight of the KNO_3 plus the NH_4NO_3 is 355 g.
9. Stir until dissolved.
10. Turn off the stirrer and remove the stir bar with long forceps, or pour the stock solution into a clean flask, saving out the stir bar.
11. Add distilled water to bring the level of solution up to 1000 ml (1 liter).

In a similar manner, make the remaining stocks. Note that cupric sulfate (in the sulfate stock) and cobalt chloride (in the halide stock) are required in amounts (2.5 mg) that are too small to weigh on many balances. To obtain this amount, weigh 25 mg of cupric sulfate and add it to 100 ml of water. Using a pipet, take 10 ml of this solution to provide 2.5 mg of cupric sulfate. Do the same for cobalt chloride.

SULFATE STOCK
1 liter at 100× concentration

Magnesium sulfate heptahydrate ($MgSO_4 \cdot 7H_2O$)	37.0 g
Manganese sulfate monohydrate ($MnSO_4 \cdot H_2O$) (if using $MnSO_4 \cdot 4H_2O$, use 2230 mg)	1.69 g (1690 mg)
Zinc sulfate ($ZnSO_4 \cdot 7H_2O$) (if using $ZnSO_4 \cdot H_2O$, use 540 mg)	0.86 g (860 mg)
Cupric sulfate ($CuSO_4 \cdot 5H_2O$)	0.0025 g (2.5 mg)
Total	39.6 g

HALIDE STOCK
1 liter at 100× concentration

Calcium chloride ($CaCl_2 \cdot 2H_2O$)	44.0 g
Potassium iodide (KI) (often omitted)	0.083 g (83 mg)
Cobalt chloride ($CoCl_2 \cdot 6H_2O$)	0.0025 g (2.5 mg)
Total	44.1 g

PHOSPHATE STOCK
1 liter at 100× concentration

Potassium phosphate (KH_2PO_4)	17.0 g
Boric acid (H_3BO_3)	0.62 g (620 mg)
Sodium molybdate ($Na_2MoO_4 \cdot 2H_2O$)	0.025 g (25 mg)
Total	17.6 g

IRON STOCK
1 liter at 100× concentration

Ferrous sulfate ($FeSO_4 \cdot 7H_2O$)	2.78 g
EDTA disodium salt (Na_2EDTA)	3.73 g
Total	6.5 g

The iron stock precipitates if not prepared correctly. Dissolve Na_2EDTA and $FeSO_4 \cdot 7H_2O$ in separate beakers with approximately 200 ml distilled water. The distilled water should have a pH of 6 or less, or the $FeSO_4$ will precipitate. Place both beakers on hot plates and heat. Slowly add the $FeSO_4$ to the Na_2EDTA (not vice versa) with continuous stirring. Cool this 400-ml solution to room temperature in the dark. After it reaches room temperature, bring the stock up to 1 liter with distilled water. Refrigerate. If this stock is cloudy and not clear yellow, then it was prepared incorrectly. Store in the dark.

Total MS salts from these 5 stock solutions: 462.8 g.

MEDIA PREPARATION

VITAMIN STOCKS

Store both of these stocks in a refrigerator.

INOSITOL/THIAMINE STOCK
1 liter

Inositol	10.0 g
Thiamine HCl	0.04 g (40 mg)
Total	10.04 g

NICOTINIC ACID/PYRIDOXINE STOCK
1 liter

Nicotinic acid	0.1 g (100 mg)
Pyridoxine HCl	0.1 g (100 mg)
Total	0.2 g (200 mg)

AUXIN STOCK

Auxin stock should be stored in a refrigerator. This stock contains 0.1 mg of 1-naphthaleneacetic acid (NAA) per milliliter of solution. Stock solutions of other auxins, such as IAA and IBA, can be made in the same way as described here for NAA.

1. Weigh 25 mg of NAA and place in a 400-ml beaker.
2. Using a dropper, slowly add, while stirring with a spatula, several drops of 1 M NaOH or KOH until the NAA crystals are dissolved.
3. Quickly add 250 ml of distilled water.
4. Transfer solution to a container and close tightly. Makes 250 ml.

CYTOKININ STOCK

Cytokinin stock should be stored in a refrigerator. This stock contains 0.1 mg of BAP per milliliter of solution. Stock solutions of other cytokinins, such as 2iP and kinetin, can be made in the same way as described here for BAP.

1. Weigh 25 mg of BAP and place in a 400-ml beaker.
2. While stirring, add 3 or 4 drops of distilled water.
3. Add 1 M HCl solution one drop at a time until the BAP is dissolved. Apply a little heat to help dissolve the crystals.
4. Quickly add 250 ml of distilled water.
5. Transfer solution to a container and close tightly. Makes 250 ml.

Figure 6-1. Prepared media, all ready to go.

Media Mixing

Having made the stock solutions, you are at last ready to mix the medium, a task much easier than mixing stocks. For propagating *Begonia rex*, you will need 2 different media: one for Stages I and II and a second one for Stage III (see Table 5).

Calculating amount of stock solution per liter of medium

Although the calculations for the amount of stock solutions needed have already been done for the Begonia media in progress (see Table 6), it would be good practice for you to figure for yourself how the amounts are determined. To determine the amount of stock solutions required for a medium, the easiest arithmetic to use is simple proportion.

You already have made a cytokinin stock solution with 25 mg of BAP in 250 ml of water. If for a particular medium you require only 0.4 mg of BAP, you need to know how many milliliters of stock solution to use to obtain that amount of BAP.

Write the original milligram amount of BAP (25 mg) over the original milliliter total of stock solution (250 ml). It does not matter if you have already used some of the stock solution; the proportion is still the same.

Next, write the milligram amount you need (0.4 mg) over the unknown milliliter amount (? ml):

$$\frac{25 \text{ mg}}{250 \text{ ml}} = \frac{0.4 \text{ mg}}{? \text{ ml}}$$

Cross multiply:

$$25 \times ? = 250 \times 0.4$$
$$25 \times ? = 100$$

Divide both sides by 25 so the unknown will be by itself:

$$\frac{25\,?}{25} = \frac{100}{25}$$
$$? = 4$$

The amount of BAP stock solution you will need is 4.0 ml.

Table 5. *Begonia rex* media formula

COMPOUND	STAGES I & II MG/LITER	STAGE III MG/LITER
MS salts	4,628	4,628
Inositol	100	100
Thiamine HCl	1.5	1.5
Nicotinic acid	0.5	0.5
Pyridoxine HCl	0.5	0.5
NAA	0.1	0.1
BA	0.1–0.5	—
Glycine	2.0	2.0
Sucrose	30,000	30,000
Agar	6,000	6,000

Table 6. **Stock solutions needed for making *Begonia rex* medium for Stages I and II**

CHEMICALS	STOCK PER LITER OF MEDIUM
Sucrose	30 g
Nitrate stock	10 ml
Sulfate stock	10 ml
Halide stock	10 ml
Phosphate stock	10 ml
Iron stock	10 ml
Inositol/thiamine stock	10 ml
Nicotinic acid/pyridoxine stock	5 ml
BA stock	4 ml
NAA stock	1 ml
Glycine (10 mg in 10 ml distilled water)	2 m
Adjust pH to 5.5	—
Agar	8 g

Source: Murashige and Skoog (1962); Mikkelsen and Sink (1978).

Checklist

When preparing stock solutions or media, a checklist like the one in the example given (see Table 7) should be filled out and followed to ensure that the right chemicals and amounts are used. If the items are checked off the moment they are added to the mixture, there should be no problem if the technician is interrupted. Including instructions insures another technician will be able to replicate the process.

Table 7. Preparation instructions and checklist for stock solution or medium

Medium: _____   Species: _____

Prepared by: _____   Date: _____

Volume to Prepare: _____   Autoclave Sterilization Time: _____

pH Desired: _____   Actual Final pH: _____

Product Number: _____ Lot Number: _____ Source: _____

Complete the table with all components and grams per batch of the stock solution or medium to be prepared, including product and lot numbers if applicable. As each component is weighed, record the actual weight. Check off each component after it is added to the solution/medium.

COMPONENT	PRODUCT NUMBER	LOT NUMBER	GRAMS/BATCH	ACTUAL WEIGHT	✓
					☐
					☐
					☐
					☐
					☐
					☐
					☐
					☐

Additional instructions and/or comments: _____
_____
_____
_____
_____

Source: Adapted with permission from *Phyto*Technology Laboratories, Inc. http://www.phytotechlab.com/pdf/MediaPreparationLog.pdf

Preparing MS-based *Begonia* medium from stock solutions

EQUIPMENT AND SUPPLIES

1 2-liter beaker (glass)
Distilled water
Hot plate/stirrer and stir bar
Sugar
2 balances
Weighing papers
1 spatula
Stock solutions (the 9 listed in Table 6)
1 10-ml graduated cylinder
2 10-ml pipets (in 0.1 increments)
1 1-ml pipet (in 0.1 increments)
pH meter
Medicine dropper
Wash bottle containing distilled water
1 M NaOH (sodium hydroxide) or KOH (potassium hydroxide) for pH adjustment
1 M HCl (hydrochloric acid) for pH adjustment
Agar
1 pitcher or other dispenser
65 culture tubes or other desired culture vessels, with caps
Test tube racks
Autoclave or pressure cooker

Procedure for making 1 liter of MS-based *Begonia* medium

1. Pour 500–600 ml of distilled water into a 2-liter beaker. Always use a container with ample room for the solution to boil. For example, use a 2-liter beaker for 1 liter of medium, a 4-liter beaker for 3 liters of medium, and so on.
2. Slide the magnetic stir bar into the beaker.
3. Place the beaker on the hot plate/stirrer and turn on the stirrer.
4. Add the sugar and stock solutions as specified in "the recipe" (see Table 6).
   a. Weigh 30 g of sugar on a balance and add to the beaker.
   b. For stock solutions requiring 10 ml—nitrate, sulfate, halide, phosphate, iron, and inositol/thiamine solutions—pour 10 ml of each solution into a 10-ml graduated cylinder and add to the flask. It is very important to clean and rinse the graduated cylinder well between measuring solutions. Measure amounts to the bottom of the meniscus (the curved surface of a liquid).
   c. Add the required amount of the remaining stock solutions using a clean 10- or 5-ml pipet for each stock solution. Always use a bulb pipetter, or pipette filler. Never mouth pipet
5. Turn off the stirrer and add distilled water to the flask up to the 1-liter mark.
6. Turn on the stirrer and adjust the pH of the medium:
   a. Calibrate the pH meter according to the manufacturer's instructions.
   b. Using a wash bottle with distilled water, rinse off the pH-meter probe.
   c. Lower the probe into the medium as it is mixing.
   d. Observe the reading on the pH meter. If the pH is below pH 5.5 (too acid), use a dropper to slowly add, one drop at a time and allowing the medium to mix after each drop, 1 M NaOH or KOH until the pH meter reads pH 5.5. If the pH is above pH 5.5 (too alkaline), then slowly add 1 M HCl, one drop at a time, until the pH meter reads pH 5.5.
7. Weigh and add 8 g of agar.
8. Turn on the hot plate/stirrer heat. Continue to heat and stir until the medium boils, but do not allow it to boil over. At this point, the agar in the medium should have changed from cloudy to clear.

9. Turn off the hot plate/stirrer and remove the flask.
10. When the medium stops boiling, add distilled water to the 1-liter mark.
11. Dispense 10–15 ml of the medium into each culture tube or vessel using a pitcher or semiautomatic pipetter. Avoid spilling any medium on the rims of the culture tubes.
12. Cap the culture tubes.
13. Place the culture tubes in culture tube racks and then place the racks in the autoclave. Alternatively, place the tubes in the wire basket of the pressure cooker; if there are too few test tubes to stand up properly in the basket, fill the gap with an empty beaker.
14. Process the test tubes for 15 minutes at 15 pounds pressure according to the manufacturer's instructions for using the pressure cooker.
15. After sterilizing, and when the cooker or autoclave pressure level is back to zero, remove the racks or basket of tubes and place them on an angle so that the medium will solidify on a slant.
16. When the agar has cooled, label and store the sterilized tubes of medium in a refrigerator or other cool, clean place.

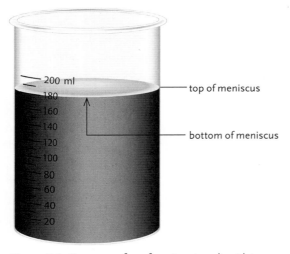

Figure 6-2. Because of surface tension, liquid in a container has a curved surface known as the meniscus. Always measure amounts to the bottom of the meniscus, not to the sides, which are higher.

Media from Prepared Mixes

For standard basal salt mixtures (without organics), 2.0 to 5.5 grams (depending on formulation) of powder are sufficient to produce 1 liter of media. A pre-mixed basal medium, for example a Murashige and Skoog (MS) modified medium that includes nutrients, vitamins, sucrose, and hormones, requires 30 to 40 grams to produce 1 liter of media. Note that powdered media are extremely hygroscopic (attracts moisture) and should be stored in a dry location. Manufacturers recommend using the entire contents once opened, thus standard packaging sizes hold amounts to make 1, 10, 50, or 100 liters of media. If storage is necessary, keep mixes refrigerated between 0° and 5°C.

Procedure for making 1 liter of medium from pre-mixed powder

1. Add 900 ml distilled water to 2-liter flask
2. Add powdered medium per manufacturer's recommendation while stirring as previously described. Be sure to rinse original container with a bit of distilled water to remove all product; add to the solution.
3. Add any other heat-stable supplements that are required such as sucrose, vitamins, hormones, and gelling agents.
4. Bring medium to final volume by adding additional distilled water.
5. While stirring, adjust pH using hydrochloric acid (HCl), sodium hydroxide (NaOH), or potassium hydroxide (KOH).
6. If a gelling agent is used, heat the medium until the solution is clear.
7. Dispense the medium into the culture vessels, place on the vessels lids, and autoclave as previously described in steps to make MS-based *Begonia* medium.

Liquid Media

A liquid medium may be prescribed in some formulas for one or more stages of growth. Liquid media are faster to make than agar media because there is no heating necessary to dissolve agar before the medium can be evenly dispensed. Changing to a liquid medium may also help solve some of the problems encountered on gelled media, such as bleeding (the exudation of substances—phenolic exudates—which discolor the agar), brown leaves, or poor growth.

Usually liquid cultures are aerated by agitation on a rotator, shaker, or rocker. Agitated liquid media will not allow waste products to build up adjacent to the culture, as can occur in gelled media. Only 5–10 ml of liquid medium should be used in a 25 × 100 mm test tube.

Some cultures in liquid medium do not seem to require aeration. Sometimes when cultures are started from explants that are 1-in. (2.5-cm) long cuttings, or longer, the base of the explant may be placed in 3–5 ml of liquid medium in a test tube and no agitation is necessary because the top of the explant is exposed to the air.

Bridges and rafts

Especially for smaller cultures (0.5–5 mm), an alternative to agitation is to insert a bridge in the test tube, flask, or other vessel to support the culture, thus allowing it to have air and convey the liquid to it (Figure 6-2). To make a bridge, cut 3 × 0.75 in. (7.5 × 2 cm) paper strips of filter paper or paper towel. Insert the strips into the test tubes with both ends down so that the middle of the strip will be above the liquid and can hold the culture. This is done prior to sterilizing the medium.

Another type of bridge can be made in the transfer hood. Place 3-in. (7.5-cm) squares of paper towel or filter paper in a beaker, cover with aluminum foil, and sterilize for 45 minutes at 15 pounds pressure in a pressure cooker or autoclave. Have sterile culture tubes ready containing 5 ml of sterilized liquid medium. In the hood, firmly grasp a piece of the toweling using sterile forceps, holding it between the forceps with the tips of the forceps at the center of the paper. With a twist of the wrist, place the paper into the test tube. If it is not grasped firmly, the forceps will simply make a hole in the paper and slide through. When the paper is partially in the liquid in the bottom of the test tube, fold over the corners to make a platform on which to place the culture. The liquid medium will wet the paper without submerging the culture. A variety of rafts, floats, bridges, or other supports are available commercially, but can be expensive and/or cumbersome.

Figure 6-3. Filter-paper bridge in liquid medium; here used for germination of *Arabidopsis* seeds.

Conclusion

Certain general guidelines for media preparation and explant pretreatment will increase the chances of starting explants successfully. If the prescribed formulas for culturing the plant at hand prove ineffective, try various standard media (both liquid and agar), starting with those that have been used successfully for related plants in the same family or genus. A useful approach for herbaceous perennials is to start them in half-strength MS salts without hormones; for woody plants, try WPM without hormones.

By now, you should have a feeling for how to mix media. If you have followed the above protocol, you are ready to find out how your plants will do in your media. We hope that most of the recipes you try will respond favorably. Let curiosity be your guide; probably some unrelated plants will respond equally well to the same formulas. Read on to learn about Stage 0, preparing the explants for the media you have ready.

7 Explants: Selecting, Managing, and Preparing Stock

An explant is the piece of a plant from which a culture is started. Theoretically, a single explant can produce an infinite number of plants. This is probably the most enlightening and the most astonishing statement that someone can make about plant tissue culture. If all goes well, one explant can produce thousands of plants. It is good to have more than one explant to achieve the desired number of plants in less time, and because typically some number of plants will be lost because of contaminants, disinfectants, or other unknown reasons. Normally, however, only a few explants are necessary, and one stock plant is usually sufficient to supply all the starts required.

Explants range in size from a microscopic 10th of a millimeter to stem pieces several centimeters long. Explants can be meristems, shoot tips, macerated stem pieces, nodes, buds, flowers, peduncle (flower stalk) pieces, anthers, petals, pieces of leaf or petiole, seeds, nucellus (the central part of an ovule) tissue, embryos, seedlings, hypocotyls, bulblets, bulb scales, corms and cormels, radicles, stolons (runners), rhizome tips, root pieces, or, though rarely, single cells or protoplasts. (Figures 2-2, 7-1, and 7-2 provide illustrations of many of these plant parts.)

While no precise instructions can guarantee successful explants, many protocols have been universally successful, and general guidelines offer a plethora of alternatives. One may well ask, When is the best time to obtain an explant? Usually, an actively growing part of the plant is the best explant for micropropagation. A stem tip may perform best if taken a little earlier in seasonal development than one would take a normal herbaceous cutting. However, the odds of success are greatest if a few explants (10 to 20) are started at weekly or monthly intervals, instead of gambling many valuable explants on only one set of conditions. For *Begonia rex*, the example for tissue culture procedures used in this book, if the plant to be used is growing well indoors, it should not need any special treatment prior to obtaining an explant.

Selecting and Preparing Stock Plants

Although the number of stock plants that are necessary for micropropagation are significantly less than those needed for propagation by vegetative cuttings, the same management principles should be observed (Bachman 2008). The assigning of a unique Stage 0 of micropropagation highlights the importance of stock plant management to optimize hygiene as well as the quality and quantity of explants. Explants that are taken from clean, insect-free, and disease-free plant materials have a greater chance of growing well *in vitro*. The stock plants from which explants will be taken need to be fertilized, growing well, and properly identified so that the large numbers of plants that are produced from them are correctly identified. The time of year when explants are removed from the stock plants may also affect

Stage 0. Differences in temperature, day length, light intensity, and water availability can affect performance from year to year. All of these are considerations in Stage 0.

Many herbaceous plants are photoperiodic; that is, day length influences whether plants are vegetative or flowering. If the desired explant comes from shoot tips or runners, environmental manipulation of the stock plant may be necessary when the season/natural day length is conducive to flowering. For example, chrysanthemum stock plants should be held under long-day conditions to maintain vegetative growth for producing shoot tip cuttings. Undesirable flowering occurs when the mum plants are grown under short days of less than 12 hours light (Dole and Wilkins 2008).

Day length can be extended with supplemental lighting. Night interruption of 1–2 hours can easily be accomplished by using a timer and also serves the same purpose—as the dark periods are shortened (what the plant actually perceives), this translates to longer days. Some species, such as carnation (*Dianthus*), flower under long days of greater than 12 hours light; to maintain vegetative shoots of such plants, day length must be shortened. This can be accomplished in a greenhouse by pulling "black cloth" (light-blocking fabric or plastic) over the bench in the late afternoon and uncovering each morning. If growing indoors, this can also be accomplished by placing a paper bag over the plant every night to prevent light from hitting the leaves. The 8-hour photoperiod is sufficient for growth but the shoots remain vegetative.

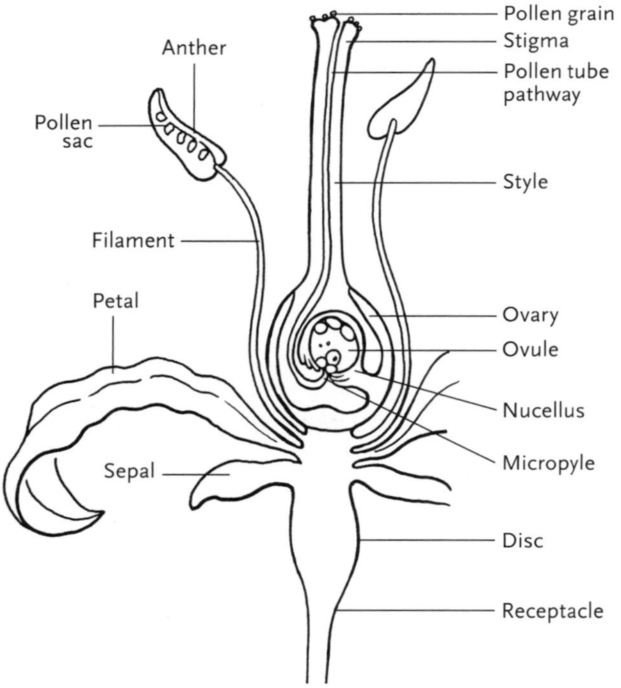

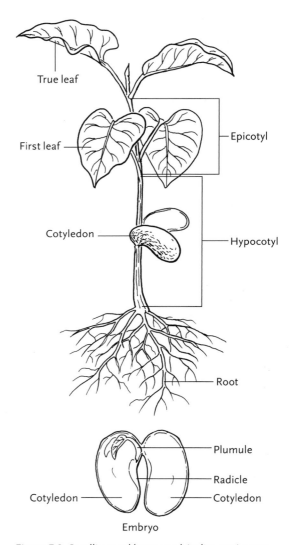

Figure 7-1. Flower parts.

Figure 7-2. Seedling and bean seed (split open) parts.

Just as juvenility is a factor in selecting material for cuttings (Hartmann et al. 2010), so it is in selecting explants for tissue culture. In general, the more juvenile the explant material, the greater the likelihood of success. With this in mind, any means by which material that is more juvenile can be taken is preferred. The juvenile leaves of *Eucalyptus* (syn. *Corymbia*), *Sequoia*, and *Vitis* are visibly different in form from their mature leaves. Mature plants of these genera sometimes produce adventitious shoots near their bases. These shoots have juvenile leaves, a fact that lends credence to the theory that the bases of plants may be more juvenile in their lower reaches than higher up. Sometimes stock plants can be repeatedly pruned back (hedged or sheared) to encourage regeneration of juvenility in new growth, which then can be taken as explants. Grafting mature shoots on juvenile rootstock can also foster some increased juvenility. Growth from rooted cuttings of mature plants is not considered juvenile material.

The number of shoots or runners a stock plant produces can also be increased by the application of plant growth regulators. Gibberellic acid (GA_3) and benzyladenine (BA) have been used to enhance shoot numbers on a number of species. Please consult the literature for effectiveness on the plant species of interest.

Obtaining Explants

When taking shoot tips, stems, buds, or flowers for explants, cut them longer than the final size that is needed. The final length may be 1 in. (2.5 cm) or more, or it may be the microscopic apical meristem. In most instances, 1 in. is a convenient size to process. Any final trimming and treatment should be done in the sterile environment of the transfer hood.

Occasionally leaves or parts of leaves are used to start tissue cultures. African violets (*Saintpaulia*), rex begonias (*Begonia rex*), snake plants (*Sansevieria trifasciata*), and a number of other plants can be propagated from leaf cuttings. Some conifer needles in culture will produce plantlets around their edges. Other leaves, after cleaning, can be punched with a cork borer and the discs placed in culture. Leaf tissue is also a common source of single cells and protoplasts, the culture of which is discussed in chapter 13.

Detecting Plant Pathogens

If possible, stock plants should be moved to a clean greenhouse, screen house, or other shelter away from dust and disease prior to taking explants. There they can be observed and given special care for a short time, usually about 2 weeks to 6 months. In controlled conditions, the plants will become as healthy as possible, and they will not be stressed, as they might in a field or garden. Wash the plants with clean water, allow the foliage to dry, and fertilize and water the plants only at the base. Subirrigation is an excellent way to deliver water and nutrients without splashing growing media and bits of organic debris onto the fresh shoots. Provide adequate fertility to insure optimum growth. The new shoots that appear will provide relatively clean explants and, in addition, may provide some increased juvenility.

Figure 7-3. This healthy African violet (*Saintpaulia*) is an excellent source of explants.

Only disease-free plants should be tissue cultured, but sometimes a plant that appears healthy is harboring a disease. If time permits, and depending on the particular plant and the budget, the plant should be tested for pathogens known to infect that species. Testing by a government or independent laboratory is usually more acceptable to a customer (due to impartiality), and government testing is required for shipping certain species across some state or national borders. Increasingly, plant propagators use pathogen detection kits, which are commercially available, to identify fungal, bacterial, and viral pathogens.

Another method of testing for disease is to graft shoots on indicator plants that are sensitive to specific diseases and will show symptoms in a matter of hours or days. Culture indexing is an effective way to test explants for internal bacterial or fungal contamination. Explant samples are placed in a special nutrient broth for up to one week; visible changes to the broth or the presence of mycelium indicate contamination. Specific information on identifying and treating for contaminants is given in chapter 11.

If observation or testing uncovers a valuable plant that is infected with a viral disease, it is

Tips for selecting explants

- The smaller the explant, the less contamination there is to remove; but the larger the explant, the more tissue there is to help establish it in culture.
- Use new, actively growing shoots whenever possible because they are cleaner than old wood.
- Plants that are grown under cover, in greenhouses, or shoots from forced, dormant branches are usually cleaner than field plants.
- Explants from younger plants often respond more quickly than those taken from older plants do.
- Plant material at the base of a plant is deemed younger than growth higher up. However, plant material at the base of a plant also can be less clean due to proximity to the ground.
- If plastic bags are fastened over actively growing shoot tips on stock plants, the new growth will be cleaner for cutting explants days or weeks later; be sure to provide shade to prevent sunburn.
- Explants may do better if taken in the morning when their water turgor is strongest, rather than later in the day. However, where heat and high temperature are not issues, cuttings may be taken at the end of the day and may perform better due to accumulated sugars (Rapaka et al. 2007).
- Leaf petiole sections 5 mm in size are recommended for explants from plants that can propagate by leaf cuttings, but a section of petiole together with a centimeter of leaf base or a node or shoot tip will also do well.
- Remove explants with tools that have been sterilized by dipping in a 10% bleach solution to reduce the spread of disease.
- Upon cutting explant material from the source plant, place the explant in a plastic bag containing a moist (not dripping wet) paper towel. The explant should remain in the bag until it is time to work on it in the laboratory.
- Keep explant material refrigerated or in a cool place if the species is not cold tolerant until it is time to work with it. For best results, process explants as soon as possible.
- If using tips or pieces of roots or rhizomes as explants, they will be easier to clean if the plant has been growing in an artificial medium, such as washed sand, pumice, or perlite. Wash these cuttings well before placing them in a plastic bag. Explants from roots and rhizomes that are removed from garden soil can be highly contaminated and are very difficult to clean.

sometimes possible to rid the plant of the disease by heat treatment (thermotherapy), followed by meristem culture. Thermotherapy can be performed *in vivo* (whole plant) or *in vitro* (explant). The length of time and the amount of heat applied vary with the tolerance of the plant, the part of the plant used, and the type of virus. A typical regimen would utilize temperatures between 96°F and 100°F (35°C–38°C) with a 16-hour photoperiod for up to 3 weeks. For example, *Pelargonium* plants have a low heat tolerance and will seriously etiolate (blanch) after about 3 weeks of heat treatment, whereas carnations (*Dianthus*) have a high tolerance and can survive heat treatment for several months. Due to the nature of bulbs and corms, they can tolerate even higher temperatures. Only healthy appearing plants with adequate energy reserves should be heat treated, because exposure to temperatures above optimal-growth temperatures is quite detrimental. Meristem tip culture would then be used to obtain hopefully virus-free explants. More on meristem culture is provided in chapter 11.

Germplasm repositories (institutional locations for collecting and preserving live-type species) are good sources of plants that are free from pathogens. At these laboratories, technicians heat treat plants, grow new plants from meristems, grow material in screen houses, test for diseases, and provide clean stock for commercial users.

Cleaning and Treating Explants

The aseptic establishment of explants is the goal of Stage I. After explants are detached from the source plant, they must be cleaned. There are no set rules for cleaning explants because the amount and kind of cleaning depends on how clean or dirty the plant material is when Stage I begins. The process that will be used also depends on how easily the explant is damaged by a cleaning agent. Disinfectants will not eliminate endogenous (internal) contaminants, and exogenous (external) contaminants will persist if the disinfectant solutions used in treatment are too weak, especially those contaminants located in plant crevices.

The most common cleaning agent used to disinfect explants is bleach in various concentrations used alone or in sequence with other disinfectants. Pure household bleach is 3%–6% sodium hypochlorite. Commonly used household bleach solutions have concentrations of 5%, 10%, and 20%. To make 100 ml of bleach solution, mix 5 ml of bleach with 95 ml of sterile water to make a 5% concentration; 10 ml bleach with 90 ml sterile water for a 10% concentration; and 20 ml bleach with 80 ml sterile water for a 20% concentration. These concentrations are often referred to as 1:20, 1:10, and 1:5, respectively. Thus 1:10, for example, means 1 part bleach in 10 parts of solution. If directions specify 0.05% sodium hypochlorite, that would indicate a 10% solution of household bleach.

Note that, over time, bleach can lose strength, and open containers also hasten volatilization of the active ingredient. Also, results may vary between laboratories depending on the concentration of sodium hypochlorite in the commercial bleach solution. For example, a 10% solution of bleach that is 5% sodium hypochlorite will not be as strong as a 10% solution of bleach that is 6% sodium hypochlorite. Be sure to stay consistent with the bleach source once a procedure is determined.

Other disinfecting agents are available and can be used either alone or after the bleach treatment. A broad-spectrum preservative and bactericide sold under the trademarked name Plant Preservative Mixture (PPM) can be used as a media cleaner to decrease contamination in cultures. Because PPM is heat stable, it can be added to media prior to autoclaving in lieu of antibiotics and/or fungicides at a rate of 0.5–2.0 ml per liter of medium. It should be noted that although PPM is bactericidal, it has to come in contact with each bacterial cell to have that effect. Because contact is required, cleaning up material that is already in tissue culture is difficult as the interior of the plant is frequently colonized. PPM will kill bacteria in/on the media and as it (the bacteria) leaches out of the plant material. As long as a medium has a sufficient concentration of PPM, no apparent contamination will result; when plantlets are subcultured onto a medium without PPM, the contaminants will often appear.

If the explant material is very dirty, it is helpful to pretreat it by placing the explant in running water for a few hours. Another approach is to agitate the material in 1:100 (1%) bleach, plus 2 drops of a surfactant such as a liquid dish detergent soap or Tween 20 per 100 ml of solution, for one or more hours; alternatively, use a higher concentration of disinfectant, say 5%–10%, for a shorter duration of 10–15 minutes. Agitate the explants with a rotary shaker, a magnetic stir plate, or by gently shaking a securely closed vessel containing the explants and disinfecting solution. Note that both a disinfectant and a wetting agent (surfactant) are necessary when washing explant material. Polysorbate 20, sold under the brand name Tween 20, is a common wetting agent, not a disinfectant, which reduces the surface tension of the material being cleaned, thus allowing the bleach to be more effective. Liquid dishwashing detergent can have the same effect as a wetting agent but should be used with care because some kinds of detergents may damage the explants. Dawn and Ivory are popular brands in many tissue culture labs.

Other disinfectants commonly used to clean explants include 70% ethyl alcohol or isopropyl alcohol; 5%–10% calcium hypochlorite; 3% hydrogen peroxide, and the pool disinfectant NaDCC (sodium dichloroisocyanurate). So-called rubbing alcohol can consist of either isopropyl alcohol or ethyl alcohol (ethanol) and the percentage of active ingredient can vary, so check the label. Ethyl alcohol is most commonly used by tissue culture labs and is available from laboratory supply vendors in larger volumes.

If there are enough explants, several small groups can be treated with various concentrations and various kinds of cleaning agents for various times of immersion in the disinfectant. Different disinfectants should never be mixed; any 2 disinfectants that are mixed in the same container can become dangerous because of the chemical reactions that can occur.

The following example of an explant cleaning treatment uses easily obtained bleach and isopropyl alcohol to prepare 10 strawberry (*Fragaria*) runner tips, each 1 in. (2.5 cm) long. The treatment is performed in a series of beakers that have been autoclaved and rinsed in 10% bleach. The recommendations for time and bleach concentration may vary according to the tolerance of the explant and the effectiveness of the treatment.

The cleaning of explants should be done in the media preparation room up until the final stages of explant preparation, both for convenience and to avoid the clutter of a stirrer/shaker, dirty rinse containers, disinfectant containers, and other paraphernalia that might contaminate the transfer hood. Transport the cleaned explants to the transfer hood while still in bleach in a covered container, and apply the final rinses in the hood using sterile technique. This final stage and sterile technique are presented in the next chapter.

Stage I explant cleaning treatment for strawberry (*Fragaria*) runner tips

EQUIPMENT AND SUPPLIES
5 100- to 250-ml beakers
1% bleach
10% bleach
Dropper bottle containing Tween 20
Magnetic stir plate and stir bar
Explant material
Distilled water
70% isopropyl alcohol (or 3% hydrogen peroxide)
Forceps
Cutters or scalpel

Cleaning procedure

1. Pour 100 ml of 1% bleach and 2 drops of Tween 20 into a sterile 150-ml beaker with the magnetic stir bar and place on the stir plate.
2. Add the explants to the beaker and stir for 10 minutes. (As an alternative to the stir plate, the explants and the cleaning materials can be shaken by hand in a tightly closed jar, or they can be agitated on a shaker.) Pour off the bleach solution, and place the explants in a clean beaker.
3. Rinse the explants with distilled water. Pour off the water, and place explants in another clean beaker. From this point on it is best to work under the laminar flow hood.
4. Rinse the explants in 70% isopropyl alcohol (or 3% hydrogen peroxide) for 5 seconds. Again, pour out the alcohol and place the explants in a clean beaker.
5. Rinse the explants in distilled water and then place them in a clean beaker.
6. Pour 100 ml of 10% bleach and 2 drops of Tween 20 into a sterile 150-ml beaker with the magnetic stir bar and place on the stir plate.
7. Add the explants to the beaker and stir for 15 minutes. If this process is not conducted under a flow hood, cover the beaker with sterile foil. Some tissue culturists use a vacuum pump at 25 mm of mercury to help the disinfectant penetrate into the crevices of the explant. Others have found that an ultrasonic cleaner is useful in cleaning explants. Still others simply shake by hand a tightly closed jar containing the explants and solution.
8. Rinse the explants 2 or 3 times in sterile water. Aseptically place the explants onto a sterile growing medium inside a culture vessel.

NOTES
If the external contaminants are not killed by the foregoing treatment, microbial growth, usually fungi, will be visible in the culture vessel within a few days. If such growth is observed, then it will be necessary to treat new explants with one or both of the following alternatives:

1. Stir the material in the 10%–20% bleach for 5–15 minutes longer.
2. Extend the alcohol dip to a minute or longer.

If any of the preceding treatments kill the explants, try weaker bleach solutions for the same or longer periods. To test the tolerance of a set of explants to 10% bleach, remove a few of the explants at 5-, 10-, and 15-minute intervals instead of soaking them all for 15 minutes. Sometimes a weaker bleach solution for a longer period is more effective. For example, try agitating explants in 5% bleach for one hour or in 1%–3% bleach overnight. Alternately, if the 10% bleach is not sterilizing the plant explant, a stronger concentration of 15% or 20% can be tried. Some plant material can tolerate the higher concentrations of bleach, but sensitive plants cannot. If the bleach solution is too strong, the explant will bleach white.

8 Sterile Technique and Aseptic Culture

Even though you may be more used to operating in gardens and greenhouses than in test tubes and jars, you will soon be at ease working with plants in miniature. You will feel the same fascination and wonder, the same concerns, cares, joys, frustrations, and curiosity that plants inspire in any setting. You will learn to detect subtle changes in growth or responses to different media, and you will recognize symptoms of stress and disease that may not now be familiar to you. Sometimes, unfortunately, you will observe pinpoints of contaminating molds, or wisps of bacteria in a medium. You will be reminded that the air is full of contaminants, that much greater care is required in the laboratory than in your greenhouse, but also that contamination will decrease as your technique and knowledge increase. Because you are familiar with the whims of plants, you know that patience and persistence are crucial for any successful crop.

Thousands of transfers are made in the transfer hood using sterile technique. The operations will go very slowly at first, but with practice, you may eventually transfer 1000 or more cultures a day, depending on the material. You will have to make many decisions based on careful observation of the cultures during transfer. Watch for plants that should be culled. Make note of what media and techniques work best, and of what changes might be made.

Sterile Technique

As we have emphasized, cleanliness is of primary importance in the transfer room, especially in the hood, where cultures are momentarily outside of their sterile protective containers—a time when they have the greatest danger of becoming contaminated. It is not easy to imagine and deal with the unseen. Microscopic organisms or contaminated particulates are everywhere, except on that which has been sterilized and subsequently protected. You have to assume that clothing, skin, counter tops, instruments, and the air are teeming with mold spores, bacteria, and a host of other invisible enemies ready to invade clean cultures. Just think where you would be if you were a bacterium.

The air that is gently streaming through the transfer hood is filtered to allow the technician to open culture vessels and make sterile transfers with reasonable assurance that the cultures will not be contaminated during the operations. Except for the HEPA filter, surfaces should be wiped down daily with a disinfectant such as 10% bleach, 70% alcohol, or a household disinfectant such as Lysol. HEPA filters are fragile and any contact with them should be avoided; take care to not push containers, instruments, or other objects into a filter. In addition, when arranging equipment in the hood, avoid any obstruction to the laminar airflow because it will change the airflow pattern and so invite contamination. Do not place any possibly contaminated items such as beakers or racks of test tubes between the HEPA filter and exposed cultures. There should be a clear path from the air flowing from the hood to the sterile cultures and equipment.

Be sure to review the section in chapter 4 on sterilizing equipment for transfer tools. The more commonly used methods include alcohol, bleach, an infrared sterilizer such as one of the Bacti-Cinerators, and glass bead sterilizers. Each method of sterilizing has its drawbacks. The alcohol flame poses a fire hazard. Bleach must be mixed fresh every 4 hours because the chlorine evaporates. The fumes that are given off from open containers of bleach can be harmful or irritating. Bleach is also a little messy to work with. When using bleach, wear gloves because it can burn your skin; it will also stain or eat holes in some fabrics. The electric sterilizers are the most expensive way to sterilize instruments. In addition, over time the forceps and scalpels will get very hot and take time to cool.

Deciding which sterilization technique is the best to use is a matter of individual preference. Whichever method you choose, it is usually best to have more than one set of instruments so that one set can be sterilized, cooled, or dried while the other is being used. Instruments should be stainless steel; otherwise, they will corrode in any of the sterilizing methods.

The procedures defined here for maintaining clean conditions in the laboratory (see below) employ bleach solutions for sterilizing instruments. Using bleach is a good method for the

General practices for maintaining a clean and safe environment

The following practices should be strictly observed at all times to maintain safe, clean conditions in the laboratory, particularly when working under the laminar flow hood:

- No eating, drinking, smoking, or storing food in the laboratory.
- Wear a protective lab coat or apron when working with cultures. Avoid wearing clothing with long, full sleeves.
- Tie-back and cover long hair. Loose hair is a potential contaminant for cultures and risks catching fire from any open flames.
- As you work under the laminar flow hood, sit straight and keep your head out of the transfer hood. Work at arm's length, as far back in the hood as is practical.
- Avoid wide sweeping arm movements over the work area inside the hood; if a hand or arm reaches from one side of the hood to the other, contaminants can fall from sleeves or arms. Keep your right hand on the right-hand side and your left hand on the left-hand side of the transfer hood work area.
- If you need to pass an object from one side of the hood to the other, use great care, keeping elbows as close to the body as possible; or simply pass the object from one hand to the other.
- Remember that the hand that holds a test tube is contaminated by it and can in turn contaminate forceps, which if not sterilized, can then contaminate the inside of a test tube during a transfer.
- Set only necessary equipment in the transfer hood. Keep the area as uncluttered as possible and do not block the flow of air coming from the hood.
- Sterilize in an autoclave or pressure cooker any jars and petri dishes containing contaminated cultures before emptying.
- Disinfect work surfaces before and after working in the hood.
- Keep transfer room doors and windows closed to avoid unnecessary drafts.

beginner, especially the beginner who wishes to spend as little capital as possible when getting started. The same general principles apply to other sterilization methods.

During Stage I, it is best to start cultures singly in culture tubes instead of in large groups within larger vessels because in this way if one start becomes contaminated it will not contaminate other explants. If a hundred or so test tubes of a particular plant are being propagated, and the loss of a few in one jar would not be a significant loss, then it is time to move them into jars.

Also, remember that, because it is difficult to distinguish one culture from another, accurate labeling is crucial. It is helpful to devise a plan for labeling and to have a second person double check your labeling. To avoid further mislabeling, it is important to work on only one clone in the hood at any one time.

Having considered some broad matters of sterile technique, we can now address the specific activities in the transfer room. In following the procedures, be sure to employ the previously mentioned general practices for maintaining a clean and safe environment. The sterile technique involved in final explant preparation, and in dividing and transferring plant material sets the pattern for all operations in the laminar airflow hood. Other procedural modifications, such as using another system for sterilizing the transfer tools, will require only slight adjustments and will be self-evident to you as you develop your technique.

Sterile technique for explant preparation

EQUIPMENT AND SUPPLIES
10% bleach in sterile plastic dish
1% bleach in sterile plastic dish
2 100-ml sterile beakers, each containing about 75 ml of sterile distilled water
1 100-ml sterile beaker containing 1% bleach solution in sterile distilled water
Plastic gloves (optional, depending on personal preference)
Forceps
Stainless steel knife (scalpel) with disposable blade
Instrument holder
Pretreated explants in beakers of 10% bleach
Sterile paper towels
Sterile test tubes of MS-based *Begonia* medium
Test tube racks for 10 and 40 tubes
Wastebasket, placed on the floor beside the technician
Labeler
70% ethanol

When preparing the bleach, disinfect 2 plastic dishes with alcohol and rinse well with distilled water before pouring in the bleach solutions.

Alternative methods for sterilizing transfer tools, such as glass bead sterilizers or infrared sterilizers, can be used in the transfer hood as well.

To sterilize the paper towels, wrap about 25 towels in aluminum foil or place in an autoclavable plastic bag and sterilize for 45 minutes at 15 pounds pressure.

A spray bottle filled with ethanol (70% v/v) is invaluable for spraying down surfaces and hands.

> Volume percent concentration (% v/v) gives the volume of the solute present in 100 ml of the solution. A 70% v/v ethanol solution has 70 ml of ethanol (ethyl alcohol) for every 100 ml of the solution.

Procedure for handling explants using sterile technique

The following procedure picks up where step 8 of explant cleaning and treatment in chapter 7 leaves off (see page 95). Left-handed people should reverse the hands indicated here. Bleach immersion is not necessary if using a glass bead sterilizer or an infrared sterilizer.

1. Turn on the transfer hood blower for at least 10 minutes before transferring (or leave it on all the time).
2. Wipe or spray inside the hood with 10% bleach, 70% ethanol (v/v), or a household disinfectant such as Lysol; do not spray the HEPA filter.
3. Wash hands thoroughly and spray them with the ethanol.
4. Immerse knife (scalpel) and forceps in 10% bleach for at least one minute and rinse in 1% bleach. Drain and place in instrument holder.
5. Using forceps, rinse the treated explants in 1% bleach, then in 2 sterile distilled water rinses. Leave the explants in the sterile water rinse while performing steps 7 and 8.
6. Using forceps and knife, lay a sterile paper towel on the counter as far back in the hood as is practical to work.
7. Return forceps and knife to 10% bleach for one minute, then rinse in 1% bleach.
8. Using forceps, place an explant from the sterile water onto the paper towel.
9. Pass the forceps to the left hand and pick up the knife with the right hand.
10. Using forceps and knife, trim the explant appropriately. (See Section Two for guidelines on trimming explants for particular plants.) For shoot tips, you can use from 0.04 in. (1 mm) to several centimeters; about 1 in. (2–3 cm) is usually a good length. Stem or petiole sections should more often be just a few millimeters. Whole young buds will naturally vary in size.
11. Return forceps and knife to 10% bleach for one minute, then rinse in 1% bleach.
12. With the left hand, grasp a culture tube containing medium for the explant. Hold the tube near its base (Figure 8-1). To prevent any microbes from falling into the tube, hold it at about a 50° angle and facing right—parallel with the front of the hood so that it is facing neither the back filter nor you.
13. Take forceps in the right hand. While still holding the culture tube and the forceps, grasp the tube's cap with the right-hand little finger and remove the cap from the tube.
14. Still holding the culture tube, forceps, and cap, use the forceps to remove an explant from the sterile paper towel and place the explant firmly on the medium in the tube. Alternatively, the explant can be transferred directly to the tube with the same knife used for trimming.
15. Still holding the forceps, replace the cap on the culture tube and place the tube in the test tube rack.
16. Return forceps to 10% bleach.

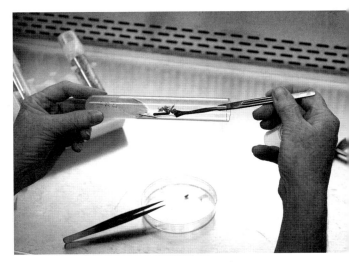

Figure 8-1. Technique for inserting a propagule (transfer) into a culture tube. The culture tube cap is held by the right-hand little finger.

Sterile technique for transferring plant materials

The sterile technique for transferring cultures is similar to that of the final stages of explant treatment just described, yet it is different enough to require redefining some of the steps.

Procedure for transferring cultures using sterile technique

1. Place the culture tubes containing cultures ready for transfer on a counter or cart outside of the transfer hood.
2. Have fresh sterile tubes containing agar medium ready to receive the transfers under the flow hood.
3. Steps 1–4 above.
4. Steps 6 and 7 above.
5. With the left hand, grasp a culture tube containing the culture to be transferred. Check the label to confirm that the tube contains the correct plant material.
6. As described in steps 12 and 13 above, hold the tube near its base and at about a 50° angle. While holding sterile forceps in the right hand, twist off the tube's cap with the right-hand little finger. Remove the cap and place it on the workbench (to be removed later), or drop it into a temporary container on the floor.
7. Using sterile forceps, remove the plant material from the culture tube and place it on the paper towel in the transfer hood.
8. Place the tube in a 40-tube test tube rack, which can go directly into the dishwasher.
9. Pass the forceps to the left hand and pick up the sterile knife with the right hand. Cut, trim, and divide the culture as necessary (see Section Two guidelines). Trim away any brown or dead material.
10. Return knife to 10% bleach.
11. Pass forceps to the right hand.
12. Using forceps, place the plant material in a culture tube with the appropriate medium, following the procedures described in steps 12–14 above. Some shoot transfers should be laid on the surface of the agar to induce bud break if there are lateral nodes. Most shoot tips should be inserted into the medium with their bases just deep enough in the medium (about one centimeter) to allow the shoot to be held upright. Transfers that are going into a rooting medium should be held upright.
13. With forceps still in the right hand, replace the cap on the tube and place the tube in a rack ready for labeling.
14. Return forceps to 10% bleach.
15. After doing about 6 culture tubes, place the towel with the disposable material on it into the wastebasket, and get a fresh sterile paper towel, again using sterile technique to lay the towel down in the transfer hood.
16. When you have finished transferring all the tubes of one variety, label each tube with a marking pen, grocery-store labeler, or hand stamp. Code the date, culture name, and your name.
17. Place the tubes in planter trays or keep in test tube racks, and move the transferred cultures to the shelves in the culture growing room.
18. Record the date, how many test tubes in, how many tubes out, multiplication rate, medium used, when next transfer is due, your initials, and any other desired information.

Other Tissue Culture Techniques

The main methods of tissue culture are classified according to the starting material, which for the home hobbyist involves obtaining an explant from a shoot tip or part of stem, bud, flower, or leaf as described in chapter 7. Several less familiar techniques are worth noting here. These include meristem culture, macerated tissue culture, nurse culture, and micrografting.

Meristem culture

The term *meristem culture* is often used incorrectly to indicate micropropagation. Literally and correctly, however, it refers to the utilization of meristematic tissue for explant material, usually the microscopic apical dome and the accompanying pair of leaf primordia. Meristemming is frequently used for establishing virus-free clones, the idea being that meristematic tissue grows faster than plant viruses can infect these new cells; therefore, the smaller (that is, younger) the meristem explant, the greater the probability that it is virus free. Many propagation laboratories use heat treatments along with meristem cultures to produce virus-free plants; they test for viruses before issuing virus-indexed plants (see chapter 11).

Strawberries (*Fragaria*) are easy to meristem compared with most plants, but the process for treating and extracting strawberry meristems is quite typical of the operation and therefore makes a good plant to use for learning the procedure (Figure 8-2). This is not an easy procedure even for the most experienced lab technicians. However, knowing precisely what to look for will come quickly with experience and will be very rewarding.

Start by selecting a young, actively growing runner before its bud leaves have expanded. Remove about 2 in. (5 cm) of the young runner tip. Because the meristems are within protective buds, initially they are clean; careful treatment should be undertaken to ensure that no contaminants are introduced.

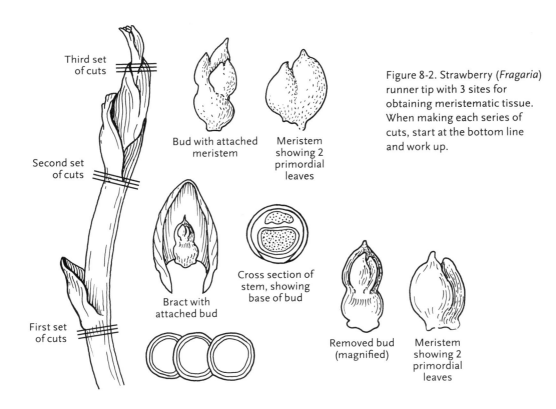

Figure 8-2. Strawberry (*Fragaria*) runner tip with 3 sites for obtaining meristematic tissue. When making each series of cuts, start at the bottom line and work up.

Meristem culture of strawberry (*Fragaria*)

EQUIPMENT AND SUPPLIES

Pretreated explant material
10% bleach in sterile beaker
70% alcohol
Dissecting microscope
Sterile water in beakers
Sterile petri dish
6-in. (15-cm) stainless steel knife with disposable blade
Small (watchmaker's) forceps or tweezers
Sterile paper towels
Culture tubes containing sterile agar-gelled medium

Procedure for meristem culture

1. Stir the explant in 10% bleach for about 10 minutes.
2. Wipe down the dissecting microscope with 70% alcohol before placing it in the transfer hood. Do not touch the microscope lenses.
3. In the hood, agitate the explants in 2 rinses of sterile water.
4. In the hood, place a sterile petri dish under the objective of the stereo dissecting microscope.
5. Using sterile technique, place a piece of sterile paper towel in the petri dish and moisten with sterile water.
6. Using sterile technique, place the runner tip on the paper towel in the petri dish under the microscope objective. Have ready a 6-in. (15-cm) knife with disposable blade and watchmaker's forceps (or tweezers).
7. Search for 2 or 3 buds on the runner tip. At some distance from the tip, you may notice a bulge in the stem, with or without an accompanying bract.
8. Using sterile technique and the sterile knife, slice across the stem just below the bulge. Make another horizontal cut at the point of attachment of the outer epidermal layer.
9. Make a vertical slit through the loosened outer layer and tear it away with the forceps. Examine the base of the stem piece to find the bud. If very young, the bud may be difficult to detect, lying flat and close to the stem. If you cannot locate the bud at the base, then examine the base of the sheath that you removed.
10. When you locate the bud (some runners may not have them in this area), dissect it by slicing away very thin pieces from the base up to the meristematic dome. The dome and 1 or 2 pairs of leaf primordia will appear clearer than surrounding tissues.
11. Touch the knife tip to this bit of tissue. Moisture will hold it to the knife tip until you place it in the culture tube and release it by lightly cutting the blade tip into the medium.
12. Cap the tube and examine it while holding it up to the light. You will barely be able to see the microscopic explant.
13. To locate the next bud, continue to slice away the stem upward toward the tip until you find another bulge.
14. Again, slice through the stem at the point of attachment of the sheathing, or outer layer. Find and dissect the bud as described in steps 10 and 11.
15. The third bud is the apical bud and will be more difficult to find because of the older leaves surrounding it. Continue to slice upward on the stem. Often a slice will reveal the base of a bud, or the dome may actually fall out.

Macerated tissue culture

Macerated tissue is one form of explant material that is used for tissue culture, particularly for growing callus cultures. Individual stem tips or meristems can be mashed with a scalpel, knife, or spatula. Sometimes cleaned explants can be homogenized in a sterilized blender. This method has worked for Boston fern (*Nephrolepis*) and for several of the other ferns, especially those that form spores on their leaves. Any macerating of explants should be done in the sterile environment of the transfer hood.

Nurse culture

Nurse culture is a technique that was developed to obtain a cloned plant from a single cell. It is a technique worthy of consideration for stimulating growth of difficult-to-grow cultures.

In this technique, a filter paper (a semipermeable barrier) is used to separate the nurse culture (usually callus tissue, in either a liquid or agar medium) from the single cell or cells. The cells receive through the filter paper nutrients and growth factors released from the callus and from the medium (Horsch et al. 1980). The growth factors provided by the nurse tissue induce growth and division of the cell or cells on the filter paper.

This technique puts to practical use the fact that a growth medium conditioned by one tissue can promote growth of a low-density culture of a different tissue. Evidence such as this suggests that metabolites critical to initiation of cell division may be the same in cells of different species.

Micrografting

Micrografting is a microscopic version of normal grafting. Although not widely practiced, in principle meristems or small shoot tips can be grafted onto virus-free rootstock to eradicate viruses from the scion plant. The practice also has potential for viral assays and as an alternative to conventional greenhouse grafting.

As is true with normal grafting, the cambium of scion and rootstock should match, but this is easy in tissue culture. The scion can either be placed on top of the rootstock or be inserted into a T epidermal incision. Rootstock can be grown in the greenhouse or tissue cultured, or it can be grown from zygotic embryos (embryos from seed).

Grape (*Vitis*) meristems have been successfully grafted on grape zygotic embryos (Zimmerman 1991). White pine (*Pinus*) meristems have also been successfully grafted on white pine zygotic embryos. In the case of the white pines, it was determined that the preferred point of graft for both juvenile and mature meristems was on the embryo hypocotyl midway between the radical and the base of the cotyledons (Goldfarb and McGill 1992).

Culture Manipulation

Once an explant culture is started, whether from a meristem, a shoot tip, a leaf stem, callus, or whatever, the question is what to do next. It is important to remember that you should never open a sterile culture tube or other container with a clean culture growing in it, other than in the transfer hood using sterile technique.

Depending on the plant and on the stage of the culture, there are numerous ways to divide and transfer growing cultures. A stereo microscope is invaluable for dividing small or complex cultures (Figure 8-3). For the micropropagation of plants, this is generally considered Stage II: multiplication. The early stage of a strawberry (*Fragaria*) meristem culture, for example, is a little green mound that is simply cut in half. On the other hand, a new culture of a *Rhododendron* shoot tip requires that new shoots be cut off and transferred and the old shoot transferred to allow other buds to elongate. An older *Rhododendron* culture will have top growth that lends itself to microcuttings. The top growth of some lily (*Lilium*) cultures is to be cut off and discarded to encourage more basal growth, which is then divided.

Frequently, cultures are divided in a manner whereby the transferred piece will have a surface area of about 1 square centimeter and have one or more shoots. Too small a division will take too long to recover and multiply. Too large a division will need to be transferred too soon for economy's sake. Section Two describes the best methods for dividing and transferring specific plants.

Do not be in a hurry to divide a new explant. Often, particularly with woody plants, it is best to leave a portion of the original stem, with new growth growing from it, for 2 or 3 transfers, gradually removing the old stem. For example, allow a new lateral shoot on a *Rhododendron* explant to grow to about 1 in. (2.5 cm) in length before removing the entire original stem. In the *Rhododendron* shoot tip culture illustrated in Figure 8-4, the top bud of the original explant is removed to help force the lateral buds. When one of the top lateral buds grows out, the explant is cut horizontally (b), on sterile paper towel in the transfer chamber, and the 2 pieces are placed in separate test tubes (c and d). When the other top lateral bud grows out on the dissected piece, a vertical cut is made (c) to produce 2 new cultures (e and f). Meanwhile, as the left lateral bud grows out on the base piece, a second horizontal cut is made (d); the excised bud, together with some of the original stem, is placed in a new tube (g). The remaining part of the original cutting (h) still contains a bud, and that may yet grow out. In the next transfer (not shown), the top buds will have developed sufficiently to be cut from the original main stem. The growth shown here would take place over several months.

Meristem explants are usually slower to develop than shoot tips. The growth from a strawberry (*Fragaria*) meristem may be readily visible within a week, but it may not be more than 2–3 mm after one month. At this size (and/or time), the meristem will benefit from being turned over. Simply move the meristem explant (using forceps and sterile technique) to a new location on the same agar slant. This serves to place the explant in fresh medium and ensures that it has good contact with the agar. In another 1–3 weeks, it may become a 6-mm green ball or base. Cut this in half and subculture each piece in a new tube.

Cultures can be considered established and in the multiplication stage (Stage II) as soon as they are successfully divided. Then the question is, How big should the transfers be, and how often should they be transferred? One rule of thumb is to transfer a piece about 0.25 in. (6 mm) in diameter when the material has primarily basal growth (mainly for herbaceous plants); or transfer a 1-in. (2.5-cm) long piece when the culture is primarily extended shoots or lateral breaks.

The usual protocol when clumps develop is to make one or more vertical cuts through the base for new transfers. In blackberries (*Rubus*), for

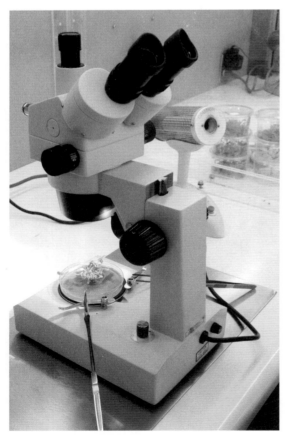

Figure 8-3. The stereo microscope as used in dividing explants.

example, if the tops are more than 1 in. (2.5 cm) long, they can be cut off and transferred in small clumps. If tall enough, 1- or 2-node cuttings can be made. Some bulbs do best if top growth is discarded and only the bases are divided and transferred. Shoots or stem pieces should be laid down on agar to provide maximum contact between the culture and the agar. However, if leaves bleed where they rest on the agar, it is better to place only the base of the stem piece in the agar.

The length of time between transfers varies according to the plant variety, culture history, size of transfer, media, and other factors. It is a sign that it is past time to transfer when the medium is discolored, leaves turn brown, growth slows or stops, or test tubes become crowded. Ideally, cultures should be transferred just prior to any of these events, but that time may be difficult to determine. Experience is probably the best instructor.

Some plants require transfer every 2 weeks, others every 6 weeks. The length of cycles will vary not only with the species but also from one stage to another. Strawberries (*Fragaria*), for example, may require transfer every 10 to 14 days after the second transfer, but after 2 or 3 months in culture the best interval may be 4 weeks (Figure 8-5). The standard transfer time for most established cultures is approximately 4 weeks; it is assumed that the nutrients and/or sugars are used up, or almost used up, by this time if the plantlets are actively growing.

The 4 stages of culture growth—establishment (Stage I), multiplication (Stage II), rooting (Stage III), and acclimatization or hardening off (Stage IV)—are not necessarily distinct. In many

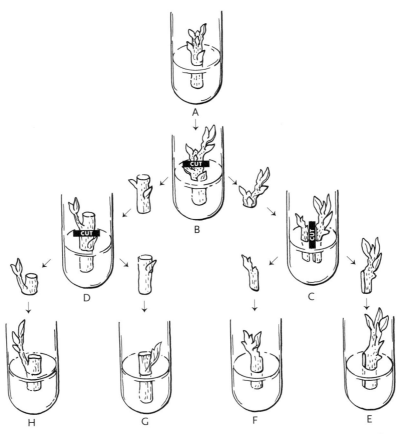

Figure 8-4. A *Rhododendron* shoot tip explant, showing growth and cuts to be made in the first 3 transfers.

instances, a different medium does not need to be used for each stage. There are 4 possible reasons for this:

1. The multiplication medium serves just as well for the establishment stage.
2. Rooting may be done *ex vitro* (out of culture).
3. A single medium may be satisfactory for the first 3 stages.
4. A prerooting medium, perhaps one with no hormones, may be applied in the late-multiplication stage.

It may be surprising to learn that cultures generally have no roots throughout Stages I and II. The absence of roots appears to help promote shoot growth and multiplication. If roots do appear, the culture will probably stop multiplying and can leave juvenility and mature.

Clean Up

Tissue culture operations produce a constant flow of used containers that require washing for reuse. For labs that have household dishwashers, the tines on the bottom shelf can be bent down to accommodate racks of test tubes, which are inverted and placed on the shelves for washing. It is not necessary to remove the agar medium from the test tubes before washing; the hot water in the dishwasher will dissolve and wash it away with no problem. To keep the test tubes in the rack, place a hardware cloth (wire mesh) cover over the tubes before turning the rack upside-down and putting it in the dishwasher. Jars can also easily be washed in the dishwasher.

Tube caps seldom need to be washed because they are sterilized with each new batch of medium. When caps appear dirty or wet, simply wash them in warm, soapy water.

Agar should be disposed of promptly so that it will not grow microorganisms. Culture media contain sugar and nutrients that are ideal for many contaminants. When cultures become visibly contaminated, the molds, yeasts, or bacteria will quickly overgrow and kill the plantlets in culture. Occasionally media containers will become visibly contaminated before they can be used for transfers. Contaminants form patches of growth in or on the medium. Liquid media become cloudy and are no longer clear when they are contaminated. If the agar is opaque, as some agars are, it can be difficult to see contaminants.

When contaminants are present, they pose a significant threat of spreading to other cultures. It is doubtful that any of the contaminants are human pathogens (most human pathogens thrive in a higher pH), but there is always this risk. The safest means of disposing of these colonies is to place containers that have the contaminants in a pressure cooker or autoclave and sterilize them before opening; then empty the contents and wash the culture vessels. If microbial contaminants appear in jars of cultures that are about ready for Stage IV acclimatization, the cultures may well be rinsed and planted.

Figure 8-5. Successful Stage III culture of strawberry (*Fragaria*).

9 The Growing Room: Plant Multiplication and Dealing with Problems

The activity in the culture growing room depends on the work in the media preparation and transfer rooms, and the success of culture growth determines the success of the entire business—it is the reason for precise media making and painstaking transferring, and is the basis by which viable plants will sell. Knowledgeable tissue culturists spend a part of each day monitoring the cultures in the growing room. The person who monitors the cultures should coordinate the media making and transfer room schedules to meet production requirements. The coordinator can also be the troubleshooter who identifies problems and tries to solve them.

Careful attention must be paid to meeting the lighting and temperature requirements of the particular plants being tissue cultured (Figure 9-1). Most cultures will do well under the conditions of light and temperature described in chapter 4, and fortunately, cultures seem to have a wide tolerance for light intensity. In general, plants in culture have low rates of photosynthesis—probably because of the sucrose in media. The fact that cultures are green indicates the synthesis of chlorophyll, which can be an indicator of adequate light. There are exceptions, of course: cultures that do better in higher or lower light intensities, or even in darkness for one or more stages of culture. Carnations (*Dianthus*) and orchids, for example, are often cultured in continuous light, whereas hosta and lily (*Lilium*) cultures can multiply in continuous darkness. *Gerbera* responds best to high light intensity (1000 f.c.) during the rooting stage.

Dealing with Problems

Many problems pop up during the operation of a tissue culture laboratory. We could try to list them all, but then something unexpected would happen to add to the list. This section covers several common problems that occur in the growing room: contaminants, media problems, air requirements, vitrification (hyperhydricity), and refrigeration.

Contaminants

Typically, the contamination rate in a well-run commercial growing room can be expected to be about 1%. Obviously, as the contamination rate increases, the cost effectiveness of the operation decreases. Contaminated cultures should be removed as soon as they are discovered. Although culture containers are almost completely sealed, contaminants can spread to neighboring vessels. Chapter 11 outlines in more detail the types and sources of contaminants commonly encountered in tissue culture, as well as methods for detecting them.

It is easy to tell when cultures are overrun by bacteria, yeasts, or molds growing on the surface of the agar (Figure 9-2). Other contaminants that grow as a cloud or haze within the medium, often surrounding the submerged base of the culture, are more difficult to see, especially if the agar is not perfectly clear. The best way to detect contaminants within a medium is to look through the medium while holding the test tube up to a light.

Culture mites are another specter; they travel invisibly on air currents, carrying with them bacteria and spores. They sometimes can be detected by tiny "footprints" across the medium. Often you will see these fungal clues lining up in a row on top of the medium from large to small—the direction that the mite is walking.

Bacteria are most easily observed especially when gellan gum or other clear agar is used because many bacteria grow below the agar or gel surface, if they are emitted from within the stem through the lower cut end. This kind of contaminant will also make the medium look cloudy or the top of the medium look slimy.

The best course of action against any contaminant is to immediately remove the culture from the growing room and sterilize the culture vessel. Re-examine sterile procedures and check for possible sources of contamination. All reasonable steps must be taken to prevent the spread of existing contaminants and the introduction of new ones. One method of protection that can be effective is to place each tray of containers within a large, sealed plastic bag. Due to the nature of the manufacturing process of plastic bags, they are sterile on the inside until opened.

In general, antibiotics are ineffective against culture contaminants. Furthermore, they can be toxic to the plants, and due to the threat of developing immunities, dangerous to humans. Occasionally, however, successes with antibiotics are reported, and some cell cultures require antibiotics. One study used a combination of 25 mg cefotaxime, 6 mg tetracycline, 6 mg rifampicin,

Figure 9-1. Growing room shelves and lighting at AG3, a commercial tissue culture lab in Eustis, Florida.

and 25 mg polymyxin B, per liter. Because these antibiotics degrade if they are autoclaved, they must be cold-filter sterilized before adding to sterilized media. They are also used as a soak in explant cleaning. Although by no means a cure-all, antibiotics are effective against at least some woody plant bacteria (Young et al. 1984), especially when used both as a presoak and in liquid media. Gentamicin sulfate, streptomycin sulfate, and penicillin-streptomycin solution have also been credited with some success. Gentamicin and streptomycin can be autoclaved and still be useful in some instances.

Some tissue culturists employ indexing, a special method of testing for the presence of contaminants. Indexing is particularly valuable to reveal systemic (internal) microbial contaminants in newly started cultures (see chapter 11 for details on indexing for biological contaminants).

Media problems

A careful technician must watch for bleeding, the phenolic exudates from the cultures that color the medium an inky purplish black. This condition can be the result of a dull knife, an old culture, or a too-liquid medium. In addition, some plant species are especially prone to produce phenolic compounds. Bleeding usually has an adverse effect on the culture, turning the base of the plantlet black and retarding its growth and multiplication. It is worthwhile to transfer such cultures as soon as the symptom is observed.

Other culture conditions to look for are changes in color or growth patterns of the plantlets. It is usually an indication that it is time to vary the formula or environment when leaves turn yellow or brown, when cultures grow too fast with a watery succulence (vitrification/hyperhydricity), when stems turn red, or when unwanted roots or callus are produced. If cultures produce roots prematurely (in Stage II), the plantlets tend to mature and will cease to multiply. This calls for quick action or else production will be curtailed. Transferring to media with higher cytokinin may be sufficient to reverse this trend and bring back multiplication. Adjusting formulas to accommodate the special needs of cultures is an art that is necessary for success.

Because of the dramatic influence hormones have on culture growth, the prime suspect to investigate when problems arise is the ratio of cytokinin to auxin in a formula. A methodical way to establish the optimum cytokinin-to-auxin ratio is to draw a grid and plot an auxin and a cytokinin on the 2 axes in increasing increments (see Table 8). For shoot cultures in the multiplication stage (Stage II), rule out ratios where the auxin content is greater than the cytokinin content (at least in the initial trials). A preliminary test might include the 6 bold-faced ratios in Table 8 that have as much or more auxin content than cytokinin content. To conduct the test, mix media of each extreme ratio, a few other ratios, and the standard formula for comparison. Grow at least

Figure 9-2. Who sneezed into the culture tube!? Three different contaminants on potato culture.

10 tubes in each variation, preferably through 2 transfer cycles. Sometimes the cultures grown in the extremes will show dramatic differences, possibly a helpful clue. For example, a 0/0 ratio (no hormones) has been known to promote elongated shoots where short shoots had been a problem. Zero hormones will not be the entire answer in that case, but it indicates that the problem is hormonal and that less cytokinin (or a different one) should be tested.

When trying to improve culture response, there are other ways to alter formulas besides varying the cytokinin-to-auxin ratio. In fact, the number of possible modifications to media is infinite. Different growth regulators can be used at different concentrations. Stock solutions can be added to media at quarter-, half-, or 2-times strength and the results compared with those of the standard strength. Individual salts or vitamins can also be varied. Try various brands, sources, quality, and quantity of agar. Agar concentration, pH, light, and air circulation are other variables to consider. Whenever a formula modification is attempted, always run a few controls (standard conditions that are causing the problem) for comparison. Trials such as these can indeed be trials—time consuming, expensive, and rarely rewarding, but oftentimes necessary.

Cultures in rooting media often benefit from the addition of activated charcoal (AC), sometimes called activated carbon, which appears to help remove residual cytokinin. The high surface area of AC also enhances its adsorbing properties of inhibitory compounds in the medium. The darkness of charcoal in the rooting area is also a beneficial factor. The presence of charcoal in rooting media makes it easy to distinguish rooting media from multiplication media, but observation of root formation is more difficult. Charcoal has also been used remedially for some problems in the multiplication stage. When using AC remember that the high surface area will affect the speed of adjust pH in the medium; purchase pH-adjusted AC to reduce this problem. Be aware that AC is very light-weight; this can be a concern when weighing it out. Finally, AC can be very beneficial for the cultures, but it is messy to clean out of culture tubes.

Air requirements

The need for air exchange in *in vitro* culture vessels is generally not a common problem. Most culture vessel systems appear to provide sufficient exchange for normal growth. Some cultures, however, such as potatoes (*Solanum tuberosum*), are quite sensitive to a shortage of air. In relatively air-tight containers, potatoes exhibit spindly growth and tiny leaves, symptoms not exhibited by rhododendrons, begonias, berries, and numerous other plants grown in the same type of containers and in the same environment.

If air exchange is a problem, an easy solution is to bore about a 0.25-in. (6-mm) hole in the polycarbonate disc covering the jar and install a piece of nonabsorbent cotton in the hole before sterilizing. This small allowance for air exchange can make a dramatic difference. Some commercially available covers that allow air exchange are Magenta B-Caps, which are vented caps that fit baby-food jars, vented closures for Magenta GA-7 vessels, and Phytacon vented lids, which fit Phytacon polypropylene culture vessels.

Another less common method to help aerate cultures involves pumping sterile air to cultures.

Table 8. **Hypothetical cytokinin-to-auxin ratios**

AUXIN (MG/LITER)	CYTOKININ (MG/LITER)					
	0	0.5	1	3	5	10
0	**0/0**	0.5/0	1/0	3/0	5/0	**10/0**
0.5	0/0.5	0.5/0.5	1/0.5	**3/0.5**	5/0.5	10/0.5
1	0/1	0.5/1	**1/1**	3/1	5/1	10/1
3	0/3	0.5/3	1/3	3/3	**5/3**	10/3
5	0/5	0.5/5	1/5	3/5	5/5	10/5
10	0/10	0.5/10	1/10	3/10	5/10	**10/10**

Note: Bold-face indicates ratios to be checked first.

Open culture containers may be placed in plastic bags into which sterile air is pumped. Alternatively, an intricate tubing system can supply fresh sterile air to each jar. The advantages of aeration are less vitrification (hyperhydricity) and greater ease of acclimatization.

Thomas and Murashige (1979) examined culture vessel air for 5 gases using gas chromatography. Culture vessels with proliferating shoots had less carbon dioxide (CO_2) than those with callus or nodal explants. The implication was that proliferating shoots had greater photosynthetic activity. Ethane quantities were usually the same as in room air. Acetylene and ethyl alcohol were virtually absent from shoot cultures. Ethylene production was quite variable, but usually higher in callus cultures.

Taiz and Zeiger (2010), in the 5th edition of the venerable text *Plant Physiology*, covered nearly every aspect of ethylene effects on plants, both at the whole plant and cellular levels. Some that might apply to shoot cultures *in vitro* are seed and bud dormancy, inhibition of shoot elongation, promotion of radial growth, respiratory changes, epinasty (leaves bend down, short nodes), root and root-hair initiation, tissue proliferation, and curtailment of apical dominance. Under certain conditions, most of these responses can also be induced by auxins; auxins, CO_2, and stress all stimulate ethylene production.

In vitro cultures can be especially sensitive to ethylene which can inhibit cell development. Studies on Habanero pepper (*Capsicum chinense*) found accumulated ethylene caused severe physiological disorders including persistence of callus in all organs, epinasty, leaf distortion, and other growth abnormalities (Santana-Buzzy et al. 2006).

Vitrification

Vitrification, also called hyperhydricity, is a physiological condition wherein tissues appear water soaked or translucent (Debergh et al. 1992). Vitrified tissues lack chlorophyll and become brittle, taking on a crystalline appearance. Other characteristics are rapid and abnormal growth, short internodes, thick curly leaves, loss of cuticle, excessive ethylene production, and death. Vitrification can be a common problem in micropropagation.

The basic cause of the condition lies in the water potential of the culture, that is, the water available to tissues. Vitrification is promoted by the type and concentration of agar and cytokinin in the medium. Low concentrations of agar and high cytokinin encourage vitrification. Chemically, the problem has been attributed to a deficiency of lignin, which is essential to the structure of cell walls and other tissues (Phan and Hagadus 1986).

Several remedies have been suggested, but the best approach is prevention. Use a minimum of cytokinin and a maximum amount of agar, depending on the extent of the problem and the requirements for optimum growth (Pasqualetto et al. 1986). Maximizing the air exchange through the container closure also helps prevent vitrification (Miles 1990). Gas permeable vessels have been shown to prevent hyperhydricity and to encourage normal shoot proliferation of potato (*Solanum tuberosum*) (Park et al. 2004).

Refrigeration

There are, of course, many other problems to contend with in addition to determining the correct medium for a given cultivar. For example, what happens when the transfer crew has not kept up with production demands and a thousand containers are behind schedule? A walk-in refrigerator will do much to alleviate the problem.

Stock plants of unused cultures can be stored in the refrigerator for several months and used as a germplasm source when needed. Chilling can ease an emergency in production. If plantlets are ready to be transferred to soil by mid-winter, their growth can be slowed down in refrigeration to delay planting until early spring, thus reducing winter greenhouse fuel costs. Some stock cultures can be held almost indefinitely in a refrigerator, with subculturing done only once or twice a year.

Certain cultures cannot tolerate cold storage, but those that can may be placed in it at any stage. Use only samples at first and check them frequently until you learn if the plant can tolerate cold storage and for how long. The temperature in refrigerators should be held at about 36°F (2°C) without light.

Cold storage is also useful for administering chilling treatments. Pear (*Pyrus*) and apple (Malus) cultivars respond with additional growth when chilled for 1000 hours prior to greenhouse planting.

Symptoms, Causes, and Solutions

From years of experience, growers know how to deal with many of the requirements and idiosyncrasies of particular plants growing in a greenhouse. Even so, plants do not always perform as expected or in the same way from one grower to another. Significantly less experience has been gained with tissue culture practices, so a grower must be prepared to vary media or other factors as problems arise.

As with most disciplines, there are many problems for which no specific solutions are known. The list of problems and possible solutions in Table 9 is not intended to say, "If you do A and B to solve C, then D will happen." This list will suggest other ideas about the causes of problems and their potential solutions. You and others will help fill in the blanks in this ever-growing, ever-challenging field. For example, the influence of container size on material is a subject of controversy among growers. It is easy enough to test the response of a culture to various containers.

In a larger laboratory, most of these problems would be referred to a research and development department. It is the duty of the person or group of research technologists in this department to troubleshoot. They are usually assigned routine tasks, such as maintaining stock cultures, quality control, sanitation, and safety.

Whether or not you have a research department, it is important to record the response of cultures to different media. The results of every trial should provide information on which future operations are based. Maintaining good records justifies the time invested and provides a key to future opportunity. Note both positive and negative results to avoid repetition of errors and to guide future protocol.

Actions to take as problems arise

- Examine laboratory records to learn when a problem started. If other cultures have exhibited the same problem, check the medium history to learn what previous responses have been. Check the original formula to see if the current ingredients agree, or if any changes were made, whether they were beneficial.
- Check all instruments to be sure they are functioning properly, particularly the pH meter, sterilizer, and water purifier.
- Try variations of medium, light, and temperature.
- Study the greenhouse and field habits and requirements of the plant in question to see if these factors yield any clues for treatment in culture.
- Consult outside sources of help, including computer networks; professional organizations; tissue culture supply companies; cooperative extension service, U.S. Department of Agriculture; other tissue culture laboratories; universities, botany and related departments; and libraries.

Table 9. **Common tissue culture problems and possible solutions**

SYMPTOM	POSSIBLE CAUSE	POSSIBLE SOLUTION
Explant dies	Disinfectants too harsh	Use weaker disinfectant
	Medium too strong	Use weaker medium (half or quarter strength)
	Timing	Obtain explants at different times
Culture blackens and dies	Contamination	Discard culture; review sterile technique and sanitation
	Bleeding	Transfer immediately and more often
	Water problem	Check water purity
	Wrong formula	Change medium
	Too hot	Reduce temperature
Explant alive, but no growth	Dormant	Chill for a month OR start explant at different time
	Needs more time	Transfer explant OR patience
	Medium too strong	Use weaker medium
	Wrong formula	Change medium OR decrease salts and hormones OR layer OR add charcoal OR add GA_3 (gibberellic acid)
	Too hot or cold	Change temperature
Growth too slow	Wrong medium	Change medium OR patience
	Too hot or cold	Change temperature
Shoots too long and spindly	Too little cytokinin	Increase air and/or cytokinin OR run cytokinin-to-auxin grid
Poor shoot multiplication	Too little cytokinin	Increase air and/or cytokinin OR run cytokinin-to-auxin grid
Shoots too short	Too much cytokinin	Decrease cytokinin OR run cytokinin-to-auxin grid
No multiplication	Too little cytokinin	Increase cytokinin OR run cytokinin-to-auxin grid
	Needs chilling	Cold store 4–8 weeks
	Too cold	Increase temperature
	Requires dormancy	Cold store 3–8 weeks
Fat stems, small leaves	Too much cytokinin	Decrease cytokinin OR run cytokinin-to-auxin grid
Unwanted callus	Wrong hormones	Decrease cytokinin OR run cytokinin-to-auxin grid OR omit auxin OR try 1–5 mg TIBA per liter of medium
Chlorotic leaves	Contamination	Check or index for contaminants
	Too hot	Reduce temperature
	Wrong formula	Change medium OR vary nitrogen
Vitrification	Osmosis upset	Reduce temperature OR increase air OR increase agar
	Too much cytokinin	Decrease cytokinin
	Lack of lignin in cell walls	Add phloroglucinol
	Wrong agar	Change agar
	Culture too old	Transfer more often
Premature rooting	Wrong hormone balance	Increase cytokinin OR run cytokinin-to-auxin grid
Red stems	Stress	Change light and temperature
	Too much sugar	Decrease sugar
	Too little nitrate	Increase nitrate
	Culture too old	Transfer more often

SYMPTOM	POSSIBLE CAUSE	POSSIBLE SOLUTION
Stage IV plants die	No roots	Grow roots in culture OR apply auxin to base
	Too hot, too dry	Adjust humidity and soil moisture
	Pathogens	Test for pathogens
	Wrong soil mix, pH, moisture, or food	Research and experiment
	Too rapid hardening	In Stage III medium: decrease sugar, open jars 1–2 weeks
Poor multiplication	Senescence	Shorten time in culture OR increase or change hormone OR add charcoal
	Pathogens	Test for pathogens
Aberrations	Medium	Change medium
	Lack of air	Provide sterile air (put filter in cap)

10 Rooting and Preparing for Transplant

Here is where we prepare our plants to go from *in vitro* to *ex vitro*—life outside the jar (Figure 10-1). Living organisms have a remarkable ability to adapt to changing environments. The ability to acclimate is especially critical to plants, where mobility is not an option. The plant must adapt to the changing environment or die. Phenotypic plasticity (the capacity to change observable traits) helps the plant alter physiological and morphological traits and processes to cope with the new environment.

Despite the delicate nature of the pampered micropropagated plant, it will adapt by reducing water loss, increasing photosynthetic capacity, and producing thicker leaves and roots—all in a matter of weeks. This is the adjustment and establishment period referred to as Stage IV.

Acclimatization can actually begin while plantlets are still *in vitro*, or it can wait until the plantlets are moved from containers to soil-less substrate. If microcuttings are to be removed in Stage II, then the hardening off will occur entirely in Stage IV (see "Omitting Stage III").

Inducing plantlets to root

If plantlets are to be rooted *in vitro* in Stage III, there are several ways of preparing for acclimatization. Container covers can be placed to fit more loosely and allow more water vapor to escape, but if left for more than a few days, contaminants may build up, depending on the purity of the room. The survival rate of carnations (*Dianthus*) approached 90% when containers were left uncapped for 9 days in a room with 50%–70% humidity before transplanting to soil (Ziv 1986). Others have used this practice successfully both indoors and in greenhouses. Depending on room humidity, it is best to remove the lids gradually.

Another effective method is to use lids with filters. This type of lid allows some desirable air exchange and can be applied much earlier than nonvented lids. Bottom cooling to reduce relative humidity in container headspace has worked effectively for *Cynara*, *Iris*, *Rosa*, and *Maranta* (Maene and Debergh 1983).

Using a stiffer agar will reduce available moisture somewhat, but it will make it more difficult to separate the roots from the agar and can cause roots to break off when they are handled. Harder agar also reduces the availability of certain nutrients. If a liquid medium is used for Stage III, many more root hairs will develop, but the roots may be less sturdy than if grown in agar and they may not adapt well to soil.

Decreasing salt levels, especially nitrates, often helps to induce rooting. The amount and type of cytokinin used in Stage II can also affect rooting. *Liriope*, *Schefflera*, and *Philodendron* have been found to survive transplanting better by 80% when 2iP or kinetin is used instead of BA in Stage II. In Stage III, cytokinin is usually eliminated and the auxin level raised. The auxin level is very important, and too much auxin can be worse than none at all. Sometimes 2 auxins used together work better than either one used independently. Lowering sucrose levels in Stage III may encourage photosynthesis, but raising it may favorably affect the water potential.

Succeeding with Stage IV

Ex vitro is often referred to as a natural environment despite the acclimatization greenhouse or growing room being a highly manipulated, very controlled environment. Growers are well acquainted with the hardening-off, or acclimatization, requirements in the greenhouse for seedlings and cuttings. Establishing cultured plantlets in greenhouse conditions is more challenging. If care is not exercised during this very critical stage, high losses can result. An abrupt change to the lower humidity and higher light levels of the greenhouse can be fatal to plantlets in mere hours. You must employ the tools available in controlled-environment production to insure survival.

Reducing water loss

A major problem faced in acclimatization is water loss. Plants have stomata (small openings located between 2 crescent-shaped sunken guard cells) that serve to control water balance and allow gas exchange for photosynthesis and respiration. They are found primarily on leaves and, to a lesser extent, on stems and other structures. Many plants have stomata only on the undersides of their leaves, but some have stomata on both surfaces.

In the process of transpiration, water vapor escapes through the stomata and cuticle (a waxy layer covering the epidermis that protects plants from excessive water loss) to the surface, where it evaporates. Normally, stomata open and close in response to turgidity, but in the high humidity of most tissue culture containers, the stomata are fixed open. When a tissue cultured plantlet is transplanted to soil, the stomata remain open until they are able adjust to the lower humidity and greater light, or until new leaves are produced. This period varies with the particular plant, its conditions in culture, and its new environment.

Electron microscopy has revealed atypical characteristics on leaves that develop in tissue culture. Often there is an abnormally high number of stomata per given surface area. The unusual

Figure 10-1. Stage III plantlets of false agave (*Furcraea*) ready for transplant into cell trays at AG3 in Eustis, Florida.

nature of epicuticular wax (a membranous covering) on leaves and stems in culture has also been observed. It was found to be abnormal in both its chemical make-up and its appearance (Preece and Sutter 1991), and the transpiration protection normally afforded by this waxy coat is depleted in cultures. Further water loss in tissue cultured plants results from guttation, the exudation of water from hydathodes (organs connected to the veins of leaves and situated at the tips of the leaves or at the vein ends on the leaf margins).

Other characteristics exhibited by tissue cultured plantlets that are deviations from normal plants can include thinner roots and stems, thinner leaves, underdeveloped palisade layers, fewer trichomes (such as hairs and scales among others), less collenchyma, reduced vascular tissue, and poorly developed and less chlorophyll.

Increasing photosynthetic capacity

In addition to these changes in morphology and transpiration, the entire photosynthetic process is confused in *in vitro* plantlets. Until Stage IV, the plantlets are heterotrophic—that is, little photosynthetic activity is required. Plantlets use very little carbon dioxide (CO_2) because they draw on the sucrose in the medium for energy instead of depending on photosynthesis. Depending on the species, some plants will exhibit the aforementioned phenotypic plasticity and the existing leaves will become photosynthetically competent. Other species remain heterotrophic until the plantlet develops new leaves in its new environment with the capacity to photosynthesize.

To encourage photosynthesis in plantlets, CO_2 can be introduced into culture vessels or the greenhouse. This is done most readily in an acclimatization chamber rather than while the plantlets are still in culture. If CO_2 is to be introduced, there must be a corresponding increase of light for the CO_2 to be effective. One study found that several woody plants, including *Kalmia*, lilacs (*Syringa*), grapes (*Vitis*), apples (*Malus*), and blackberries (*Rubus*), benefited significantly when CO_2 was introduced into an acclimatizing chamber (Mudge et al. 1992). In this study, optimum irradiance from fluorescent fixtures was at about 130 µmol m^{-2} sec^{-1} (about 1000 f.c. or 10,000 lux) for *Kalmia* and slightly higher for *Syringa*. Carbon dioxide levels were at 1200 parts per million, monitored with a CO_2 monitor/controller. A household-type humidifier provided humidity. A more practical way for the average laboratory to increase CO_2 levels is to loosen container covers during the first day of the Stage IV process.

Omitting Stage III

For some species, Stage III (the rooting stage *in vitro*) can be omitted. Plantlets are transplanted and rooted into soil-less substrate directly from Stage II (Figure 10-2). The success of this practice depends on the particular plant, its ease of rooting, and on the grower's level of control over the environment. Sometimes the roots that develop in Stage III are not functional outside of the culturing media, in which case the plantlets

Figure 10-2. Stage II plantlets of dwarf brush cherry (*Eugenia mytrifolia* 'Globulus', showing roots starting to emerge from callus.

may as well be transplanted from Stage II (Figure 10-3). The advantages of skipping Stage III are less labor costs, reduced transfer room use, decreased expenses, and more space in the culture growing room.

In one report (Mikkelsen and Sink 1978), the production cost of *Begonia rex* was cut by 50% when direct rooting was employed. However, the plantlets of most species require Stage III for more rapid rooting, better height, more sturdiness, or better conditioning. Ultimately, the quality of Stage IV (*in vivo*) plantlets depends on the quality of the plantlets that come from the growing room.

Transplanting

Common procedure for transferring plantlets from the culture growing room to the greenhouse is to plant the plantlets in soil-less substrate in cell trays, and place them in high humidity in tunnels or tents on benches in shaded greenhouses. Over a 2- to 4-week period, the bench cover should be gradually opened or pulled back and the amount of mist gradually reduced to lower the humidity, thus allowing the existing leaves to adjust and/or assisting new leaves to grow. These enclosures are variously equipped with mist, bottom heat, exhaust fans, cooling pads, and lighting with appropriate timing and light-sensing devices. An entire greenhouse might be similarly equipped in lieu of a tent, but the larger the greenhouse, the more difficult it is to control moisture and humidity.

A growing room or growth chamber, equipped with fluorescent lights over shelves, similar to the culture growing room is a good option for establishing tissue cultured plantlets (from either Stage II or Stage III) into soil-less substrate. Humidifiers and fogging systems can do the job well, especially in smaller areas. Heat and light are easily controlled in this scenario, but space is limited.

As plantlets are moved from Stage II or III to soil-based or soil-less mixes, it is desirable to wash off as much of the agar from the roots as possible because molds, yeasts, bacteria, and insects thrive on the nutritious agar. Plantlets coming out of Stage III are rarely uniform in size within the container. Sort plantlets by size before transplanting to ensure uniform trays (Figure 10-4). Plants of the same size will take up water and nutrients at similar rates. The smaller plantlets may need to be "babied" along; the larger the plantlet, the greater the chance for survival. Various fungicides should be tested on a few plantlets to learn which ones can be used safely; always carefully follow the instructions on fungicide labels. Some growers use a commercial rooting-hormone (auxin) powder or dip at this point.

Figure 10-3. New roots emerge from a bitter switchgrass cultivar (*Panicum amarum* 'Dewey Blue'), transplanted at Stage II. North Creek Nursery, Oxford, Pennsylvania.

Growing media and cell trays

Soil-less substrates (growing media) are the best choice for transplanting because they are clean and easy-to-use. Most consist of varying ratios of sphagnum peat and aged bark with perlite or vermiculite as aggregates. Sand, coconut fiber (coir), and rice hulls can also be substituted or included in the mix. Substrates specific to orchids and other epiphytes include redwood or fir bark, lava rock, tree fern fiber, charcoal, and perlite. There are many commercial growing substrates that work well. Good aeration and pore space is critical to root development.

The water holding capacity (WHC) of the substrate along with cell size impact the scheduling of mist for proper management of root zone aeration. Polystyrene plug or cell trays are usually 20 × 10 in. (50 × 25 cm) with numerous cell depths, shapes (square, round, hexagonal) and counts per tray (38, 50, 78, and so on). Foam trays are also available. When selecting a cell size, consider not only the newly transplanted plantlet but also the leaf span after several weeks' growth. If reusing trays, be sure they have been properly cleaned and disinfected prior to receiving the new transplants.

Maintaining high humidity

Maintaining 100% humidity is crucial at transplant. The period for gradually reducing the volume and frequency of irrigation/mist varies by genera, but a good rule is 2–4 weeks from "intensive care" to "outpatient" for most herbaceous material. On a smaller scale, cell trays can be covered individually with plastic bags (tents) or humidity domes (clear plastic, raised covers that fit over standard cell trays and help maintain moisture). Be vigilant in checking under the covers as temperatures can rise rapidly under the clear plastic on a sunny day, causing heat stress and water loss. Condensation can also be a problem, so consider propping up a corner, perforating the cover, or otherwise ensure some kind of air exchange.

For larger scale operations, a greenhouse mist bench can be used. Utilize static (micro) misters and avoid large droplets common to standard overhead irrigation heads. Check misters periodically; tiny particles or carbonate build-up can clog them. Misters are commonly used to

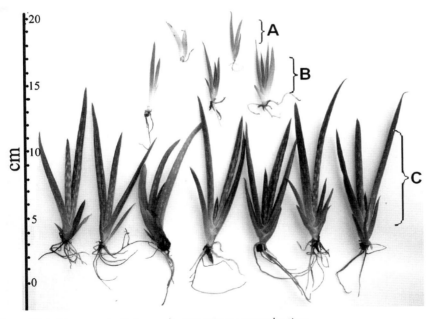

Figure 10-4. *Aloe vera* plantlets vary by size prior to transplanting.

maintain high humidity—the finer the mist, the better. Misters need to be controlled by a timer and watched carefully so that the root zone does not remain saturated.

Fog systems are also available and do a good job of maintaining high humidity without overwatering plants that are being hardened off. These systems are expensive and prone to clogging by carbonates and other minerals in the tap water. Whether mist or fog is utilized, it can be difficult to maintain humidity when exhaust fans or vents are working to reduce heat. If this is a problem, plastic tents can be used within the high humidity greenhouse.

Light requirements

Light management is critical in Stage IV. Light levels should be increased gradually if possible over the first few weeks of acclimatization. Although lighting requirements will vary widely, many cultures root better in increased light (350 to 600 f.c.) because of the increased carbohydrates that are generated as the plants photosynthesize. Other cultures, such as some bulbous and rosaceous plantlets, form root initials more readily in darkness than in light. Depending on geography and/or the season, efforts should be made to reduce light intensity by covering the bench with shade cloth or with spun poly row cover, such as Reemay (Figure 10-5). Aluminized mesh is another choice to diffuse light.

The duration of light (photoperiod) can influence the rate and type of development of many species. Photoperiod is a misnomer, for in reality, duration of darkness triggers growth and reproduction in daylength-sensitive species. For

Figure 10-5. Greenhouse bench with spun poly cover to maintain humidity and diffuse light; also shade cloth to reduce light intensity and temperature.

example, hybrid *Echinacea* Stage IV transplants benefit from 24-hour lighting: it keeps the plant actively growing and prevents the plant from becoming reproductive. The plant's energy is directed to root and shoot growth, not flower formation (Pilon 2011).

Temperature control

In addition to humidity, light, and growing media, temperature control during Stage IV is also important. Air temperature is not the only concern. The temperature of the root zone can greatly effect root development. Many growers utilize bottom heat, especially in early spring (Figure 10-6). At Terra Nova Nurseries, bottom heat is maintained at an even 72°F (22°C) during the day and turned down to 66°F (19°C) at night (Heims 2010). The little plantlets are transitioning from a comfortable, controlled temperature to one that fluctuates and may not be as ideal as was in Stages II and III.

If Stage IV occurs in the greenhouse, temperatures will fluctuate with the season that they are removed with hot greenhouses in the summer and cool greenhouses in winter. A good way to prepare the plantlets for this change is to adjust their temperature prior to removal from the culture vessel. This process may be as easy as placing the Stage III culture vessels in the new temperature environment a few days to a week before Stage IV.

Timing

If the micropropagator sells to (or is) the finishing grower, the timing of Stage IV needs to be considered in the scope of the greater production cycle. For example, most perennial species require vernalization (cold period of 6 to 12 weeks) to bloom and/or be ready for market in the spring. The hardened-off liner should be transplantable early enough that the plant has sufficient root growth and bulk to ensure survival as it goes into the cold period (overwintering).

The number of species and cultivars available of micropropagated hardy perennials, tropicals, and succulents has increased tremendously since 2000. Greenhouse or nursery finishing growers often have the option to let the original propagator do the work of rooting and acclimatization for them, and receive the finished product much like any other vegetatively propagated liner. But if the grower has the proper acclimatization facility, there can be cost savings in ordering in Stage III material and producing their own liners. Reduced shipping charges and per-liner costs can be significantly reduced.

Figure 10-6. Monitoring growing media temperature to optimize root growth of *Hosta* transplants at Battlefield Farms, Rapidan, Virginia.

11 Dealing with Biological Contaminants

This chapter describes some of the primary methods used to detect and identify microbial and viral contaminants. The topics related to contamination that will be discussed are as follows:

- Laboratory sanitation as it relates to solid surfaces, air, and humans.
- Biological quality-control procedures for monitoring sanitation.
- Detection and partial identification of external and systemic bacterial and fungal contaminants using plant tissue indexing procedures.
- Detection and identification of viral contaminants using ELISA (enzyme-linked immunosorbent assay), an immunological procedure.

The term *biological contaminants* refers primarily to bacteria and fungi found on or within explants or in the laboratory. When the term *aseptic* is used, it means that the plant material is free from bacteria and fungi; it does not necessarily mean that it is free from viruses, which are not easily seen. More on this is explained later.

Sources of contamination in the laboratory include laboratory air or solid surfaces, humans, and improperly prepared tissue culture media. Biological contamination of explant tissue is thought to originate either primarily from a systemic contaminant or from improper use of aseptic laboratory procedures (Cassells 1991). Occasionally, external contaminants on explant tissue, such as a bacterial endospore, may be resistant to chemical sterilization. An endospore is a highly resistant, dormant asexual cell that forms in some microbes. When placed in a favorable environment, it germinates into an actively growing vegetative cell.

Plants normally excrete substances from their roots that stimulate microbial growth of both pathogens (disease-causing agents) and beneficial microflora in the rhizosphere (the soil surrounding plant roots). Bacteria can enter plant tissues through natural openings or wounds, from which entry points they begin to colonize the plant's tissues. Fungi may in some cases penetrate plants directly, but often utilize natural openings or wounds.

Alternatively, bacteria, fungi, or viruses can enter plant tissues with the assistance of vectors (disease carriers), including certain insects (Tarr 1972, Matthews 1981). Plants carry a wide range of microflora, including prokaryotes (bacteria and bacterium-like organisms), fungi and algae, as well as viruses and viroids (infectious agents smaller than viruses and composed only of RNA) (Campbell 1985). Viral contaminants do not pose a problem when located outside of a plant; only after they infect the plant can viruses use the plant's biochemical systems for viral reproduction.

Contamination in tissue culture operations can be costly, particularly if the contamination rate is greater than 1%, or if it is a pathogen such as tobacco mosaic virus (TMV) or *Agrobacterium tumefaciens*, the agent of crown gall disease. If contaminants are observed in test tubes or jars, the containers should not be opened and should be sterilized in an autoclave or equivalent before discarding the contents.

Some plant tissue culture managers do nothing more than make a casual observation of contaminants. If the agar is clear and no foreign growth (such as bacteria, molds, or yeasts) is observed, and if the plants look healthy, they are presumed healthy. All too often, no further quality control steps are taken due to the time and expense involved, lack of interest or lack of information. There is sometimes a willingness to gamble that unseen contaminants either do not exist or will not cause any problems. But if the propagator is seriously concerned and is dedicated to having the cleanest possible plants for sale or for in-house use, then additional testing is necessary to determine the presence, origin, and nature of microbial contaminants and to better avoid any such problems in the future.

Determining the source of microbial contamination in tissue culture containers is no small task. Visible contaminants are usually first noticed after some days or weeks of incubation in the growing room, when containers are being handled as they are moved from the transfer room to the growing room. Contaminants have a much faster rate of growth than plant tissues, so when they populate a container they will typically overrun and destroy the culture, even if they are not actually pathogenic.

Laboratory Sanitation

Sanitation for the laboratory, as previously discussed, entails first substantially reducing, and then keeping, the microbial population in the air and on surfaces to acceptable levels. This includes cleaning surfaces by using water, soap, or detergent. *Disinfestation* implies the use of chemical agents to kill pathogenic microbes on the surface of plants or objects without sterilizing the plants or objects—sterilization would destroy plant viability. Disinfectants are a class of chemicals used for disinfestation. *Sterilization* is the destruction of all living matter. Sinks are cleaned or disinfected, explants are disinfected, and media and transfer tools are sterilized before use.

A good sanitation program for media preparation, culture, and transfer operations, and the growing rooms themselves minimizes the need for monitoring quality control. If an enviable 1% or less of containers in the culture growing room has obvious microbial contaminants, it is an indication that the laboratory has good sanitation procedures. If contamination is significantly higher than 1%, then it is desirable to initiate procedures to analyze systematically the sources of the problem, whether they are microbes that are systemic in the plants or merely surface contaminants from the laboratory. Isolation and identification of contaminants may be needed.

Possible sources of contamination are the air from the transfer hood HEPA filter, various solid surfaces, such as tools and containers present in the transfer hood, and flaws in the sterile technique used by the person transferring cultures. An exercise in the proper use of sterile technique is detailed in chapter 8. Anyone not already competent in sterile technique will benefit by first practicing these procedures and testing themselves on the effectiveness of their technique before doing critical work. To do so will save time and money.

Practical knowledge of sanitary technique is necessary for establishing a clean lab environment that does not have a high potential for contributing contaminants to cultures. A daily or frequent washing and rinsing of laboratory floors and bench, cart, and table surfaces is essential. Sweeping the floors, other than for spillage reasons, should be avoided because it stirs up dust particles. A detergent solution containing a surface tension depressing agent (such as 1% Tween 20) is adequate for washing laboratory surfaces. After washing, surfaces should be rinsed and mopped with clean water to speed drying. The walls should be washed at least once every 2 weeks, and the ceiling at least once per month.

Human element

Studies have shown that the number of bacterial colonies found per liter of air rises in direct proportion to the number of people in a room (Nester et al. 1995). This would suggest that there are advantages to having a laboratory with few personnel.

Some individuals tend to "shed" more hair (and/or microorganisms) than others. Such shedders, especially, should wear gloves, head covers, and perhaps masks. If properly positioned, faces and hair should not be over containers or even in the hood. A shield or other barrier at forehead height can serve as a reminder to keep well back from open containers. Coughing or sneezing results in the release of many infectious particles. Swab samples taken from hands and arms can be grown as described earlier in the chapter for solid surfaces. One test that can illuminate the importance of cleanliness is to place your fingers on agar media in petri dishes before and after hand washing. After incubation, compare the colonies growing in each of the dishes. Careful prewashing of hands and exposed arms with disinfectant soap or spraying down with alcohol is recommended for all those involved in tissue culture transfer.

It should be noted that most contaminants encountered in tissue culture operations are of little or no risk to healthy people, provided standard microbiological procedures of containment are employed. The few organisms posing a slight risk are usually normal body-inhabitant species. If you have a special health problem, you should be especially diligent in following the precautions listed. Examples of such conditions are diabetes, allergies (especially to fungi), and disorders of the immune system. To maintain safe conditions in a tissue culture transfer room, common sense and strict personal hygiene should prevail.

Biological Quality-Control Procedures for Monitoring Sanitation

Two different types of procedures are used in the laboratory to monitor sanitation. The first detects contaminants on solid surfaces, the second detects contaminants in the air.

Testing for solid-surface biological contaminants using streak plates

A quality-control test for detecting and enumerating microbial contaminants on solid surfaces is the cotton swab–streak plate procedure. In this test, swab samples are taken from various solid surfaces in the laboratory and tested on microbial-enrichment agar media that encourage any microbes present to form colonies, thereby enabling their enumeration and subsequent identification. (A colony is a visible population of cells that arises from a single cell).

Identifying at least the categories of contaminants is often helpful for tracing the source of contamination. The number of colonies found provides an approximation of the number of microorganisms present per unit of surface area. Regular testing will allow the lab supervisor to note any changes in the cleanliness of the lab over time. It is advisable to repeat the test after attempting to correct the problem.

The common microbial-enrichment agar media for detection of contaminants are trypticase soy agar (TSA) and Sabouraud's dextrose agar (SDA). TSA medium is used to detect bacteria, and SDA medium is used to detect fungal contaminants.

The streak-plate procedure shown in Figure 11-1 should give ample dilution for seeing individual colonies following incubation. If not, you may wish to use an alternative streaking procedure (Figure 11-2), or one of your own modification. Modified procedures may allow for better isolation of single colonies, particularly if the swab sample contains a high concentration of

Streak plate testing of solid surfaces for contaminants

EQUIPMENT AND SUPPLIES
Microbial-enrichment agar media
Balance for weighing media
1-liter Erlenmeyer flasks
Distilled water
Aluminum foil
Autoclave or pressure cooker
Sterile petri dishes (plates), 100 × 10 mm
Labeler
Sterile, wooden-stick swabs with 1-cm-long non-absorbent cotton or polyester (Dacron) tips, in capped test tubes
Test tube rack and basket
Test tubes containing sterile water, capped
Glass-marking pencil
Inoculating loop and handle, approximately 8 in. (20 cm) long
Bunsen burner or infrared sterilizer such as the Bacti-Cinerator
77°F and 98.6°F (25°C–37°C) incubators, if available
Wastebasket

Place the individual cotton swabs in test tubes, cap, and sterilize in autoclave or pressure cooker. Various cap types, including nonabsorbent cotton plugs, screw caps, and push-on caps, are available.

Preparing microbial-enrichment agar plates

1. Prepare TSA medium (from a commercial dehydrated mix) in a flask using 45 g mix per liter of distilled water. For SDA medium, prepare using 65 g per liter.
2. Cover the flasks with aluminum foil and sterilize in an autoclave for 15 minutes at 250°F (121°C).
3. After processing, cool the freshly autoclaved agar to approximately 115°F (46°C)—cool enough that the flask of agar can be held to your cheek without burning your cheek. Agar will solidify at approximately 105°F (40°C). After cooling to room temperature, sterilized agar may be kept in the flask for a short time under refrigeration and remelted prior to use.
4. When ready to pour the agar from the flask, use sterile technique to partially lift the petri dish cover with one hand (see Figure 11-1b), and pour approximately 20 ml of the agar into the plate with the other hand. If the agar is too hot when poured, moisture will condense on the underside of the lid. If it is too cool, lumps will form in the agar. If the sterilized medium hardens before it can be poured, it will require remelting in a steamer or on a hot plate. Use extreme caution because the agar may burn. A magnetic stirrer is helpful here. Sterilized medium can also be remelted in a microwave, again using extreme caution because the agar may quickly boil over.

Preparing swab samples from a solid surface

1. Clean your hands or use gloves.
2. Delimit a solid surface area (bench top, ceiling, or wall) of 3 in.^{2} (19 cm^2).
3. Label each test tube containing a sterile swab to identify it with its corresponding surface sample location.
4. Carefully remove a sterile swab from a test tube, holding the test tube cap as illustrated in Figure 8-1. Replace the cap and return the empty test tube to the test tube rack.
5. Moisten the swab by dipping it in a test tube of sterile water and squeeze out the excess moisture by pushing against the inside wall of the test tube.
6. Swab the area to be sampled using a back-and-forth motion and, without touching any other surfaces, return the swab to the test tube from which it came.
7. Replace the cap and return the test tube containing the swab to the laminar flow hood for streaking on the agar plate.

8. Using fresh swabs for each surface, repeat the swabbing procedure for other surface areas.

Preparing an agar streak plate

The purpose of streaking is to dilute the contaminants sufficiently to enable the formation of individual colonies, each one presumably arising from a single contaminated cell.

1. With a marking pencil, divide the underside of the petri dish containing the agar (as prepared above) into 4 quadrants (Figure 11-1a). On the underside of the petri dish, label the date, the area to be sampled, and the type of medium—TSA or SDA.
2. Aseptically remove a processed swab from a test tube with one hand. With the other hand, lift the petri dish cover above the plate and hold it at an angle such that it shields the agar surface from possible contamination during the streaking process (Figure 11-1b).
3. Still holding the petri dish lid slightly open, gently touch the swab containing the inoculum to the upper-right surface of quadrant one (Figure 11-1c).
4. Discard the swab into the wastebasket.
5. Using a fresh sterile swab, streak the inoculum by gently rubbing the inoculum back and forth approximately 20 times over the whole area of quadrant one (Figure 11-1d).
6. Replace the petri dish lid and discard the swab.
7. Sterilize the inoculating loop by heating it in a Bunsen burner flame or infrared sterilizer (Figure 11-1e).

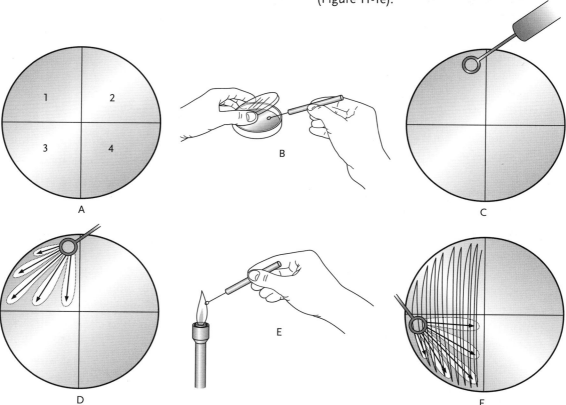

Figure 11-1. Procedure for streaking an agar streak plate: (a) divide the petri dish into 4 quadrants; (b) lift the petri dish cover; (c) touch the swab to one quadrant; (d) streak the inoculum; (e) sterilize the loop; (f) rotate the petri dish and streak the inoculum from the first quadrant into the second. Repeat steps d through f until all 4 quadrants are streaked.

8. Lift the petri dish lid and cool the inoculating loop by touching it to any uninoculated portion of the agar surface. Wait for the sizzling sound to stop.
9. Rotate the petri dish 90° counterclockwise.
10. Streak the inoculating loop back and forth from the first quadrant to the second quadrant of the agar plate with about 25 streaks (Figure 11-1f). Streak quickly, letting the loop ride easily on the agar surface. It is all right if the streaks overlap. Too much pressure will break the agar surface, and too little will result in irregular streaks.
11. Streak the contents of the second quadrant into the third quadrant using the same procedure (steps 7–10).
12. Streak the contents of third quadrant into the 4th quadrant using the same procedure (steps 7–10).

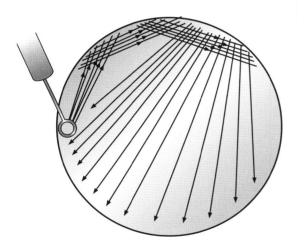

Figure 11-2. Alternate streaking procedure for streak plates.

contaminants. One occasional problem is the presence of *Proteus* spp. or other motile bacteria, which rapidly spread over the entire plate, resulting in only a few or no individual colonies.

After the streaking procedure, the petri dish containing the inoculum must be incubated. Incubate TSA plates at 98.6°F (37°C) for 2–3 days and SDA plates at 77°F (25°C) for 5–7 days. The plates should be inverted to prevent any moisture inside the lid from falling on the inoculum. If no incubator is available, store the plates at room temperature for 1 or 2 weeks.

Following incubation, observe the petri dishes for the number and kinds of microbial colonies. If very few colonies appear on the plates, it is safe to conclude that the surfaces tested had only a few organisms able to grow on the selected enrichment media.

A clean, well-maintained lab should result in a low number of microbial colonies. Over time, keeping a log of your results will give additional meaning to such studies. For example, you may find that at certain times of the year there is a heavier load of contaminants than at other times. This may also suggest increasing the frequency of the sanitation program during such times, which often correspond with increased human traffic entering and leaving the lab.

Observing contaminants on streak plates. For preliminary characterization of microbial colonies that may have developed on your streak plates, observe the following characteristics:

- Color: green, golden, tan, gray, black, white, pink, red
- Size: pinpoint, small, medium, large, spreading
- Shape: round, irregular
- Elevation: flat, convex, craterlike
- Margin: smooth, irregular
- Texture: smooth, rough, woolly, filamentous
- Consistency: mucoid, moist, glistening, dry

A combination of morphological and physiological studies is required for accurate identification of bacteria and yeasts. Bergey's *Manual of Determinative Bacteriology* (currently in its ninth edition, see Holt 1994) is a useful key for identifying bacteria; *Yeasts: Characteristics and Identification* (Barnett et al. 2000) is a key for identifying yeasts. Filamentous fungi, such as molds, can sometimes be identified by making a study of their colonial (Figure 11-3) and cellular morphology (Figure 11-4) using a conventional 40× dissecting microscope and comparing what you find with a key, such as Barnett and Hunter's *Illustrated Genera of Imperfect Fungi* (4th edition, 1998). Dissecting microscopes with higher magnification (to 200×) are also available. These microscopes have an inverted stage, with the illumination source directed from above the stage. They can also be used for observing tissue cultures and microscope slide preparations. A covered slide preparation is sometimes necessary to better observe details of the cellular morphology of filamentous fungi. An introductory laboratory manual, *Microbiology Experiments: A Health Science Perspective* (Kleyn et al. 1995), contains exercises related to all of the above identification procedures.

For normal production quality control, the above procedures for detection and enumeration of contaminants should prove adequate. Further identification of colonies to the genus and species level requires additional training. A good introductory course in microbiology usually includes methods for bacterial identification and sometimes methods for yeast and mold identification. Photographs of commonly seen contaminants may be useful for training new lab employees.

Using a wet mount slide to observe contaminants. It is easy to determine if a non-filamentous colony is a bacterium or yeast using a wet mount slide preparation for observation under a compound microscope.

Using microbiological stain to observe contaminants. Bacteria are often difficult to observe with a wet mount because their optical density (the degree of opacity of a translucent medium) is similar to that of water. Staining will increase the optical density of the bacteria so that their shape and arrangement can be distinguished more easily. Note their shape (rod, coccus, or spiral), their arrangement (cocci may be single, double, tetrad, or cluster), and their size, all of which are important factors for identification.

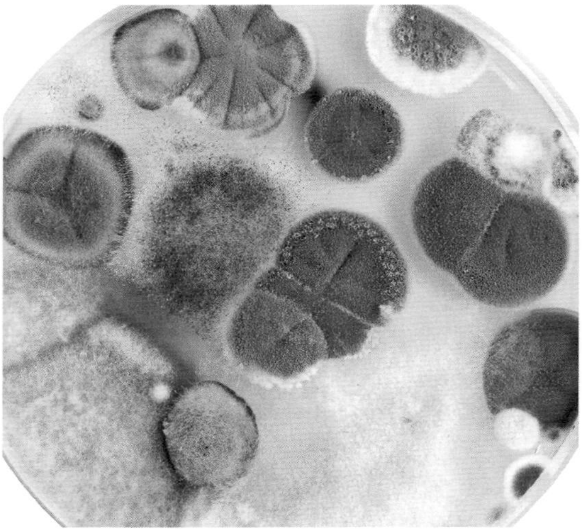

Figure 11-3. Airborne mold and other fungal spores can come from old building materials and improperly filtered ventilation systems.

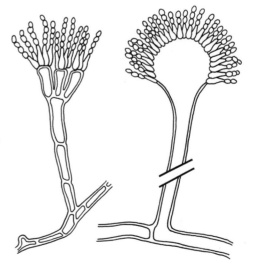

Figure 11-4. Some molds, such as *Penicillium* (left) and *Aspergillus* (right), can be identified by the cellular structure of their fruiting bodies.

Preparing wet mount slides

EQUIPMENT AND SUPPLIES
Microscope slides (75 × 25 mm), 0.13–0.16 mm thick
Coverslips (22 × 22 mm), 0.13–0.16 mm thick
Dropper bottle containing distilled water
Inoculating needle
Compound light microscope
Lens immersion oil (high viscosity)
Microscope lens paper and lens cleaner

Procedure

1. Place a small drop of distilled water on a clean microscope slide.
2. Place a minute amount of the microbial colony on the tip of a sterile inoculating needle and mix into the drop of water. For best results, a very slight clouding should occur after mixing with the water. The needle must be wiped off and resterilized after every use.
3. Cover the drop with a clean coverslip. If the coverslip floats on the drop of water, the droplet is too large.
4. Place the slide on the microscope stage and examine it initially with the low-power (10×) objective. When the object is in focus, you should see single cells, and perhaps you will be able to determine whether they are bacteria, which appear very tiny at this magnification, or yeasts, which are much larger than bacteria and may show some buds attached to the mother cell (Figure 11-5). Sometimes it is necessary to use a stain to see bacteria or yeasts.
5. If your microscope is parfocal, switch to the high-power (45×) objective. When properly in focus, you can readily determine if the cells are bacteria or yeasts. If they are bacteria, place a drop of immersion oil on the surface of the coverslip and carefully switch to the oil immersion objective (95×). Now you may see one or more bacterial forms: coccus (spherical; Figure 11-6), rods (cylinders, bacillus), or spirals (regularly curved rods). Actinomycetes (a group of microscopic bacteria, often pathogenic, that resemble fungi) may exhibit filaments as well as individual cells (see Figure 11-8).
6. When finished, carefully wipe off the excess oil from the oil immersion objective with a piece of clean lens paper.

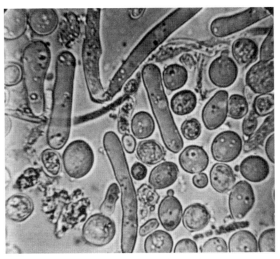

Figure 11-5. Two yeasts and a mold: *Saccharomyces cerevisiae* (baker's yeast) with oval cells, some with buds, and *S. pastorianus* (lager yeast) with cigar-shaped cells.

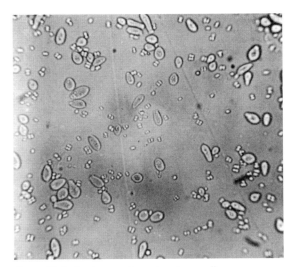

Figure 11-6. A mixture of yeast and a small coccus (spherical) bacterium, *Sarcina*.

One disadvantage of staining is that it kills the bacteria, making it impossible to determine motility. Some bacteria have flagella (long, whip-like appendages), which enable motility. Movement is sometimes difficult to determine under the microscope because bacteria also exhibit Brownian motion (the irregular motion of small particles suspended in a liquid or a gas, caused by the bombardment of the particles by molecules of the medium). The flagella of bacteria are very thin and require special stains to be visible. Some common bacterial stains are methylene blue (also used to stain yeast cells) and crystal violet.

Staining bacteria and yeasts for observation

EQUIPMENT AND SUPPLIES
Microscope slides (75 × 25 mm), 0.13–0.16 mm thick
Coverslips (22 × 22 mm), 0.13–0.16 mm thick
Glass-marking pencil
Dropper bottle containing distilled water
Inoculating needle
Bunsen burner or infrared sterilizer such as the Bacti-Cinerator
Forceps or clothespin
Prepared crystal violet and methylene blue stains
Blotting paper or paper towel
Compound light microscope
Lens immersion oil
Microscope lens paper

Procedure

1. Prepare a clean slide by washing with a detergent solution. Rinse and dry with a paper towel.
2. Using a glass-marking pencil, inscribe a circle on the center of the slide.
3. Place a small drop of distilled water inside the circle.
4. Using a sterile inoculating needle, remove a small amount of a colony and mix it in the water until the drop becomes slightly cloudy. If the broth is too cloudy, uneven staining will occur; if too clear, it may be difficult to find the bacteria. Allow the slide to dry for 10–15 minutes.
5. Attach the bacteria to the microscope slide by heat fixing. This is done by passing the slide through a Bunsen burner flame 2–4 times, until the bottom of the slide is lukewarm. An infrared sterilizer may also be used. In gently heating the slide, most of the bacteria will adhere to the slide without causing serious distortion of cellular morphology.
6. Hold the slide level over the sink using forceps or a clothespin. Add a drop or 2 of stain inside the circle. Let stand for 20–30 seconds.
7. Wash off the excess stain with tap water and dry the stained slide with blotting paper or a paper towel, by dabbing at it gently. Take care not to rub the stained area sideways, which could remove the bacteria.
8. Observe the slide using the oil immersion objective of the microscope, following the same focusing procedure outlined above in steps 4 and 5 for wet mount slides (that is, start with the low-power objective).
9. When finished, carefully wipe off the excess oil from the oil immersion objective with a piece of clean lens paper.

Procedure for observing yeasts

1. Place a small drop of methylene blue on a clean slide.
2. Using a sterile inoculating needle, remove some growth from a yeast colony and mix it in the methylene blue until the drop is slightly cloudy. Place a clean coverslip over the drop.
3. Observe with the 10× and 45× microscope objectives.

Testing Laboratory Air

Properly filtered inside air should not pose a contamination threat in the laboratory. The threat comes largely from people and their activities within the laboratory. The air in the media preparation room need not be as clean as the air in the transfer and growth rooms. However, to maintain the highest cleanliness standards, the air in all 3 rooms should have some positive pressure. The laboratory's ventilation system should be constantly removing old air and introducing new, fresh air. Periodic testing will ensure that air cleanliness remains high.

Using petri dishes for air sampling

This method is especially useful for periodic assessment of the microbial content in the laminar flow hood while it is operating.

Usually each colony will originate from a single cell, which multiplies to form a colony large enough to be seen with the unaided eye. The number of colonies found in the laminar flow hood should be minimal. If the numbers increase significantly, that would suggest a faulty HEPA filter.

Using a membrane filter apparatus for air sampling

A membrane filter apparatus attached to a portable battery-operated vacuum pump is sometimes used to assess air quality in the transfer room or the greenhouse. The pump has a vacuum source to collect air contaminants on a membrane filter disk. A direct microscopic examination of the membrane provides a total count of both viable and nonviable particles (Kleyn et al. 1981).

This method of testing air is more expensive than the petri-dish method. A portable air-sampling pump from Mine Safety Appliances with an adjustable airflow rate of 0.5–3.5 liters per minute costs about US $675. A pump from SKC (224-44XR) costs US $750 and is adjustable from 0.005 to 5 liters of air per minute.

A membrane filter holder assembly, such as the 47-mm polycarbonate in-line filter holder by Pall Life Sciences (#1119) runs about US $100, and the corresponding filters (0.45 µm porosity, 47-mm diameter with grid; Pall Life Sciences GN-6 Metricel #66278) cost US $130 for a package of 200).

Collecting air contaminants in a petri dish

EQUIPMENT AND SUPPLIES

12 covered petri dishes containing sterile agar media, either MS or SDA media for testing fungi, or TSA media for testing bacteria

Tape (such as Parafilm) for sealing the petri dishes

Labels and pen

Procedure

1. Remove the lids from the petri dishes and place the lids on a sterile surface with the inside down so that no contaminants will drift into them.
2. Place the open petri dishes in various locations throughout the room for various times, such as one at each location for 1 minute, 5 minutes, 15 minutes, and 30 minutes.
3. Replace the lids and seal the dishes with tape.
4. Label each dish to identify the medium, exposure time, and location.
5. Invert and incubate in a warm place at 77°F–86°F (25°C–30°C) for a week or 2. Petri dishes should be inverted during incubation to prevent any moisture present on the lids from falling onto the agar surface, and thus interfering with colony formation.
6. Examine the dishes daily for numbers and kinds of colonies growing on the media.

Testing laboratory air with a membrane filter aparatus

EQUIPMENT AND SUPPLIES
Autoclave or pressure cooker
Membrane filter holder assembly
Sterile membrane filters
Tweezers or forceps
Battery-operated pump, with vacuum source
Tygon hose, 0.25 in. (6 mm)
Petri dish
Scalpel or knife blade
Microscope slides 3 × 1 in. (7.5 × 2.5 cm)
Acetone or lens immersion oil
Compound light microscope

Procedure for enumerating and differentiating total particles in air

1. Autoclave a wrapped membrane filter holder assembly outlet with attached hose barb adapter, support screen, and inlet.
2. Using sterile tweezers, insert the support screen in the membrane filter holder outlet.
3. Aseptically remove a sterile membrane filter from the package and place it on top of the support screen.
4. Aseptically screw the filter inlet onto the top of the filter outlet.
5. Attach the hose barb adapter to one end of the Tygon hose. Attach the other end of the hose to the inlet fitting on top of the pump (see Figure 11-9).
6. Turn on the pump and adjust to the desired air-flow rate. After a prescribed period of time, turn off the pump. The desirable airflow rate and duration time of sampling depend on the contamination level of the air sample.
7. Using sterile tweezers, aseptically remove the membrane filter from the membrane filter holder and place the filter in a sterile petri dish.
8. With a sterile scalpel, cut off a small portion (about 10 × 10 mm) of the membrane and transfer it aseptically to the surface of a clean microscope slide.
9. Add a drop of lens immersion oil to clear the membrane surface, and place the slide on the microscope stage.
10. Using the low- and high-power objectives, examine the slide for particle types (amorphous dust, bacteria, and spores) and numbers relative to one another (Figures 11-7 and 11-8). By examining a number of microscopic fields, you can determine the average number of particles per field. This in turn can be related to the number of particles per field per liter of air.

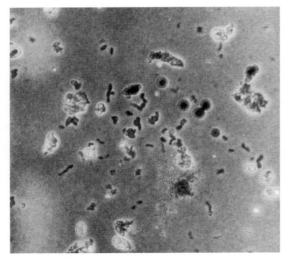

Figure 11-7. Direct microscopic examination (×400) of dust on the surface of a membrane filter, showing bacteria, fungal spores, and inanimate matter.

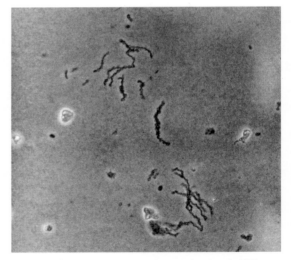

Figure 11-8. Direct microscopic examination (×400) of dust on the surface of a membrane filter, showing filamentous actinomycete cells.

Procedure for Enumerating and Differentiating Viable Particles in Air

The procedure for determining viable (capable of germinating, living, growing) particles in the air differs from that of determining the total number of particles in that the viable particles are collected in a container of sterile water, rather than on the surface of a membrane filter. This is done because many viable particles will become dehydrated and die when collected on a membrane surface.

In this process, the container for the sterile water has 2 outlets: one serving as a hose attachment to the pump, and the other serving as an air inlet. After collection of the air sample, the container is removed from the pump, and the water is poured through a membrane filter. The outlet of the membrane filter is attached to a vacuum flask for filtration. The membrane filter is then removed and transferred aseptically to a suitable agar medium. After incubation, the membrane is inspected for viable colonies.

Detection and Elimination of Systemic Contaminants—Indexing

The detection of pathogens in culture material is affected by the growing environment. Cool and dry conditions will reduce pathogen activity, thus masking their presence and causing growers unknowingly to propagate symptomless but infected stock. Plant disease is often an economically limiting factor in the production of ornamental plants by cuttings. Systemic diseases are caused by various plant pathogens, such as the bacteria *Dickeya chrysanthemi* and *Xanthomonas axonopodis* pv. *dieffenbachiae*, which are highly pathogenic to several foliage plants (Knauss 1976).

Of great concern are bacteria that attack vegetable crops, especially those crops that are propagated vegetatively, such as potatoes (*Solanum tuberosum*). When plants are infected, crop yield can be reduced by as much as 50% or more. Particularly offensive are *Pectobacterium carotovorum* var. *carotovora* (synonym *Erwinia carotovora*), which causes potato soft rot, *Pectobacterium carotovorum* var. *atrosepticum*, which causes blackleg, and *Clavibacter michiganensis* subsp. *sepedonicus*, which causes bacterial ring rot. Some pathogenic fungi include *Verticillium albo-atrum* in chrysanthemums (*Dendranthema* ×*grandiflora*); and *Fusarium oxysporum* in cuttings of carnations (*Dianthus caryophyllus*), among others. A wide range of viruses also attacks a great many plants of nutritious and ornamental value.

Because of such concerns, most states involved in potato production have programs for producing certified "seed" potatoes (vegetatively produced potato tubers that are used in place of true seeds). In these programs, seed potatoes are certified as being free of soft- and ring-rot bacterial infections, as well as having a low level (3% or less) of certain viral infections (leaf roll, mosaics, and spindle tuber viruses). After replanting 3–5 successive crops, propagating from tubers each time, the incidence of infection will increase, thereby decreasing yield to the point where new certified seed potato stock becomes necessary. New seed potato stock initiation and multiplication is done via test tube culture.

Certification programs have also been undertaken for several major fruit crops. For example, some states provide disease-free foundation stock of strawberry (*Fragaria* ×*ananassa*) plants. Usually these have been tissue cultured from heat-treated, virus-tested stock plants.

Systemic contaminants can be detected by various indexing methods. Indexing refers to any of several procedures for determining the presence of systemic microbial and viral contaminants in suspect plants. Many indexing methods are now available for detecting systemic pathogens in a variety of plants.

Virus Detection Methods

Early knowledge of viral plant diseases dates back to written records of several hundred years ago. Notable examples are potato leaf roll and tulip break, a color variegation found in infected tulips (*Tulipa*). Although it, like most viral infections, is detrimental to plant health, tulip break was once considered highly desirable by humans because of the ornamental appearance of the colorful streaks caused by the virus.

Many plant viral infections can be recognized by visible symptoms and further diagnosed by inoculation of an indicator plant with an extract from the infected plant tissue. Unfortunately for detection purposes, latency (the nonexpression of visible virus symptoms) is often a feature of virus infection in plants growing *in vivo*, and it appears to be the norm for plants growing *in vitro* (Cassells 1992). Some of the more common visible viral lesions are mosaic and mottle patterns on leaves, chlorotic leaf spots, and yellow coloring. Other aberrations include dwarfism, ring spots on leaves, leaf curl, and leaf wilt (Hill 1984). Another common symptom of a virus is the formation of inclusion bodies in plant tissue. Their presence can be determined by preparing stained slides of the plant material and observing them with a microscope.

Viruses are obligate parasites consisting primarily of DNA or RNA with a protein coat. They depend on the metabolism of the host cell for their reproduction. Unlike bacteria and fungi, viruses cannot reproduce outside of a cell or in a nutrient medium. Once inside the host tissue, they reproduce and often spread systemically within the host. In some cases, the virus can remain latent and escape detection.

Viruses in some virus families are able to integrate into the host genome. A similar phenomenon occurs in bacteria infected with a bacterial virus (bacteriophage), whereby the virus becomes integrated into the bacterial chromosome and replicates only when the bacterial chromosome replicates. A repressor protein produced by the bacteriophage prevents it from being released from the bacterial DNA. When the repressor protein is no longer synthesized or is inactivated, the bacteriophage is released from the host DNA. Once released, large quantities of viral DNA are rapidly synthesized, eventually bursting the bacterial cell.

Plant viruses may be filamentous, rod shaped, spheres, or brick shaped (Figure 11-9). Some viruses that attack bacteria look like spiders out of science fiction. Viruses are so small that an electron microscope is necessary to view them. A few diagnostic labs offer electron microscope testing or bioassays on indicator plants, and these techniques will detect unknown or unsuspected viruses.

Serological tests are used most often today for the detection of plant viruses. A serological test is the use of serum reactions that occur between certain proteins (antigens, in this case the virus coat protein) and antibodies (serum globulins). Antibodies are produced by inoculating an animal with an antigen preparation. When sap from a plant suspected of being infected with a

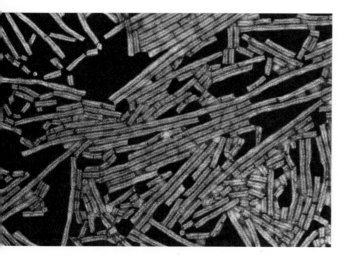

Figure 11-9. Scanning electron micrograph (SEM) of tobacco mosaic (TMV), a filamentous virus containing an RNA core and a protein coat. TMV infects many other plants besides tobacco (*Nicotiana* spp.).

particular virus is mixed with serum containing a specific antibody to that virus, an antigen-antibody reaction occurs. By using specific indicator chemicals, the antigen-antibody reaction can be detected colorimetrically (by observing colors).

ELISA testing techniques

For routine detection and identification of viral contaminants, an ELISA (enzyme-linked immunosorbent assay) procedure is commonly used. Until the development of this serological technique, routine detection of viruses was impossible other than by employment of a large staff, a greenhouse, and the use of many indicator plants. Routine ELISA testing can reassure both grower and customer that plants are free from viruses of particular concern.

At least 6 companies worldwide offer reagents and accessories for ELISA detection of about 100 different plant viruses (Gugerli 1992). The ELISA tests are simple to use, reliable, and relatively inexpensive. They are sold as individual reagents or as kits, which include all the necessary reagents including positive and negative control samples. Modern developments include the use of immunostrips and lateral flow devices that make identification of many common viruses possible without the use of lab equipment.

When using prepared kits, you do not need to determine the optimal dilution except for the test sample of suspect virus antigens. If you prepare your own coating antibody and related enzyme-labeled antibody, you must determine the optimum working dilutions. Working dilutions for the coating antibody and enzyme-labeled antibody that give the greatest amount of color in conjunction with the least amount of color in the control wells must be chosen for use with the double antibody sandwich (DAS) technique; a scheme for this purpose is shown in Figure 11-10. For many plant viruses, a coating antibody suspension of 1 mg per ml of water may be further diluted to 1–2 micrograms (µg/ml). Working concentrations greater than 10 µg/ml can reduce the intensity of the virus-binding reaction. Optimum enzyme-labeled antibody dilutions range between 1:5000 and 1:10,000 (Hill 1984).

The DAS ELISA technique is the most common among several modifications of the ELISA technique that have been developed. The ELISA test is conducted on a polystyrene microplate that has a number of flat-bottomed wells for reactions; each well has a capacity of approximately 400 microliters (µl). The 4 major reactions are illustrated in Figure 11-11.

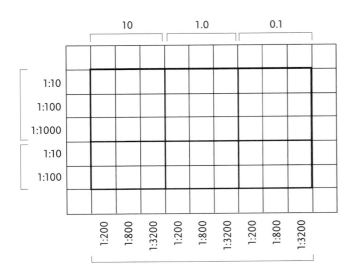

Figure 11-10. Scheme for preparing working dilutions of antigen and antibody used in the DAS ELISA test. Each square represents a reaction well with a particular dilution ratio.

ELISA test reactions

Reaction 1: Applying the coating antibody

1. In each reaction well, add the specific virus antibody desired. It will become adsorbed on the surface of the plastic. (Some labs use prepared microplates, such as those manufactured by Agdia, whereas other laboratories prepare their own antiserum, which they add to the plates after preparing the proper working dilution.)
2. After adsorption (2–4 hours), wash the microplate with a buffer solution to remove any excess serum. Shake microplate until dry. The phosphate buffer solution (0.03 M, pH 7) contains 2% polyvinylpyrrolidone (PVP), 0.2% ovalbumin (prepared on the day of use), and 0.5 ml Tween 20.

Reaction 2: Addition of the antigen

1. Prepare a sap extract from a virus-suspect plant by mixing equal volumes of plant leaves and phosphate buffer, plus 0.5 ml of Tween 20 per liter.
2. Pulverize the mixture with a mortar and pestle or use a hand-held homogenizer, available with easy-to-roll bearings that glide over the surface of plastic bags containing the leaf/buffer sample.
3. Prepare a working dilution of buffered antigen.
4. Add 200 µl of diluted antigen to several wells of the microplate, where it will become trapped and adsorbed on the surface of the specific antibody film.
5. For controls, fill wells with a sample of the buffer used for extraction and a positive control and a negative control. A positive control is an antigen that is able to attach to the surface of the coating antibody. It can attach because it has a physical configuration complementary to the physical configuration of the coating antibody. A negative control is an antigen with a physical configuration that is not complementary to that of the coating antibody.
6. After adsorption (2–4 hours), wash the microplate with the buffer solution to remove any excess antigen. Shake microplate until dry.

Reaction 3: Addition of the enzyme-labeled antibody

1. Add to each well 200 µl of a working dilution of the specific antibody complexed (labeled) with the enzyme alkaline phosphatase. In the wells, the antibody will attach to another available reaction site on the antigen molecule.
2. After incubation (2–4 hours), wash the microplate with the buffer solution to remove any excess enzyme-labeled specific antibody. Shake microplate until dry.

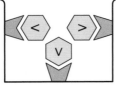
Reaction 1

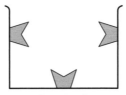
Reaction 2

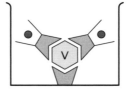
Reaction 3

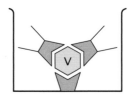
Reaction 4

Figure 11-11. Four major reactions of the DAS ELISA test.

Reaction 4: Addition of the enzyme substrate

1. Add 300 µl of freshly prepared substrate (para-nitrophenyl phosphate) to each well. If the antigen containing a bound antibody-alkaline phosphatase complex is present, a reaction will occur between the enzyme and the para-nitrophenyl phosphate to form a colored product. The time required for development of the color complex is usually less than 30 minutes.
2. If a series of microplates need to be examined, the color reactions can be stabilized by adding 50 ml of 3 M NaOH to each well, followed by slight agitation of the microplate.
3. Cover the microplates with a clinging plastic wrap and store in the refrigerator for several hours.
4. The results can be determined either by visual examination or by inserting the microplate in a plate reader for colorimetric determination. Virus quantification is also possible by measuring light absorption with a colorimeter.

Confirmatory test using indicator plants for virus detection

1. Prepare sap extract (as described in reaction 2 above) from a plant that tested positive with an ELISA test.
2. Dust carborundum (with a grit particle size of 600 mesh) or diatomaceous earth on one or more leaves of a known indicator plant for the virus. With a gauze pad, gently rub the sap extract into the dusted leaves. The purpose of the abrasive particles is to provide wounds that allow the virus to infect.
3. Store the plants in an isolated area and examine in 1–3 weeks for characteristic virus symptoms.

12 Business

Although this book is devoted to tissue culture and not to business management, certain important economic considerations must be addressed. As with any business, it is desirable to predict the financial success of a commercial laboratory, record it in progress, and periodically evaluate it in retrospect. Most operations cannot afford to operate without a tentative plan and a projected budget. Good management involves constant planning, study, observation, review, and adjustment.

Marketing

Good marketing practices are essential for success. Marketing is an art and a science that must interact with management, technology, and common sense. It must equate with salesmanship, for without sales there is no commercial enterprise. If you have a superior product, you can probably find a market for it; but if you first find a need and then grow to fill that need, you will have an even better chance of selling your product.

Perhaps the most appropriate circumstance for establishing a tissue culture laboratory is as an adjunct to an existing nursery, which then becomes a built-in customer for the tissue culture operation. If the product does not sell as a microcutting, rooted plantlet, or liner (small, established plant), then there is room to grow it on within the nursery until the market is ready.

Sometimes an independent laboratory will form an association with a nursery that will serve as the desired cushion between production and market. The nursery can grow-on the plantlets on a consignment basis, or simply be an ongoing source of orders. An independent laboratory can act as a custom grower for various nurseries, or if a laboratory is tied to a nursery, it can custom grow for outside nurseries in addition to supplying in-house needs. Custom orders are a more comfortable approach than speculative growing, which has much uncertainty and risk. In fact, this is the best way to succeed: before you begin to micropropagate a new plant, make sure that you have a buyer who is willing to take your product when it is produced.

Orders for woody plants or bulbs generally need to be placed a year in advance. Foliage plants require less time. If the laboratory anticipates certain orders, an inventory of stock cultures can be maintained to help cut down the lead time. As stated previously, stocks can be maintained on a minimal medium and refrigerated for slower growth to lengthen the time required between transfers.

It is important to have money up front for custom orders. Unless you have previous experience with a plant in culture or it is commonly micropropagated, you should charge a fee for the establishment of a new plant *in vitro*—US $500 to $1000 is not an unusual amount to charge for this activity. If you are unable to find an established procedure for tissue culturing a desired plant, the labor required for getting it started and determining optimum nutrients and conditions can be quite costly. The agreement with the customer should state that all or part of the fee is at risk because you may not succeed in attempting to establish the plant in culture. Several laboratories might be growing for a single buyer if the customer has been shopping around; without any financial agreement, you could be left with a crop and no market.

Certain large operators offer a brokerage service. With this service, brokers will buy and consolidate material from several small micropropagation facilities to fill large orders or special needs. In addition to the large brokerage companies, individual brokers are constantly on the road to accomplish the same end. Brokers will buy plants from small producers at about 20% of the price. This may not offset production cost, but it is better than destroying unsold products. Occasionally larger nurseries resort to dumping, the practice of unloading an over-stocked crop at less than cost. Such action may be deemed necessary by the nursery, but it promotes an unstable market and places the smaller producers at risk.

Sometimes other laboratories will order a few plants *in vitro*, which is a way for them to bulk up stock and avoid development time and expense. Fulfilling such small orders for other laboratories can be a risk because those laboratories well might become your competitor. For this reason many laboratories will not sell axenic (noncontaminated) or *in vitro* plants. An alternative is to sell culture test tubes for their real worth, the cost of development—asking US $500 or $1000 per test tube will quickly discourage any such attempt to undermine your market.

A catalog and website are excellent ways of announcing the products or services you have to offer. The catalog and website need not be elaborate, but they should be neat, easy to read, inviting, and factual. Provide information on common and scientific names, sizes, availability, prices, quantity discounts, and shipping charges. Be sure to include an order form, full address, phone and fax numbers, and updated availability lists. Other desirable but optional details are illustrations, directions for planting and care, other relevant information on plants, or company history.

It is good business practice to study the markets. Find out what other laboratories are doing—if they will tell you. When a laboratory has invested time and money into research and operation, it is easy to understand why they might withhold information and hope to profit by their advantage. The other side of the argument is that the free exchange of knowledge pushes technology forward and more people will benefit in the end. At the very least, by communicating with other laboratories and consulting their catalogs you can find out what they are growing and how much they are charging. Visit nurseries and wholesale florists to learn what is in demand. Read grower journals such as *GrowerTalks*, and *American Nurseryman*; also gardening magazines like *Fine Gardening* and *Horticulture* to learn what is being marketed.

As you study the market, try to find out if there will be more profit in selling plantlets, liners, or gallons. What are the going prices? How much would each of these products cost you and return in profit? Is it really cheaper to tissue culture the varieties in question than it is to maintain stock plants and grow from cuttings? How long will it take to establish a plant in culture? Boston ferns (*Nephrolepis*) may take one month, strawberries (*Fragaria*) 2 months, and *Rhododendron* 6 months. What is the rate of multiplication? How long is the rooting period? Is it cheaper to build a laboratory and produce the plantlets, or to have them custom micropropagated in an existing laboratory? Answers to these questions will help to answer the main underlying question of whether there will be a profit.

Worldwide Microplant Production

During the late 1970s and early 1980s, there was a surge of new laboratories built in the United States. A 1986 survey indicated over 250 laboratories operating in the United States (Jones 1986). By 1990 competition, mergers, fear of mutations *in vitro*, and buy-outs had succeeded in closing a number of doors. Surprisingly, according to a survey presented in newsletters and journals, there were only 125 in 1992 (Bridgen 1992). An accurate number of laboratories is difficult to determine, but now there are probably less than 100 commercial laboratories in the United States. Some are minor adjuncts to various nurseries large and small. Others are large enterprises with or without nursery facilities. There are mergers, failures, and takeovers, and those institutions in support of research.

With increasing worldwide trade, the tissue culturist will be competing with markets both at home and abroad. No comprehensive survey can be found on the worldwide volume of plants produced from tissue culture; however, the highest concentrations of tissue culture laboratories can be found in Southeast Asia, Europe, and North America. China and India by themselves produce hundreds of millions of plantlets annually. Annual production of ornamental foliage plants alone was estimated at 254 million worldwide in 2006, accounting for a wholesale value of US $219 million (Chen and Henny 2007). Another rapidly growing sector is tropical fruit crops, led by banana (*Musa*) as the world's 4th most important food crop.

A limited but intensive survey of 46 members of the Tissue Culture Association was conducted by Dennis Yanny (1988). The 30 members who responded (10 were no longer in business and 6 did not reply) reported 701 full-time employees and 412 part-time employees, none of whom had a college education. The average annual production was 1.45 million plants, mostly foliage plants. Of these, 24% sold as plantlets, 44% as liners, and 32% as finished plants. The average initial investment was US $36,700 (minimum was US $10,000). The average current investment (at the time of the questionnaire) was US $236,000 (minimum US $20,000). Laboratory space averaged 2700 ft.^{2} (250 m^2); the smallest measured 1200 ft.^{2} (111 m^2).

One study has placed total micropropagated plant production in North America at 84.7 million per year (Jones and Sluis 1991). No production number was reported for orchids, but foliage plants dominated with 32.5 million per year. *Syngonium* had an output of 16 million plants per year, woody ornamentals were reported in excess of 13 million per year, and ferns were estimated at 12 million. *Gerbera* may be losing ground due to some seed now coming true to color. Other major micropropagated plants included bulbs, fruits, vegetables, and trees.

With increasing worldwide trade, the tissue culturist will be competing with markets both at home and abroad. Again, little current information is available, but looking back, Western Europe and Israel combined accounted for an annual total of 212.5 million micropropagated plants in 1988. The major products reported were 92 million pot plants, 38 million plants for cut flowers, and 19.4 million fruit trees. The Netherlands alone produced 16 million *Gerbera*, 13 million bulbs/corms, 12.8 million ferns, 2.4 million orchids, and 4 million African violets (*Saintpaulia*) (Debergh and Zimmerman 1991).

Eastern Europe is credited with producing 55 million micropropagated plants per year. A limited report from Asia estimates annual production at 74 million, which includes 44 million orchids. Production numbers from South and Central America are also difficult to find. Orchids, bananas (*Musa*), *Eucalyptus*, carnations (*Dianthus*), *Anthurium*, and potatoes (*Solanum tuberosum*) are the most common products of South and Central American tissue culture operations. One Venezuelan laboratory produces 1.5 million bananas per year. Statistics on micropropagated products from Australia and New Zealand are incomplete, but 20 laboratories reported a total of 25 million plants. Except for a few ornamentals, grown primarily in Egypt, most plant tissue culture in Africa is dedicated to breeding and improving food crops (Debergh and Zimmerman 1991).

With this review of worldwide activity in the field, a person contemplating starting a tissue culture business may be discouraged. One might conclude that there is a minimum of profit to be had from standard tissue cultured products, such as foliage houseplants, rhododendrons, and some perennials. If that is true, then the greatest profit lies in being the first with plants newly issued, patented plants (especially if it is your own patent), plants that are difficult to propagate, and unique plants.

Many foreign markets are not satisfied by internal production. The best way to find these markets is to travel to the countries that have potential buyers. There are special hurdles in foreign markets, primary among these being customs. There was a time when plants *in vitro* were considered free from disease as long as no contaminants were visible. It has become common knowledge, however, that *in vitro* plants may have invisible endogenous pathogens such as virus, the effects of which will not be visible until long after the plants have left the culture jars. Consequently, customs requirements vary with time, country, state, and the type of plant. Often a phytosanitary certificate is required, which involves a government inspection and an inspection fee before material can be shipped. Where there are regulations, the minimum requirement is that the plants have never been in soil (as opposed to soil mix or artificial soil) or, sometimes, that the growing medium be washed off. Everyone loses when shipments are delayed or destroyed, so check with appropriate government agents to avoid potential problems.

Costs

Because a successful operation usually will expand, the original planning of the physical structure of the facility should provide for potential growth; otherwise, the result will be a piecemeal, expensive, disorganized assortment of additions. A local contractor can help determine facility cost, but shopping for yourself with a detailed plan in hand can save considerable expense.

Equipment can be changed more readily than a building. Scientific equipment is fine, but oftentimes, cheaper household equivalents can simply be adapted, for example, a kitchen dishwasher instead of a laboratory glasswasher, or a pressure cooker instead of an autoclave. Alternatives should be studied and balanced against the size and nature of the operation being considered. In other words, do not build a lab to produce 150,000 plants annually and then buy equipment for a million plants, nor should you build a lab for a million plants and try to get by with household equipment.

Detailed studies of tissue culture costs are available in the literature to determine profit or loss (de Fossard 1993, Standaert-de Metsenaere 1991). Records should be kept of overhead costs, purchases, labor costs, and losses. Ideally, you should itemize overhead, investment, depreciation, chemicals, transfer rate, multiplication rate, growing-on costs, labor, losses, and shipping costs. Then determine a profitable price. If you have not yet tissue cultured anything, you will have to base your figures on a survey of the market and your best educated guess. If you are operating an efficient laboratory and can afford cost analysis to determine your pricing, your prices will probably be in line with demand-oriented or competition-oriented pricing. If you omit the analysis and simply use competition-oriented pricing, you will quickly know whether you are making any money. If you are selling at a loss, an analysis might help you find where you can save money.

Labor is the greatest cost in a laboratory. Labor costs, which involve media making, transferring, culture surveillance, and benefits, can account for as much as 80% of operating costs (Jones and Sluis 1991), so management must constantly watch for ways to maximize labor efficiency. The relative cost of premixed or partially premixed media should be studied and weighed against the cost of labor to mix it; however, the desirability, or necessity, of having the flexibility of varying the ingredients in media should not be underrated.

The preliminary cost projection should stipulate the number of employees required and their qualifications. Personnel should already have some knowledge of tissue culture, sterile technique, chemistry, and plants, or these will have to be taught. Moreover, like so many other positions, common sense, cleanliness, a sense of humor, and the ability to get along with others is of more importance than any technical background, which can be learned on the job.

Regardless of background, an employee must have a clear understanding of the difference between research and production. The inquisitive person will want to experiment, the overly precise will want to pick at transfer material or dawdle with labeling or records. There is no faster way to lose profit. These kinds of individuals need to understand that the laboratory is a business, not a university for training. Problem material will require special treatment and will slow production, and these matters should not be left to the amateur. Efficient mainstream production is at the heart of successful nursery tissue culture.

Most larger plant tissue culture laboratories have a research and development (R & D) department that may be responsible for studying problems in production, new plant varieties, sanitation, safety, and quality control. Production problems include media formulation, multiplication rate, root initiation, and detection and elimination of microbial and viral contaminants. Other functions of the R & D group may be maintaining cultures in storage and training new staff.

Genetic engineering is another function of R & D. The department may be assigned to develop new hybrids using transformation (introduction of foreign genes into a plant), protoplast fusion, or somaclonal variation. They may be asked to fingerprint plant material (identify DNA or proteins using electrophoresis) or to investigate the production of secondary products *in vitro*. (See chapter 13 for a discussion of these and other biotechnological processes.)

If no one within an existing nursery organization is qualified or available for training to direct the tissue culture operation, advertising in technical and trade journals or the local newspaper frequently can turn up the right person for the job. College placement offices can assist in identifying technicians. Oftentimes competent personnel can be located through meetings of the International Plant Propagators' Society (IPPS), the International Association for Plant Biotechnology (IAPB), or the Tissue Culture Association.

The period during which transfer activity is reduced is an ideal opportunity for the established grower to rotate personnel from laboratory to nursery. Cultures targeted for spring out-plant make winter the busiest time of the year inside, but it is usually a slow time outside in the nursery. The advantages of retaining year-round employees are evident to employers who must hire, train, and evaluate new employees.

Laborsaving devices are generally a good investment. Usually the more automated an operation is, the lower the labor costs; on the other hand, more automated equipment is also more expensive, such as using an autoclave instead of a pressure cooker, a semi-automatic media dispenser instead of a pitcher, an electronic balance, or an automatic labeler. If capital is available for laborsaving equipment, the investment will pay for itself in some defined period.

Travel time within the laboratory translates into dollars—whether this is the time it takes to walk from one room to another or the time used by the employee's hands to make a transfer—it costs money. Pass-through windows in a small operation save time by eliminating the need to carry or cart cultures and materials through doorways from room to room. Similarly, careful organization and positioning of equipment around the transfer technician or in the media preparation room contributes to the most efficient use of labor.

The size and shape of culture containers is

another economic concern. Test tubes are valuable for starting cultures and restricting contaminants; larger containers are more economical in terms of transfer time. The adept technician can transfer approximately 130 tubes per hour when dealing with good material and a 3- to 6-times multiplication rate. However, transferring cultures into pint jars or squared-shaped jars (such at Magentas) at a rate of 20 containers per hour, each with 16 propagules, is equivalent to 320 culture tubes per hour.

Losses from contamination of tissue cultures, especially from disease in Stage IV, can be devastating. The sooner contamination is detected, disposed of and/or treated, and the sooner appropriate prevention measures are taken, the more cultures and plants will be saved. Contamination can wreak havoc on a good plan. A realistic plan will take into account some overage for losses, which should not exceed 2%. Any greater amount calls for serious investigation. Money spent for air filtering, sterile transfer, and containers that prevent contamination is money well spent, as is time spent cleaning surfaces.

The fear of contamination is well founded, but operations can spend initially an excessive amount attempting to meet hospital-quality sterility. Purchasing, replacing, and cleaning special clothing, such as gowns, goggles, masks, gloves, and slippers, can be very expensive without being effective. Gloves are advisable primarily to protect the hands of workers from harsh disinfectants, and secondly to protect the cultures. The effectiveness of laboratory coats is questionable because they may be no cleaner than normal clean clothing. If a problem with contamination emerges, track down the source and then buy what is needed to combat it. Experience, coupled with a contaminant quality-control program, will demonstrate which sanitary practices are most effective for you.

The use of refrigeration for some cultures will slow down contamination as well as culture growth if the cultures cannot be cared for on time. Refrigerated storage of cultures is a tool that can be incorporated either deliberately in original plans or as recourse to ease workloads when there is too much to transfer at one time. If cultures become aggressive and are ready for transfer ahead of schedule, it is possible to refrigerate some species until it is time to transfer them (Figure 12-1). Strawberries (*Fragaria*), *Rhododendron*, carnations (*Dianthus*), and chrysanthemums are some of the plants known to tolerate several weeks or months of storage at temperatures of 34°F to 36°F (1°C–2°C), usually without any light.

Occasionally natural lighting can be used in tissue culture growing rooms to conserve energy consumption. Because natural light fluctuates significantly and is difficult to control, however, it is not recommended, and power usually is not a major expense. Fluorescent lighting is constant, relatively inexpensive, and usually provides more than enough heat required in the growing room. It is economical to schedule the

Figure 12-1. A crowded test tube (foreground) indicates that it is past time to transfer the plantlets. Ideally, plantlets are transferred before they become overgrown.

dark cycle during the day, which saves on cooling costs, and maintain the light period during the night to save on heating costs, especially if the heat from the lights is circulated throughout the laboratory, minimizing the need for other heat sources. Even when ballasts from fluorescent fixtures are relocated outside the growing room to prevent overheating and reduce cooling requirements, they still provide a heat source that can be tapped by the air circulation system. Turning off lights and appliances not in use and providing good insulation are basic to energy conservation and sustainability.

Saving on media chemicals by buying inferior quality is not a wise practice because it will have a high impact on plant performance for a relatively small percentage savings. Most often, the extra cost of buying quality chemicals and supplies will pay off in the long term.

A tissue culture operation offers savings in space compared to a conventional nursery operation. Less space is required to carry stock plants, to start plants, and to multiply them. With careful scheduling of laboratory and growing-on space, the grower need allocate fewer dollars to space and will realize a higher return on the dollars invested in land and buildings.

Deciding What to Grow

When contemplating a product mix, an obvious first step is to identify plants that respond to micropropagation. Section Two of this book includes tissue culture protocols for a number of species. Additional resources are available in the form of journal articles. Online and searchable journals such as *In vitro*, *HortScience*, and *Journal of the American Society of Horticultural Science* hold countless articles detailing micropropagation procedures that have proved successful for the authors. However, many "recipes" are top secret; once a company's R&D cracks the code, that information often becomes proprietary and is not made public. One of the best resources for those interested in propagation of any form is the International Plant Propagators' Society. Members have access to proceedings (papers presented by scientists and commercial propagators at IPPS conferences the world over), including a veritable treasure trove of micropropagation protocols.

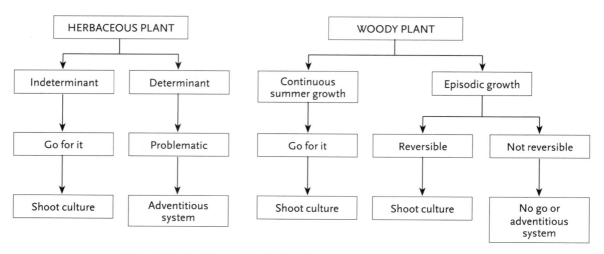

Figure 12-2. Decision-making ladder for micropropagation of herbaceous and woody plants. In the absence of an established protocol for a particular plant, the plant's growth habit can help predict the ease or difficulty of micropropagation.

So what if you can't find a protocol for a particular species? Observing the growth habit of the plant can give you some clues as to a species' potential for successful micropropagation. Brent and Deborah McCown (1999), both academics and nursery owners, developed a very logical decision ladder for determining the potential successful micropropagation of both herbaceous and woody plants (Figure 12-2).

Herbaceous plant growth can be determinant or indeterminant—think of tomato vines as an example. *Helleborus* ×*orientalis* is a good example of a determinant plant—the foliage and flowering stems arise from the crown separately; there is no branching, no leaf nodes. *Helleborus* has only recently been successfully tissue cultured; however, the protocols remain mostly proprietary. Indeterminant plants such as perennial salvia or the houseplant *Pothos* continue growing from apical meristem. These are generally more amenable to shoot culture.

Woody plants, such as willow (*Salix*), that continue to grow throughout the season are usually good candidates. Those expressing "episodic" growth (once-per-season flushes) can be much more difficult to culture. *Rhododendron* and lilacs are good examples. As with determinant herbaceous plants, alternatives to shoot culture such as somatic embryogenesis or organ culture may be the only option.

Plant Patents and Trademarks

The existence of trademarks and patents on plants presents a challenge for the propagator. Know the source of your stock and any restrictions associated with use of the cultivar and/or brand/marketed name. Trademarks and patents are applied for and issued through the U.S. Patent and Trademark Office (USPTO) in the United States. Outside the U.S., various other proprietary designations are available including Plant Breeders Rights.

Trademarks do not restrict propagation of a plant variety, but protect the name under which a plant is marketed. True cultivar names are not valid for trademarking, rather trademarked names are used in branding and promotion. Anyone can use the TM designation. The symbol ® designates that the trademark is federally registered with the USPTO.

Patents provide a time-limited right to exclude others from use of plants and plant products. Plants that are patented are not allowed to be propagated without a license. In the United States, new plant varieties created by breeding or by a laboratory process are patentable. Protection is extended to plants produced by either sexual or asexual reproduction and includes propagules such as seeds or tissue culture plantlets. Natural-source materials are not patentable. In addition, processes to create certain plants and plant products are patentable, including tissue culture protocols.

The interest in patents has skyrocketed, especially with ornamental plants (see Table 10). The ability to collect royalties on novel varieties combined with aggressive marketing by third-party agents representing multiple plant breeders has fueled this increase. It is worth mentioning that often the costs of obtaining a patent (US $2000 and above) for a new ornamental plant may be

Table 10. **Plant patents applied for and granted in the United States**

DECADE	PLANT PATENTS APPLICATIONS	GRANTS	INCREASE IN GRANTS FROM PREVIOUS DECADE
2000–2009	10,696	9,436	132%
1990–1999	5,376	4,057	57%
1980–1989	2,790	2,589	69%
1970–1979	1,654	1,531	—

Source: Compiled from U.S. Patent and Trademark Office (USPTO) submissions to World Intellectual Property Organization (WIPO). Original data available at www.uspto.gov/web/offices/ac/ido/oeip/taf/us_stat.htm.

more than the royalties that will be received from licensing that plant.

Crop Planning

One of the biggest problems in commercial tissue culture laboratories is overproduction, that is, producing too many plants for a limited market. The propagation of hundreds or thousands of plants that are just going to be destroyed is not only wasteful, but it is very costly. A production plan for each of the crops being micropropagated must be a high priority. Table 11 shows a sample crop production plan for preparing 12,000 field-ready 'Earliglow' strawberry plants for shipping out on 1 April 2014. A manager should develop a time frame, especially for crops that must multiply to numbers in the thousands by spring delivery. A good plan starts with the number of plants ordered and the scheduled delivery date. Working backward toward the explants, target numbers are established and the container requirements noted for specific dates.

After a crop plan has been established for each variety of plant, an overall space schedule must be made to assure adequate space and product flow. Usually, as numbers increase, cultures are relocated. When these moves are planned ahead of time, there will be a minimum of confusion and wasted motion. In addition, you should devise a plan for *in vitro* media and an estimate of the labor hours.

When you have about enough plantlets to fill your order, you can create a table to help you determine how many plantlets to put into rooting and how many into multiplying in the last 2 transfer cycles (see Table 12 for an example). For example, say you need 25,000 plants. At 30 plants per rooting jar, you will need 833 rooting jars to end up with 25,000 plants. You have on hand 200 multiplying jars. The plants have been multiplying at the rate of 2 to one. You judge that each multiplying jar will be divided to fill 3 rooting jars. Thus, the plantlets in your 200 initial multiplying jars will require 400 multiplying jars or 600 rooting jars after the first transfer. This gives you too many multipliers (requiring 1200 rooting jars next transfer) but not enough rooters this transfer,

Table 11. **Sample crop production plan for 12,000 field-ready 'Earliglow' strawberry plants**

DATE	NO. OF UNITS	(OUT TO)	TRAYS	6-PACKS	JARS	TUBES
04/01/14	12,000	Ship	—	—	—	—
02/01/14	12,250	Greenhouse	—	—	—	—
12/15/13	12,500	Acclimatizing (2)	174	72/tray	—	—
12/01/13	12,800	Acclimatizing (1)	64	—	—	—
11/01/13	12,800	Rooting, singles	—	—	512	—
10/15/13	3,200	Multiply	—	—	67	—
10/01/13	1,067	Multiply	—	—	23	—
09/15/13	350	Multiply	—	—	7	—
09/01/13	128	Multiply	—	—	—	128
08/15/13	60	Multiply	—	—	—	60
08/01/13	30	Multiply	—	—	—	30
07/15/13	20	Transfer	—	—	—	20
07/01/13	25	Explants	—	—	—	25

so you need to determine how many should be turned out to rooting this transfer in order to have 833 rooting jars after the next transfer.

Make a table with jars to go to multiplying on one axis and jars to go to rooting on the other—the sum where the 2 axes meet is the 200 jars you have on hand. The table could run from 0 to 200 on both axes, but a likely range is illustrated Table 12. The table shows that if plants from 80 of the 200 jars are transferred to 160 multiplying jars, the next transfer will produce 480 rooting jars (that is, the 80 multiply to 160, which are then transferred to 3 times as many rooting jars, 480). The remaining 120 of the original 200 multiplying jars will fill 360 rooting jars this transfer. With this combination—80 jars to multiplying and 120 to rooting—840 rooting jars will be produced by the end of the next transfer.

Planning may appear to be a futile exercise, but as you plan, you will find it enlightening to make up a set of schedules. It is also essential for successful production. Before you finish, you will realize that you are able to visualize the total operation much more clearly.

Keeping Records

Most nurseries and tissue culture operations utilize software programs for their record keeping. Options can be as simple as a spreadsheet program or one of many available nursery- and/or greenhouse-specific inventory systems. Adapt the programs of your choice for keeping records on production, performance, and finances. Technicians should record their hours working under the hood, *in vitro* observations, and production records (number of jars in and out) to help with crop planning. The careful records that are required for research projects are usually not economically justifiable for basic production purposes.

Records are useful only if they are referred to, and not just buried in a drawer or in a file on the computer. The object of record keeping in production is to provide sufficient cross-checks for peace of mind, without creating costly busywork. At one end of the spectrum of record keeping are the identification of each explant and its source plant, and the maintenance of this identity throughout the history of the clone. Where a number of clones of the same cultivar are involved, such meticulous record keeping facilitates the elimination of a clone if contamination or other problems develop. At the other end of the spectrum is a running file of each crop, which provides the customer's name, address, and telephone number, the name and code of the cultivar, date started, transfer dates, number transferred, media used, location (on numbered growing room shelves), culture conditions, tests performed, and results.

Another invaluable record is a chemical inventory, noting when a new container of chemical is opened, when and where it was purchased, the cost, and where it is stored in the laboratory. Today, prepackaged media commonly being used, it is insurance to record the batch number

Table 12. **Final transfers to achieve required stage III plantlets**

JARS TO ROOTING (R)	JARS TO MULTIPLYING (M)			
	60	70	80	90
140	120 M × 3 = 360 R + 420 R 780 R			
130		140 M × 3 = 420 R + 390 R 810 R		
120			160 M × 3 = 480 R + 360 R 840 R	
110				180 M × 3 = 540 R + 330 R 870 R

when it arrives. A reliable system for reordering chemicals will avoid the problem of suddenly being without one when it is needed.

A carefully established and maintained record of general financial accounts is also essential because it is easy to lose money quickly on a tissue culture operation. It can be a year or more before any money comes in and several years before you know if you have been successful. It is a good strategy to work with your customers and request 25% or 50% down at the beginning with the remainder due at delivery. This process not only assists with cash flow, but it gives you a guarantee that they will not cancel the order. At some point you must answer the question, Are we making any money? A good set of financial records, including a quarterly or annual profit and loss statement, meticulously maintained, will provide answers to this question and is the key to successful financial management.

The foregoing suggestions, coupled with professional advice on business planning and projections, should provide a firm foundation on which to answer some of the questions posed earlier and to decide whether you want to pursue starting a tissue culture business.

Shipping

The best method of packing tissue culture materials for shipping depends on the stage at which the micropropagated plants are shipped. Rarely is it necessary or desirable to ship plantlets in a sterile condition. Shipping plantlets in autoclavable plastic containers or foam trays involves less cost and risk than glass. Preferably, they should be shipped without agar because it is likely that a container will not remain upright throughout shipment.

If the order is for microcuttings, place them in a clean, firm plastic storage dish with a tight cover. If the microcuttings are padded with damp paper toweling, they will not have room to tumble around in the container. Sometimes they can be placed in plastic bags and heat-sealed to maintain high moisture levels. In either case, the container should be placed in a box and surrounded with packing peanuts.

If rooted plantlets are to be shipped, a good method is to jelly roll them in moistened (not dripping wet) paper towels. Remove the rooted plantlets from the jars or rooting trays, and roll about 50 of them in moistened paper towels. Pack 10 rolls in a plastic bag, and place 6–12 plastic bags in a small carton lined with sheets of foam to protect against excess heat or cold during shipment.

Rooted liners (Stage IV) are shipped in the same way that other rooted propagules are shipped (Figure 12-3). Rooted liners will survive packed in a box for 3–4 days. Most often, they remain in trays and packed in water-resistant boxes, often with netting or some other method to hold them in place during the rigors of shipping. Be sure foliage is dry before packing to prevent fungal diseases. During colder weather, foam panels help insulate the young plants.

Figure 12-3. Packing system for shipping trays of rooted liners to customers. Terra Nova Nurseries, Canby, Oregon.

13 Plant Biotechnology

Plant biotechnology—especially in vitro regeneration and cell biology, DNA manipulation and biochemical engineering—is already changing the agricultural scene in 3 major areas: control of plant growth, protecting plants against abiotic and biotic stresses, and production of value-added foods, metabolites and pharmaceuticals. Plant biotechnology faces several major challenges in the coming decades: alleviating the hazards of abiotic stress, improving pest control, maintenance and improvement of the environment, improvement of food quality and design of 'specialty food' using biochemical engineering, and production of biomaterials. Two parallel research approaches will most likely exist simultaneously in the near future: the transgenic approach (expression of unique genes and specific promoters and transcription factors), and the non-transgenic approach (genomics-assisted gene discovery, marker-assisted selection, efficient mutations, and clonal agriculture).

—Arie Altman (2003)

Plant biotechnology encompasses a vast array of disciplines and techniques. These include traditional culture methods, germplasm storage, genetic improvement, and methods of generating useful products from plants. It is beyond the scope of this book to explore these techniques in depth; however, we will attempt to present a brief overview of these areas.

Plant biotechnology has been and will continue to be useful in producing plants from test tubes. The converse is also true: manipulating plant tissues in test tubes is an important part of plant biotechnology. Plant pathogens have been genetically engineering plants by inserting their DNA and hijacking the plants' metabolic processes for their own use. Researchers use plant pathogenic organisms and other biotechnologies to insert gene(s) of choice (not human DNA) into plants for a variety of economically important purposes. Biotechnology has also been important in generating other plants that are useful to us:

- Plants that can produce elevated levels of pharmaceutically useful compounds.
- Haploid plants through anther culture for genetic studies and breeding material.
- Doubled haploid plants by arresting mitosis that are completely homozygous.
- Somaclonal variation in tissue culture.
- Advances in cryopreservation that have enabled us to store and manage collections of plants more efficiently, especially those that cannot be conserved by seed.

Plant biotechnology can produce useful plants, but its use can also produce ethical, legal, and environmental issues that need careful thought and regulation.

Genetic Engineering

In a broad sense, both tissue culture and genetic engineering can be defined as biotechnology—given an understanding of biotechnology as the human manipulation of genetic material—but the distinction between the 2 fields is as important as their parallels. *Tissue culture* or *micropropagation* does not involve the transfer of exogenous/

foreign genetic material of an organism, but rather multiplying the genetic material present in the plant's cells. *Genetic engineering* can be defined (in part) as the purposeful altering of genetic material. Tissue culture can aid or complement genetic engineering by providing means for readily multiplying plant material.

Humans have been altering plants for centuries through selective harvesting, hand pollinating or cross pollinating of plants, and multiplying or inducing mutants. Almost all commercially available foods and most breeds of domesticated animals are the result of human selection and breeding, and advances in the field of molecular biology have helped to further improve and fine-tune genetic manipulations. Most genetically engineered medicines are accepted without question, as are most "natural" mutants such as the navel orange. But often genetically engineered food products are met with opposition and rejected as "Frankenfoods."

There is an increasing public outcry against genetically modified organisms (GMOs), especially with respect to the safety of our food supply. A major concern for many people is the increasing use of genetically modified seed for many of the major food crops including soybean, corn, and canola. It is estimated that greater than 90% of U.S. soybean acreage is genetically modified (NASS 2010). The biggest fears are monopoly of seed controlled by one or several multinational companies; aggressive pursuit of GMO intellectual property by these companies; very limited genetic diversity of major food crops such as corn; and transgene flow to non-GMO plants.

Expanding populations will require more food grown in less space with less water, with greater tolerance to weather and pests, and with better nutrients and keeping quality. Fortunately, industries and governments are working together to provide such food products, developed with sensible controls, and to educate the public regarding genetically engineered products. There is no way to halt progress, but it is possible to harness and guide it into safe, positive, constructive pathways. The field of biotechnology is undergoing an explosion of new research and new results, and the potential for further advances is remarkable.

Some purposeful applications of genetic engineering include providing a better understanding of cellular functions, learning how the genetic code directs the formation and maintenance of an organism, and improving cultivated plants by introducing traits of nutritional, aesthetic, utilitarian, pharmaceutical, or horticultural value, such as insect and disease resistance (Greenberg and Glick 1993). Some of the plants, mostly food crops, that have been transformed by inserting foreign genetic material include soybean (*Glycine max*), rice (*Oryza*), corn (*Zea mays*), potato (*Solanum tuberosum*), lettuce (*Lactuca sativa*), cotton (*Gossypium*), and tomato (*Lycopersicon*).

The expanding role of plant tissue culture as an accompanying tool in genetic engineering and other plant biotechnology warrants a discussion of the molecular basis of life. Even a superficial concept of what is happening in the world beyond our vision increases our understanding of new products in the markets, of research articles appearing in newspapers or journals, or of events in our own laboratories and greenhouses.

Plant DNA

Plant characteristics are passed from generation to generation in genes that are encoded by a chemical compound known as deoxyribonucleic acid or DNA. Various metaphors have been used to describe the double-helix structure of DNA (Figure 13-1). It has been compared to a circular stairway, a ladder, or a double strand of beads. If you were to unwind a spool of thread and form many small loops, which you then wad up in your hand, you could begin to appreciate the mass and proportion of the DNA threads that make up the chromosomes within the nucleus of a cell. Consider each of the interlocking pieces of the zipper as representing one of the 4 nuclear bases—adenine (A), thymine (T), guanine (G), or cytosine (C)—that code for all the processes of living organisms. A and T are always paired across the strand from each other, as are C and G. A base pair is held together with hydrogen bonds and, with its contingent sugar and phosphate, is called a nucleotide. On the outer edges of the long molecule is the sugar-phosphate backbone, the handrails of the metaphoric DNA staircase. These 4 bases encode the vast complexity of genes and their expression in all living things. Exactly how this is done is explained later where DNA is transcribed into mRNA and later translated to proteins.

Mitosis and meiosis

Plants grow by creating more cells. To do this, each cell must first double its genetic material and other cellular components. When a cell is ready to divide (multiply), the folded mass of DNA further condenses into chromosomes. The DNA is "unzipped" by a polymerase enzyme that travels down the double strand, separating the 2 strands. As a double strand is separated, a new complementary strand forms, so that once again a double strand is formed. The new pairs move to opposite sides of the cell via tubule formation so that a new cell wall can form between them and divide the cell into 2 distinct cells (Figure 13-2).

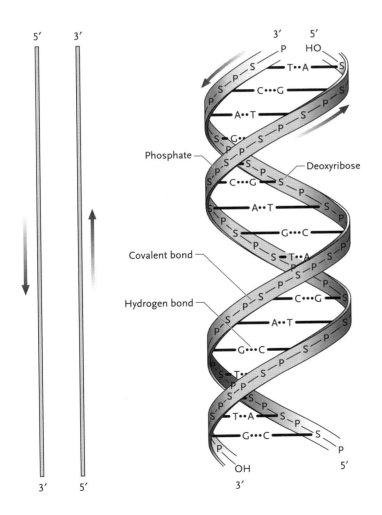

Figure 13-1. Two different ways of representing the DNA molecule. The 2 complementary strands run in opposite directions (antiparallel), with the 5-prime end containing a phosphate bonded to the sugar and the 3-prime end having a hydroxyl group (OH) on the sugar. The 2 dots between A and T indicate that 2 hydrogen bonds exist between these molecules; the 3 dots between G and C indicate that there are 3 hydrogen bonds. The purines (A and G) and pyrimidines (C and T) are on the inside of the molecule.

The new cells contain nuclear material that is identical to that of the previous cell and to that of all the other cells in that organism. In this vegetative process of cell multiplication called mitosis, the diploid complement ($2n$) of chromosomes remains the same.

The sexual process of cell reproduction, called meiosis, is one in which the diploid complement of chromosomes is reduced from $2n$ to n; it is a reduction division. The products of meiotic division are the reproductive cells, the gametes or germ cells, also known as a microspore or pollen (male) and a megaspore or ovule (female). In the final step of the sexual process, the diploid chromosome number is reconstituted when the male and female gametes unite to form the fertilized egg, which then grows into an embryo. The random separation of chromosomes in meiosis accounts for the variability in offspring. In a diploid plant, there are 2 copies (alleles) of each gene; both can be dominant or recessive, or one dominant and one recessive. Each pollen or egg randomly receives one of these alleles. Additional variation occurs when chromosomes are paired, at which time they may exchange genetic material; this is called chromosomal crossover. This process and mutation of genes (by the environment, other organisms, or humans) account for most of the genetic evolution of plants.

Ploidy and plant improvement

Plant DNA is organized into tightly wound chromosomes within each plant cell nucleus. The amount of genetic material can vary in different cells and life stages. In general, most plants and other living organisms have 2 sets of each chromosome (diploid = $2n$) during the major part of their life cycle. Many flowering plants are diploid organisms and have between 14 and 22 ($2n$) chromosomes. However, some diploid plants have many more chromosomes; adder's tongue fern (*Ophioglossum vulgatum*) has approximately 1200 chromosomes!

The number of chromosomes and resultant genetic material are not directly correlated with genetic complexity. For example, ferns have considerably more genetic material than tomatoes. This is known as the C-Value paradox: the amount of DNA in the haploid ($1n$) cell of an organism is not related to its evolutionary or biological complexity.

Additionally, some plant species have multiple sets of chromosomes or a condition known as polyploidy. This condition is naturally occurring and fairly common among our produce (for example, potatoes, tomatoes, strawberries, and blueberries), some ornamentals, and grasses, but is very rare in mammals. An estimated 30%–70% of flowering plants probably have polyploid origins. Triploid plants have 3 sets of chromosomes in somatic cells, tetraploids have 4 sets, hexaploids have 6 sets, and octoploids have 8 sets. For example, wild wheat is diploid ($2n = 2x = 14$), wild emmer wheat is tetraploid ($2n = 4x = 28$), and commercial durum wheat used to make bread is hexaploid ($2n = 6x = 42$). Strawberry species can

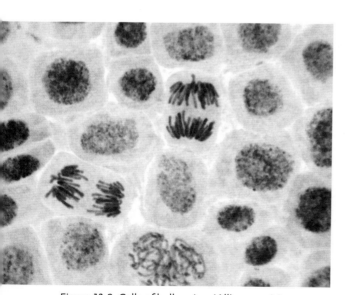

Figure 13-2. Cells of bulb onion (*Allium cepa*) in various stages of mitosis. After the chromosomes in a cell duplicate, they move to opposite sides before the cell divides.

be diploid, tetraploid, pentaploid, and octoploid. The woodland strawberry (*Fragaria vesca*) is diploid ($2n = 2x = 14$) while commercial strawberry (*F. ×ananassa*) and native *F. virginiana* are octoploid ($2n = 8x = 48$).

Characteristics of polyploid plants include larger organs: thicker leaves, flowers, and fruits. Shoots and stems are often thicker with shorter internodes and tend to be more heterozygous, vigorous, and adaptable due to multiple sets of chromosomes, especially if the sets of chromosomes come from 2 different ancestors.

Polyploidy can be induced by the application of colchicine, a plant extract originally derived from autumn crocus (*Colchicum autumnale*), which inhibits formation of spindle tubers during mitosis. This results in duplication of the genetic material/chromosomes without subsequent cell division, yielding these aberrant numbers of chromosomes. Other inhibitors of mitosis include oryzalin, trifluralin, and nitrous oxide gas (Bouvier et al. 1994, Eeckhaut et al. 2004). Typically, seedlings with actively growing meristems are sprayed or soaked with the compound. Polyploidism can result in sterile plants, enhanced flower and fruit size, increased vigor, pest resistance, and stress tolerance. Another commercial use of inducing polyploids is in creating triploid watermelons and citrus. Because of triploidy, the embryos are not fertile and thus the fruit are seedless.

From DNA to proteins

Heritable plant genetic material, encoded as DNA within the nucleus, serves as a template for RNA. When the RNA polymerase enzyme "unzips" the double strands of DNA, a single-stranded complementary messenger RNA (mRNA) is formed according to the coded fragments of the DNA. This mRNA is transported from the nucleus into the cytoplasm of the cell to the ribosomes. The mRNA is then translated to produce proteins.

Over 90% of DNA is presumed nonfunctional because it does not "read" the same as that which is known to be functional. Intermixed in sequences of nucleotides are nucleotides that do not code for anything, as far as is currently known. These are called *introns*. The nucleotides that are known to be productive are called *exons*. After the DNA is transcribed, producing the single-stranded complementary RNA, the RNA removes the introns with restriction enzymes. The remaining functional sequences, the exons, are spliced together to form mRNA.

A codon is a group of 3 nucleotides that code for a particular amino acid (Figure 13-3). A gene is a section of DNA, made up of several hundred nucleotides, that codes for a protein. There are 20 amino acids that are combined during translation of the genetic code to form proteins with various chemical modifications (see Table 13). Hundreds of proteins can be found in a single cell. Many proteins serve as enzymes, organic catalysts that assist cellular reactions but are not used up in the processes. Proteins also play a structural role in the make-up of cell components such as cell walls, membranes, and chromosomes.

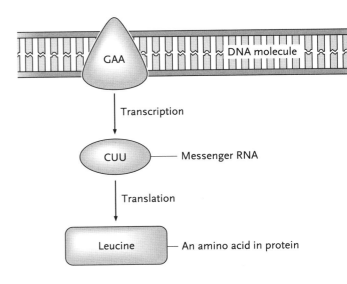

Figure 13-3. The codon GAA codes for the amino acid leucine via the messenger RNA CUU.

Table 13. **Amino acids and corresponding DNA codons**

AMINO ACID	SYMBOL	SLC	DNA CODONS
Alanine	Ala	A	GCT, GCC, GCA, GCG
Arginine	Arg	R	CGT, CGC, CGA, CGG, AGA, AGG
Asparagine	Asn	N	AAT, AAC
Aspartic acid	Asp	D	GAT, GAC
Cysteine	Cys	C	TGT, TGC
Glutamic acid	Glu	E	GAA, GAG
Glutamine	Gln	Q	CAA, CAG
Glycine	Gly	G	GGT, GGC, GGA, GGG
Histidine	His	H	CAT, CAC
Isoleucine	Ile	I	ATT, ATC, ATA
Leucine	Leu	L	CTT, CTC, CTA, CTG, TTA, TTG
Lysine	Lys	K	AAA, AAG
Methionine	Me	M	ATG
Phenylalanine	Phe	F	TTT, TTC
Proline	Pro	P	CCT, CCC, CCA, CCG
Serine	Ser	S	TCT, TCC, TCA, TCG, AGT, AGC
Threonine	Thr	T	ACT, ACC, ACA, ACG
Tryptophan	Trp	W	TGG
Tyrosine	Tyr	Y	TAT, TAC
Valine	Va	V	GTT, GTC, GTA, GTG
Stop codons	Stop	—	TAA, TAG, TGA

SLC = Single-letter database codes.

Inserting foreign DNA into plants

Genetic material in plants has been modified in many ways both naturally and through other means. Some organelles in the plant cell, such as mitochondria, which generate the cellular energy molecules in the form of ATP (adenosine triphosphate), are thought to be prokaryotic bacteria in origin and were incorporated into plants several billions of years ago. Chloroplasts, which harvest light energy from the environment into the plant cells, are also thought to have been incorporated into plants by endosymbiosis from cyanobacteria. These 2 very essential organelles are examples of "foreign" DNA that became parts of plants' cellular components and function over time, or horizontal gene transfer. Plant improvement is still mainly dependent upon conventional breeding, selection, and hybrid seed production. Genetic engineering does offer some unique advantages including extremely directed insertion of one or several biologically important traits (for example, pest resistance, flower color, cold tolerance) into an otherwise perfectly suitable plant and overcoming biological/incompatibility barriers.

Bacteria and viruses. Bacteria and viruses are particularly adept at inserting their DNA into plants, sometimes in a symbiotic relationship but oftentimes for their own gain. *Agrobacterium tumefaciens*, a plant pathogenic bacterium know for causing crown gall, is perhaps the most famous and widely used example today. This bacterium contains a plasmid (a small, extra-chromosomal, circular piece of DNA that replicates independently of the chromosomal DNA). When introduced into the plant, the plasmid hijacks the plant cell genetic machinery and metabolism, incorporating its genetic material, resulting in the formation of a tumor known as crown gall.

It is now possible to remove tumor-producing genes from bacteria like *Agrobacterium tumefaciens* and substitute genes perceived as useful, such as genes for herbicide resistance (Gasser and Fraley 1992). With such plants, herbicides can be used to kill weeds without affecting the genetically engineered plants expressing the gene that conveys herbicide resistance. To determine if the foreign gene DNA of interest (T-DNA) has been successfully incorporated into the plant DNA and is being transcribed and translated, or "expressed," the T-DNA typically also contains genetic code for a selectable marker. The selectable markers are usually either antibiotic resistance (for example, neomycin and kanamycin), where only plant cells expressing the resistance gene from the inserted T-DNA will survive, or fluorescent proteins, modified genes that code for a protein from jellyfish that make cells glow under specific spectra of light and can be visually selected for genetic transformation (Chalfie et al. 1994). Other traits that are introduced in such a way into plants are insect and viral resistances, stress tolerance, altered fruit ripening, and altered flower color (Greenberg and Glick 1993).

The closely related *Agrobacterium rhizogenes* carries a root-inducing plasmid (Ri-plasmid) that causes root tumors. This bacterium has not been as widely used for genetic engineering as *A. tumefaciens* except in the production of hairy roots that are valuable for the production of phytochemicals, enzymes, and foreign proteins (Chandra 2012). It should be noted that both bacteria are considerably more adept at transforming dicotyledonous plants but monocots such as rice have been transformed (Toki 1997). Efficient transformation of plants is dependent upon virulence of the *Agrobacterium* strain upon the intended plant host, plant response, and selection/regeneration of transformants (Li et al. 2011).

Genetic engineers have used *Agrobacterium tumefaciens* as well as many other bacteria and yeast to insert foreign DNA. They integrate pieces of foreign DNA into pieces of plasmid DNA, which when inserted into a suitable host bacterial cell, replicate and produce useful products, such as insulin, interferon, and certain antibiotics.

Though it certainly is incredible that, with nuclear material so infinitely small, scientists can cut out genes and move them into other organisms, any study of nature reveals that the combining of genes is nothing new. Consider, for example, a virus. These amazing bits of DNA or RNA reside in a protein coating. When separated from living material, viruses are virtually inert chemicals; however, if a virus enters a living cell, whether bacterium, animal, or plant, the virus suddenly comes to life. Settled on the wall of an organism, it shoots its own DNA out of its coating through the organism's cell wall to merge with the organism's DNA, which then becomes recombinant DNA (DNA combined from different sources). In such an instance, the viral DNA replicates only when the host cell's DNA multiplies. Sometimes a virulent virus can infect the host cell, producing a diseased state in which the virus replicates itself until the cell wall bursts and a myriad of new viruses spread out to find other cells to victimize. The only defense an organism has is its restriction endonuclease enzymes, which act as scissors to cut out unwanted viral DNA. Scientists have isolated numerous restriction endonucleases, enzymes that cut different chromosomal DNA base sequences.

The insertion of foreign (desired) DNA into a plasmid requires 2 different enzymes (Figure 13-4, steps 1 and 2). The first enzyme is a restriction endonuclease-type enzyme. Such enzymes recognize and become attracted to specific base sequences in double-stranded plasmid DNA, where they cut the DNA. The same restriction endonuclease is used to prepare the desired DNA. The net result is overlapping cuts with sticky ends on both the plasmid and desired DNA molecules, which enables the desired DNA molecule to fit into the plasmid DNA. The second enzyme, ligase, sews these parts together, once again forming a circular plasmid molecule.

The next step in this process of genetic engineering is the introduction of the recombinant plasmid molecule into the host bacterium, or "transformation" of the bacterium to have it replicated or cloned—many identical copies are made (Figure 13-4, step 3). The desired DNA must permeate, or pass through the bacterial cell's membrane. This can occur in nature, but most commonly the host bacterial cells have been selected for this ability, or artificially induced by chemical treatment to be more amenable to uptake of desired DNA. Special techniques are used to introduce the recombinant plasmid DNA into the bacterial cells. One of these techniques is electroporation, whereby the host bacterium and the recombinant plasmid DNA are mixed together and treated with an electrical current. The electrical current weakens the bacterial cell

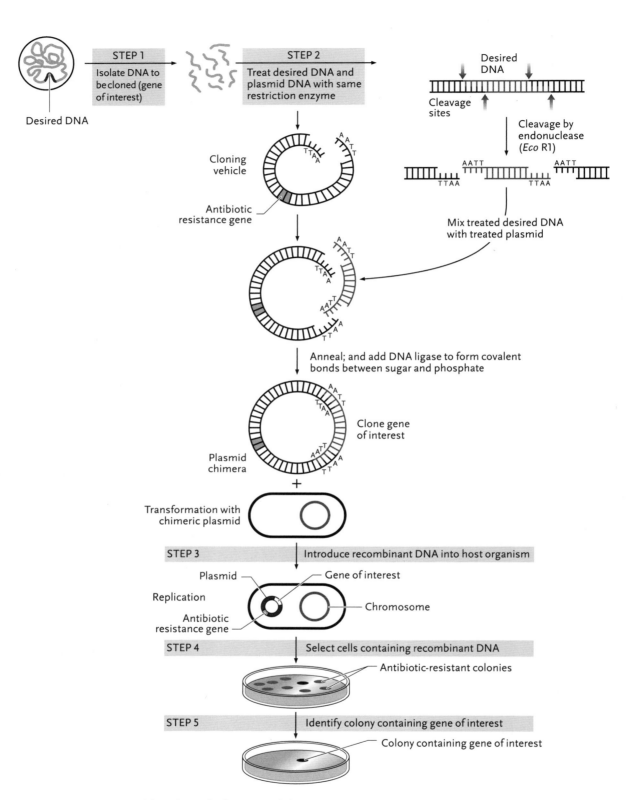

Figure 13-4. The scheme for forming and cloning recombinant DNA using a bacterial host cell.

membrane, enlarging "pores," thereby enabling the fragmented DNA to enter the host cell and be replicated.

To identify the recombinant DNA clone, the plasmid used for cloning the desired DNA must contain a marker gene, for example, a gene for resistance to a certain antibiotic. Such a marker enables selection of the host cells when they are placed on a growth medium containing the antibiotic (Figure 13-4, step 4).

The bacterial cells containing the recombinant DNA plasmid are now grown in selection media such that the plasmid is replicated rapidly, usually overnight is sufficient to generate enough plasmid DNA. The plasmid DNA is then extracted back from the bacterial cells and separated from the bacterial chromosomal DNA based upon size, usually using a filtration step. This now-concentrated-plasmid DNA can be used for *Agrobacterium*-mediated plant transformation after it is again electroporated into an *Agrobacterium* strain of interest. The *Agrobacterium* cells containing the disarmed tumor-inducing plasmid can now use their former cancer-inducing machinery to incorporate the gene of interest and selectable marker(s) of the recombinant plasmid into the plant cells' DNA.

Other methods of inserting foreign DNA. While Agrobacterium-mediated gene transfer is applicable to many dicotyledonous plants, some plants, like soybean, are not amenable to the process, nor are monocots. For these crops, researchers have developed a few other strategies for transferring genes of interest into plants.

Another method for plant gene transfer is the use of polyethylene glycol (PEG) for direct intake of DNA by plant protoplasts. Protoplasts are plant cells with cell walls removed and only the plasma membrane to hinder ingress of DNA inside. However, lacking a cell wall can often render them unstable and fragile. Young leaves can be used to make protoplasts by surface sterilizing the leaves and then enzymatically digesting the cell wall. Suspension cell cultures are often used as the cells are actively growing and typically synchronized with respect to cell division cycle.

Electroporation can also be used to move foreign DNA into protoplasts. An electrical pulse of about 750 volts/cm is briefly applied (5 to 10 milliseconds) to the protoplasts (Bates 1999). The pulse opens small holes or "pores" in the plasma membrane. The major drawback to using protoplasts is the long waiting period for the single cells to grow into plants of significant size.

Biolistic methods (particle guns or "gene guns") now allow direct transfer of DNA into plants that are recalcitrant/nonresponsive to *Agrobacterium*-mediated transformation or are monocots. Tungsten or gold microparticles are coated with the DNA to be injected, and they are then shot into plants at a speed of 1000 miles per hour (about 1600 kilometers per hour). This system was first successfully demonstrated in 1987 (Sanford et al. 1987). The projectile of the "bioblaster" is driven by an exploded .22-caliber blank cartridge. Thousands of cells are pierced and inoculated at once by this method (Glick and Thompson 1993, Wood 1989).

BioRad (Hercules, California) markets a commercial system called the Biolistic PDS-1000/He, which uses helium gas as the propellant. The company has also developed a hand-held Helios® Gene Gun system that can be used on plant tissue that cannot be placed in the vacuum chamber. A further modification of this system is the double-barreled gene gun where 2 simultaneous bombardments with control and test DNA preps onto the same leaf allow significantly fewer bombardments while providing statistical measures of transformation efficiency (Kale and Tyler 2011).

Other biolistic systems have been developed. One such device was the Particle Inflow Gun (PIG) in which helium was used to transport microcarriers into the vacuum chamber instead of rupture discs and macrocarriers in the original BioRad design (Gray et al. 1994). This system is cheaper than the BioRad system and has been successfully used to transform a number of crop plants. The ACCEL system uses high voltage shock waves to propel microcarriers with genetic material into plants (McCabe and Christou 1993). The major advantage of this system is reduced tissue damage compared to gunpowder or helium.

DNA fragments are commonly analyzed by electrophoresis (often called fingerprinting), a system whereby substances are separated by size and charge, which determines the distance they travel on a gel within an electrical field. The smaller fragments will move faster and farther than the larger ones. After the fragments have separated, the current is shut off and the separated fragments are photographed, measured, compared with controls, identified, and sampled for further analysis, treatment, or transfer. Segments of the gel can be used as probes.

Electrophoresis can be used to separate DNA fragments, proteins, and enzymes. It has been used in the past to separate isozymes (isoforms of a given enzyme) to differentiate between species or even cultivars such as those of red raspberry (*Rubus idaeus*) (Cousineau et al. 1993). Since the 1990s, many plant species and other organisms are differentiated by sequencing of DNA fragments rather than electrophoresis. One example is the use of the highly variable ITS (internal transcribed spacer) region of nuclear ribosomal DNA to differentiate among species in the Asteraceae (Schmidt and Schilling 2000) rather than electrophoresis.

Mouse ear cress (*Arabidopsis thaliana*), a common weed in the mustard family (Brassicaceae), is a favorite experimental plant for genetic engineering because its genome has been sequenced, it grows well in culture, can flower in 5 weeks, can produce thousands of progeny in 10 weeks, and has the smallest known genome of any higher plant—10 chromosomes (= $2n$) and 70 million base pairs. It has been used extensively to study gene function through mutagenesis and the selection of mutants (Figure 13-6).

The ability to selectively insert new genes into plants is possible because of the rapidly evolving ability to sequence the genomes of many

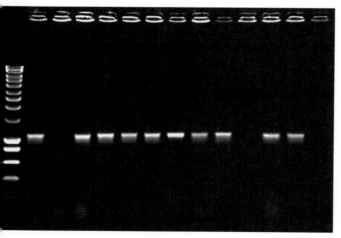

Figure 13-5. In electrophoresis, which helps researchers visualize DNA as separated by size, the DNA is negatively charged and will migrate through a polymer gel toward a positive charge. The DNA fragments are visualized by adding ethidium bromide as a dye. As the DNA fragments move through the gel, they pick up the dye, which intercalates between the double strands, or "ladder," and are visible when illuminated by a UV light.

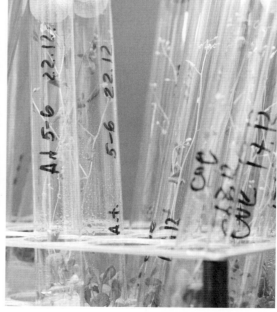

Figure 13-6. *Arabidopsis* is a very small plant that has had a huge impact on biotechnology.

plants. The cost and time to sequence genomes has exponentially decreased over the years. Several of the major crop species have been completely sequenced (see Table 14). As previously noted, the first was *Arabidopsis*, sequenced in 2000. Currently, a project to sequence one thousand plant genomes is underway. The purpose of sequencing is to provide a gene map to relate various traits to specific pieces or fragments of DNA. This knowledge has tremendous potential for assisting the cure of certain diseases and enhancing our understanding of human nutrition and body physiology. In the realm of plants, this knowledge can mean broadening the gene pool for hybridization, developing healthier and more nutritious plants, creating more beautiful surroundings, or averting worldwide famine. The future of biotechnology is in our hands, and it is our obligation to see that every reasonable precaution is exercised so that the products of this new science only benefit society.

Table 14. **Plant genomes sequenced through 2011**

SCIENTIFIC NAME	COMMON NAME	PUBLICATION	YEAR
Arabidopsis lyrata	—	*Nature Genetics*	2011
Arabidopsis thaliana	Arabidopsis	*Nature*	2000
Brachypodium distachyon	Brachy	*Nature*	2010
Brassica rapa	Turnip	*Nature Genetics*	2011
Cajanus cajan	Pigeon pea	*Nature Biotechnology*	2011
Cannabis sativa	Cannabis	*Genome Biology*	2011
Carica papaya	Papaya	*Nature*	2008
Cucumis sativus	Cucumber	*Nature Genetics*	2009
Fragaria vesca	Strawberry	*Nature Genetics*	2010
Glycine max	Soybean	*Nature*	2010
Lotus japonicus	Lotus	*DNA Research*	2008
Malus ×domestica	Apple	*Nature Genetics*	2010
Oryza sativa L. subsp. *indica*	Rice	*Science*	2002
Oryza sativa L. subsp. *japonica*	Rice	*Science*	2002
Phoenix dactylifera	Date palm	*Nature Biotechnology*	2011
Physcomitrella patens	Moss	*Science*	2008
Populus trichocarpa	Poplar	*Science*	2006
Ricinus communis	Castor bean	*Nature Biotechnology*	2010
Selaginella moellendorffii	Selaginella	*Science*	2011
Solanum tuberosum	Potato	*Nature*	2011
Sorghum bicolor	Sorghum	*Nature*	2009
Thellungiella parvula	—	*Nature Genetics*	2011
Theobroma cacao	Chocolate	*Nature Genetics*	2010
Vitis vinifera	Grape	*Nature*	2007
Zea mays	Maize	*Science*	2009

Culture Methods

The following are culture methods that not only have the potential for plant regeneration but also have biotechnological applications in such areas as physiology, nutrition, disease and herbicide resistance, and the production of secondary products (Figure 13-7). The requirements for these procedures are generally more sophisticated and demanding than for basic tissue culture and may not be realized by even the most astute students or skilled laboratory scientists.

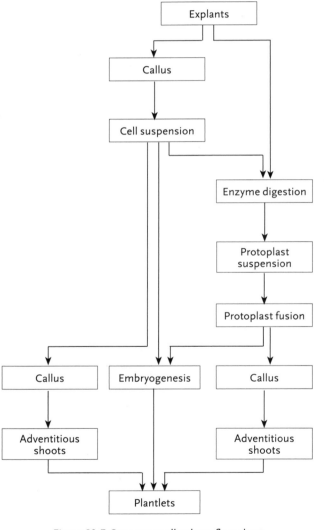

Figure 13-7. Summary cell culture flow chart.

Callus culture

Callus is a disorganized proliferated mass of actively dividing cells, often thin-walled parenchyma cells. Historically, carrot (*Daucus carota*) and tobacco (*Nicotiana*) callus cultures have been the classic models for plant tissue culture. Regeneration of whole plants from callus was standard procedure until the merits of shoot tip culture were realized; a callus phase usually was induced before obtaining shoots and roots. For some plants, the callus stage is still required, particularly in difficult-to-grow plants like the grass family (Poaceae).

Callus is a source of cells for liquid cell suspension and embryogenic cultures—the ultimate form for mass production, as millions of plants can be started in a very small space. Callus can be friable or firm textured and it can be green, cream, yellowish, or brown colored.

Callus culture will produce more mutant plants than does shoot culture, but sometimes mutations are deliberately induced for genetic variability (see "Somaclonal Variation" elsewhere in this chapter). Careful management can minimize the potential for mutations in callus, primarily by limiting the length of time as callus and by frequent transfer.

Protocol for producing carrot (*Daucus carota*) callus

1. Clean viable seeds in alcohol and bleach.
2. Apply several sterile water rinses.
3. Place on half-strength MS medium in test tubes or petri dishes.
4. After germination, place 5- to 10-in. (cm) sections of hypocotyl on Gamborg's B5 medium (see *Catharanthus* in Section Two) with 0.1 mg/liter of the auxin 2,4-D.
5. Place in the dark at 77°F to 85°F (25°C–28°C).
6. Transfer every 4 weeks until there is sufficient growth for the particular study.

Cell culture

Cell culture technology has many objectives, among them being the study of cell nutrition and metabolism, disease and pesticide resistance, the production of secondary products, somatic hybrids, haploid culture, protoplast isolation and fusion, genetic engineering, mutagenesis, embryogenesis, and other phenomena. Cell suspension cultures are a convenient means of studying cell utilization of and response to different nitrogen and carbon sources. The data gathered from such studies are important not only for cell cultures but also for application to plants in the field or greenhouse. Cell cultures are used for screening cell resistance to introduced pathogens, herbicides, fungicides, salinity, and other factors. The cells that do not succumb to the introduction of such factors can develop to be whole plants possessing corresponding resistance.

Cell culture is a productive and efficient method of multiplication (barring mutations). The cells and/or embryos are ultimately plated out (spread on agar medium) and grown out, with or without a callus stage. Some cells will give rise to somatic embryos; in other cases, plantlets are obtained only by way of organogenesis, or from callus-derived adventitious shoots.

Cell and cell suspension cultures are derived from callus cultures. Cell suspension cultures are more readily obtained if the callus is friable. Such cultures require a rotator or open-platform orbital shaker. Gamborg's B5 is a satisfactory medium for carrot cell suspension culture; 2,4-D is added and agar is omitted. Antibiotics are often necessary.

Somatic embryogenesis

Somatic (or nonzygotic) embryogenesis is the formation of embryos from somatic (vegetative) cells, not zygotic (sexual) cells. This must not be confused with embryo rescue, which is the removal of an embryo from a seed and using that embryo as an explant for micropropagation. Dividing cells undergoing embryogenesis first form a globular stage, then a heart stage, and then a torpedo stage, from which plantlets develop. Factors that affect nonzygotic embryogenesis include the type of explant (immature embryo explants usually), and certain genotypes are more prone to somatic embryogenesis and this ability is often heritable. Embryogenic cells are typical smaller in size, isodiametric in shape, and the cytoplasm appears dense. The success rate of obtaining plants from embryogenic cells is rather low, occasionally as much as 50% but typically only a few percent.

Although somatic embryo cultures and cell cultures tend to produce abnormal plantlets, an increasing number of species are being produced successfully by these methods. A promising potential application for somatic embryogenesis is encapsulated artificial seed (Redenbaugh 1990). Synthetic seeds are useful when asexual propagation *in vitro* is desirable, but seedlike propagules are more convenient than plantlets. Examples may be when there is a need for large numbers of trees for reforestation or hybrid seed that is difficult or expensive to produce. Embryos can be encapsulated in a gel to make artificial

Protocol for producing a cell suspension culture

1. Cut 4 to 6 callus pieces, each about 0.5 in (1 cm) square.
2. Transfer the callus pieces to 125-ml Erlenmeyer flasks containing modified Gamborg's B5 medium.
3. Stopper the flasks with cotton and cover tops with foil.
4. Place the flasks on a shaker in continuous light.
5. Subculture every week by allowing the cells to settle and decanting part of the medium, which is then replaced with fresh medium. A cell turbid suspension usually develops within 6 weeks.

seeds, then field planted by fluid drilling (field planting of seeds in water). Celery (*Apium graveolens*), carrots (*Daucus carota*) (Nag and Street 1973), and bananas (*Musa*) are among the crops that have been planted using encapsulated embryos, with varied success.

Protocol for conditioning cells in suspension for embryo development

1. Decant the cell suspension medium and replace with fresh Gamborg's B5 liquid medium.
2. Continue agitation on the shaker.
3. In 2 weeks, decant the medium again.
4. Spread the cells on half-strength Gamborg's B5 or MS medium with agar and without hormones in test tubes or petri dishes.

Protoplasts

Protoplasts, which are cells that lack their normal rigid cellulose walls but retain their plasma membranes, are a valuable research tool in biotechnology and are useful for plant breeders and hybridizers. Because protoplasts lack walls, many substances are readily taken up through the fragile membranes, including viruses, nuclei, chloroplasts, DNA, proteins, other macromolecules, and other protoplasts. Protoplasts are used for genetic manipulations because isolated DNA and viruses can be incorporated or protoplasts of 2 different genotypes can be fused into a novel hybrid.

Starting material for protoplast cultures may be any rapidly growing tissue cultured callus or shoots, cell culture, or leaf mesophyll from young, fully expanded leaves. Most growers probably will not want to attempt protoplast culture themselves because it is a very difficult procedure, but it is important to understand the principles in order to appreciate their potential.

Protoplasts are useful to the scientist because they are single cells, can be obtained from diverse plant tissues, lack walls and thus can be manipulated, and eventually can be grown back into entire plants. Various enzymes are used to remove cell walls and produce protoplasts. Both cellulases and pectinases are needed. Fresh plant parts and cultured cells (callus or suspensions) can be used to generate protoplasts. The plant cells need to be incubated in the enzymes plus an osmoticum like sorbitol, mannitol, sucrose, or glucose. The osmoticum will pull the cell away from the cell wall. Then filtering is needed to separate protoplasts from debris and enzymes.

Protoplasts obtained from cell suspension cultures of callus are colored off-white. Protoplasts from cell suspension cultures of leaf mesophyll cells are green. When fusing protoplasts from 2

Two methods of isolating protoplasts

Protocol for using *in vitro* starting material
1. Pretreat cells for 1–2 days in the dark.
2. Centrifuge cells in Gamborg's B5 medium.
3. Incubate with filter-sterilized enzyme solution (cellulase and Macerozyme) for 1 to 12 hours.
4. Filter through a 60-micron (μ) screen.
5. Centrifuge.
6. Using a microscope, observe for protoplast release.
7. Add to appropriate culture-growing medium.

Protocol for using a leaf as starting material
1. Clean leaf in bleach and rinse.
2. Peel off the epidermis and re-apply the enzyme solution.
3. Incubate with filter-sterilized enzyme solution (cellulase and Macerozyme) for 1 to 12 hours.
4. Filter through a 60-micron (μ) screen.
5. Centrifuge.
6. Using a microscope, observe for protoplast release.
7. Add to appropriate culture-growing medium.

different source plants, it may be useful to prepare the protoplasts by these different methods—from callus and from leaf mesophyll cells—to facilitate observation of the fusion and distinction of the dissimilar cells that are being fused. Protoplasts from 2 different plants can also be distinguished from one another microscopically by staining with different colored nontoxic vital stains.

Haploid culture

Haploid cultures are usually derived from pollen via anther culture and rarely through microspores. The purpose of this procedure is to convert polyploids to lower ploidy levels, produce homozygous lines, and to select desirable combinations of traits or genes. Haploid plants are valuable to plant breeders and geneticists because there is only a single set of chromosomes to determine inheritance and expression of various traits. Haploid plants are often distinguishable by their smaller size, and they are often less vigorous than their counterpart diploid plants especially if there are lethal recessive alleles present.

Haploid plants can be rapidly converted to fertile homozygous diploids (doubled haploids) that have 2 sets of identical chromosomes; these are equivalent to inbred lines obtained through conventional breeding, but usually obtained much faster. Homozygous diploids can occur when chromosomes of haploid plants spontaneously double or they can be induced to double using colchicine (an alkaloid extract from the autumn crocus [*Colchicum autumnale*] that arrests mitosis by preventing the separation of duplicated chromosomes to daughter cells). Doubled haploids allow for the selection of recessive mutations.

Haploid plants are recovered *in vitro* when plants are regenerated from haploid cells that are formed during meiosis. Haploid cells from males are the microspores (androgenesis) and haploid cells from females are the egg cells (gynogenesis). Usually the males are used through anther culture because there are many anthers per plant, they are easily accessible, and more successful. Many factors affect success in anther culture:

- The developmental stage of the flower buds, anthers, and microspores.
- Pretreatments to the anthers (cold shock or heat treatments).
- Media.
- The condition of the plant from which anthers are taken.

Isolated microspores from the anther can also be used to produce haploid plants, but this procedure is more complicated and less likely to succeed.

Once plants are produced, they can be determined to be haploids by counting chromosomes under a microscope, measuring DNA content by flow cytometry, using epidermis peels to count the number of chloroplasts in guard cells, or examining size and fertility of the plants. If the plants are self-pollinated, the progeny will be identical to the regenerated plant.

Anther culture procedures vary significantly from one species to another, and the failure rate is generally high. Many factors contribute to success including genotype, stage of microspore development, components in the culture medium, and anther pretreatment involving a cold shock. More detailed information on producing haploid cultures and, more specifically, haploid tobacco and potato plants, can be found in Hadziabdic et al. (2011), Wadl et al. (2011), and Bhojwani and Dantu (2010).

Somaclonal variation

Somaclonal variation is variation in plant cells that is generated by the use of tissue culture; it can be spontaneous or it can be induced. For the micropropagator who wants to produce clones, somaclonal variation is a nightmare; for a plant breeder, this process is a wonder. This phenomenon can be a genetic change and can be sexually transmitted to the progeny. Sometimes variation *in vitro* can be epigenetic, which is not genetic change. Epigenetic changes are the result of environmental pressures and cannot be transferred to the next generation by seed. They can, however, sometimes be stable if propagated vegetatively.

The most easily observed somaclonal variations are gross morphological changes including leaf color and shape and growth habit. Somaclonal variation can be used to direct genetic changes in plants to develop environmental adaptation to cold or salt tolerance, insect resistance (Hadi and Bridgen 1996), and new plants, especially ornamentals (Brand and Bridgen 1989). If this variation arises from sporophytic tissue, it is called somaclonal variation; if it arises from gametophytic tissue, it is called gametoclonal variation.

There are ways to discourage somaclonal variation in plant tissue cultures, and these procedures are important to remember when producing clones. Organized growing points like the shoot tips or axillary buds have the least chance of mutation. Adventitious shoot formation (as opposed to axillary shoot formation), callus, and cell suspensions have the greatest chance for genetic mutations because they are not genetically stable. Long-term cultures and cultures that are forced to grow quickly have greater probabilities of mutation. Media with high levels of cytokinins and auxins, especially like 2,4-dichlorophenoxyacetic acid (2,4-D), are more prone to problems with genetic stability. Unfortunately, it is not always easy to tell if cells have been mutated because the expression of mutations is often strongly dependent on the environment and micromutations often occur that are difficult to identify.

Embryo culture

Embryo culture, also known as embryo rescue, is the aseptic removal of an immature or mature zygotic embryo from a seed and growing it *in vitro* with the goal of obtaining a viable plant (Bridgen 1994). This technique has been used for more than half a century to save the hybrid products of fertilization when they might abort or otherwise degenerate.

The major applications of embryo culture have been for plant breeding especially for interspecific hybridization because of embryo abortion. By aseptically culturing the embryo in a nutrient medium, this problem may be overcome. Embryo culture can shorten the breeding cycle by overcoming dormancy in seeds. Dormancy may be caused by endogenous inhibitors or specific environmental factors like light, temperature, and dry storage requirements. Embryo culture can also be used to propagate plants vegetatively because embryos are responsive as juvenile tissue. In addition to the applied uses of embryo culture, the procedure is useful in basic studies to study nutritional and growth requirements of the embryo as well as precocious germination, to determine seed viability, or to produce haploid plants following distant hybridization.

The process of embryo culture can be difficult due to the tedious dissection that is necessary and the complex nutrient medium requirements. Embryos are located within the sterile environment of the ovule, so surface sterilization of the embryos is not necessary. Instead, entire ovules or ovaries are surface-sterilized, and then the embryos are removed aseptically. The removal of small embryos is difficult and requires the use of a dissecting microscope and microdissecting tools; large embryos are not difficult to excise. Successful growth of embryo cultures depends on many factors such as the media, light (or the absence of light), and temperature.

New Forests

The demand for timber, pulp, paper, and wood derivatives is increasing as populations grow, but forests the world over are declining. Reforestation has not kept pace with the steadily growing demand for wood products. With the urgent worldwide need for reforestation seedlings (preferably from superior trees), tissue culture has done much toward helping the situation. Progress is due in part to the millions of dollars that has been spent on research to solve the many problems of propagating superior forest trees, particularly conifers, via tissue culture.

Embryos, cotyledons, juvenile buds, needles, and fascicles have responded to culturing far more readily than have explants from mature trees. Research has been plagued by slow growth, aberrations, poor rooting, vitrification, plagiotropic growth, and genetic instability. Both the difficulty in cleaning older material and the lack of response by mature tissues have repeatedly blocked the way to micropropagation of select conifer trees. However, the production of high-quality somatic seedlings via somatic embryogenesis has been successful on a commercial scale for several economically important conifers (Grossnickle et al. 2011).

Growers are well acquainted with the problems related to lack of juvenility in rooting cuttings. In many cases, the methods applied to retain juvenility or induce rejuvenation for purposes of cutting wood also apply to obtaining juvenile material for explants for tissue culture. Hedging or shearing has long been practiced to retain trees of a particular species in a juvenile state in order to provide cuttings that will root.

Alternatively, rejuvenation of mature stock of some species is sometimes accomplished through sequential cuttings, grafts, or hedging. Another method is to apply repeatedly a cytokinin to appropriate nodal areas. This technique has been known to produce witches'-broom, the shoots of which display some degree of juvenility and consequently respond as tissue culture explants.

One approach employed by timber companies is to multiply full-sib (where both parents are known) seedlings *in vitro* for plantlets to increase the orchard, and then use the seed from that orchard for reforestation. The cost effectiveness of reforesting directly with the tissue cultured plantlets is questionable.

Mature trees are occasionally tissue cultured by the induction of somatic embryogenesis from cells of the nucellus. In theory, if the embryo is removed from a seed, the tissues that remain will have the same genetic make-up as that of the female parent plant, not of the embryo. Because of the ease of cleaning seeds and the juvenility of nucellus tissue, this is a viable approach to micropropagation of mature trees. The subsequent cell culture with ensuing somatic embryogenesis is classic procedure.

With respect to hardwood forest trees, the first organogenesis in woody tissue culture was observed by Roger J. Gautheret in 1940 when he induced buds from cambial tissues of elm (*Ulmus campestris*). The next hardwood milestone was not reached until 1968, when Lawson Winton reported the first true plantlets from aspen (*Populus*). To date, other micropropagated forest hardwood trees include *Acacia*, birch (*Betula*), sweet gum (*Liquidambar*), *Paulownia*, *Santalum*, teak (*Tectona*), and *Eucalyptus*. Most of these have been successfully cultured from mature trees, in addition to more routine success from juvenile sources.

Germplasm Storage

Germplasm storage is important because it preserves valuable stock plants and in a manner that keeps it disease- and insect-free. The world's plant germplasm historically has been confined to "living gene banks"—the collections of plants in plantations, orchards, arboreta, and botanic gardens—and to seed gene banks for storing seeds that lend themselves to drying and maintenance at low temperatures, such as at the U.S. Department of Agriculture's Seed Storage Facility at Fort Collins, Colorado, part of the National Center for Genetic Resources Preservation (NCGRP) run by USDA/ARS. Many other collections exist around the world. The concern among conservationists has been that such collections are inadequate, seeds do not store well, the viability of seeds decreases with time, and this is not suitable for asexually propagated plants.

Plant tissue culture offers an opportunity to store plant materials *in vitro*. It can be as easy as placing cultures in refrigeration at 40°F–43°F (4°C–6°C) for short-term storage or as high tech as cryopreservation for long-term storage. The ultimate goal is to slow down or stop growth without damaging the plants or changing the genetics, while saving effort, expense, and space.

Cryopreservation is a procedure to freeze plant material to stop growth; it is used extensively with animals. Plants can be cooled to –321°F (–196°C) with liquid nitrogen to stop all metabolic activity so that plants can be stored indefinitely. Small, actively dividing nonvacuolated cells, such as shoot meristems, cell cultures, pollen, and embryos are best to use (Panis and Lambardi 2006).

One factor that is critical for successful cryopreservation is that water in tissues must not form ice. One way to do this is by increasing osmotic tolerance by growth on highly osmotic media, sequential transfer to increasing sucrose concentration, growth on abscisic acid (ABA), or glycerol. Cryoprotectants that have been used include dimethyl sulfoxide (5%–10% solution DMSO), sucrose (0.75 M), glycerol (10%–20%), and various combinations of DMSO, glycerol, and sucrose.

Tissue also undergoes cold acclimation with long cold nights and short warm days. The 3 most common techniques include controlled cooling, vitrification, and desiccation. Controlled cooling involves collecting dormant wood in the winter, providing additional chilling, slowly dehydrating, and then storage in liquid nitrogen. Vitrification (not to be confused with hyperhydricity, which refers to unhealthy plants; see chapter 9) involves encapsulation of tissue in vitrifying solution so that the cytoplasm becomes glasslike when flash-frozen. Desiccation involves removal of water by drying (typically with silica gel) and then storage in liquid nitrogen. Seeds are often cryopreserved in this manner as they typically contain very little water.

The procedures for thawing cryopreserved materials are as important as the protocols for freezing. Rapid thawing is often successful if components like polyoxyethylene-polyoxypropylene are added. Testing for survival after freezing includes biochemical tests, but the best evidence of survival is regrowth.

The International Board for Plant Genetic Resources (IBPGR) and cooperating laboratories have established a database for research, sources of germplasm, and other information. They stress the importance of disease indexing and eradication and of the characterization of clones in the various germplasm repositories. Cryopreservation is not yet used very extensively for the preservation of valuable plant materials in spite of its potential.

Secondary Products

Plants are not as complacent and vulnerable as they may appear. They respond to elicitors (stimulants) by producing secondary metabolites (secondary products). Metabolites are chemicals essential to metabolism, but secondary metabolites are not necessarily required for life. They include the alkaloids, terpenoids, and glycosides. Secondary metabolites often provide the plant's defense and individual response to the environment. Some of these products are allelopathic, referring to the phenomenon of chemicals emitted by plants that influence or inhibit the growth of other plants. Perhaps best known is juglone, a toxin produced by black walnut trees (*Juglans nigra*) that is leached from the trees into the ground, where it prevents the growth of some broad-leaved plants.

Many secondary metabolites are of research value or commercial importance, such as pharmaceuticals, medicinals, dyes, food additives, natural flavors, perfume fragrances, gums, and pesticides. The products are conventionally extracted from whole plants or parts of plants. Traditionally the plants are field grown, or they are collected in the wild where they are often in short supply, limited by season and weather, of unreliable yield, and of questionable quality. The collection and removal of plants from the wild is of some concern because of the depletion and potential extinction of certain species and because of the effect of removing native plant species on the natural habitat and environment. The application of plant tissue culture to growing cells, callus, or plantlets for extracting secondary products is the object of intensive research.

Tissue culture production provides the opportunity to develop higher yields than is possible from field-grown plants. Higher yields in cell cultures are achieved by adjusting media, selecting high-yield cell lines, genetic engineering, or applying elicitors, factors that help induce the sought-after chemicals.

The first step toward obtaining secondary products *in vitro* is to grow explants on a solid medium to produce callus. The second phase is to use cells from the callus to proliferate in liquid cell cultures. Although yields from cell cultures can be higher than yields directly from plants, they are, at best, still low. It is necessary to grow large quantities of cells in tanks (reactors or fermentors) to produce profitable volumes. The optimal medium, environment, treatment, and reactor must be carefully determined. Extraction involves equipment for chemical separation as well as technical ability, but the premium prices that are paid for secondary metabolites are an attraction; estimates range from hundreds to thousands of dollars per kilogram for certain products, but cost-effective procedures for mass production of these products are costly to develop.

One of the first secondary products to be successfully produced commercially by cell culture was shikonin (shikon), a secondary metabolite from the root of *Lithodora zahnii* that is used to remedy inflammation, to treat burns and hemorrhoids, as an antibiotic, and as a red dye (Curtin 1983). Mitsui Petrochemical Industries in Japan succeeded in producing this medicinal, which is widely used in Japan but also of interest elsewhere. Because *Lithodora* cannot be grown commercially in Japan, the company was determined to avoid the need for importing shikonin from China and Korea. By conventional methods, *Lithodora* requires at least 5 years to where the roots will produce a mere 2% shikonin. With careful selection of cell lines, Mitsui was able to obtain 15% of dry weight within a matter of months; its production was estimated at 65 kilograms (143 pounds) per year.

Because of the red color of shikonin, it was relatively easy to select for higher-yielding cell lines—the more red color, the greater the product. Mitsui determined that modifying the media of Elfriede Linsmaier and Folke Skoog and of P. R. White would increase shikonin yield significantly. This increase in shikonin production from *Lithodora* is achieved by raising the copper levels in the medium (Fujita et al. 1981), by treatment with a fungal homogenate, and by using a special extraction treatment (Chang and Sim 1994). Perhaps the biggest challenge has been to increase the scale of production from agitated liquid media in flasks to larger containers that could still be agitated without injuring the delicate cells.

Three types of bioreactors have been developed that allow for larger scale agitation of media: airlift fermentors, rotating drums, and stir tanks. An airlift fermentor provides agitation by introducing filtered air from below, which gently agitates the medium as the bubbles rise to the surface. The stir tanks appear to be the favored method for most large-scale operations (Singh and Curtis 1994). The powerful agitation mechanisms that are used for microbial reactors are too strong for plant cells, so impeller speed must be reduced yet still provide adequate aeration.

E. Ritterhaus and colleagues demonstrated the capability of stir tanks for effective production of secondary metabolites. Using a 75,000-liter tank with an undivided impeller for low shear, they grew coneflower (*Echinacea purpurea*), which has a cell concentration of over 200 grams fresh weight per liter (Singh and Curtis 1994), and thus has a greater output of secondary products. *Echinacea* species contain inulin, betaine, resins, sugars, mineral salts, fatty acids, and echinacein. Echinacin, a commercial extract of *Echinacea*, inhibits the activity of hyaluronidase, a spreading factor enzyme found in snake venom, in some streptococci, in heads of leeches, and in sperm (O'Neil 2006).

Asian ginseng (*Panax ginseng*) and American ginseng (*P. quinquefolius*) are widely used in health foods, cosmetics, pharmaceuticals, and other products, especially in Asia. Seed germination in these species is poor and plants require 5 years to produce seed; however, explants from roots, seeds, or flowers can produce plantlets *in vitro* in 3 months. With support from the Nitto Deako Corporation in Ibarak, Japan, Tsutomu Furuya and Keiichi Ushiyama (1994) were able to produce 500 mg of dried ginseng per day, using 2- and 20-ton bioreactors. Tissue cultured ginseng compared favorably to conventionally produced ginseng with respect to treating gastric ailments, hypoglycemia, and blood flow when tested in rats. The 2 researchers state that tissue cultured ginseng inhibits ulcer formation, although ginseng from cultivated plants does not. Some dispute the claims of the medicinal value of ginseng, however. Nevertheless, American ginseng is becoming exceedingly rare in the wild, having been so intensively collected, and both ginseng species are grown commercially.

Taxol, the most-discussed drug for cancer treatment, traditionally is extracted from the bark of yew (*Taxus*) trees, thus thousands of regal giants from Pacific Northwest old growth forests have been destroyed. Following great political and environmentalist pressure, several alternatives have been developed. Other parts of the trees have been found to produce taxol as well as the bark. Several laboratories are now tissue culturing needles to produce callus for cell culture production of taxol. Meanwhile, other laboratories are striving to synthesize part or all of the complex compound.

When certain cell cultures are treated with elicitors, especially in combination, the yield of secondary metabolites can be increased significantly (Chang and Sim 1994). Among the elicitors that have been discovered to serve this purpose are fungi, ultraviolet radiation, hormone deletion, heavy metal ions, detergents, agar, and heat. Madagascar periwinkle (*Catharanthus roseus*), is of particular interest because it is a source of vinblastine and vincristine, alkaloids used in cancer treatment. When a fungal homogenate was added to cell cultures of *C. roseus* the production of alkaloids was three times that of cultures lacking the fungal elicitor.

A filtered extract from the fungus *Pythium* added to a cell culture of the forb *Bidens pilosa* induced the cells to produce secondary products—certain antibiotic phytochemicals—that were not produced when the filtrate was absent. This phenomenon invites speculation on the incredible potential of cells and on the infinite number of microbial and other inducing agents that exist and the diversity of biochemical products they might produce.

Researchers are turning to plant genetic engineering to help convert cellulose-rich biomass into alternative fuel sources such as ethanol (Sticklen 2008) One of the hurdles of producing cellulosic ethanol is the costly process of breaking down the cellulose and removing the lignin. Enhancing the ability of plants like switchgrass (*Panicum virgatum*) to self-produce enzymes (cellulases) reduces costs associated with the process.

These brief examples of applying tissue culture to obtain secondary products barely touch the surface of this exciting subject. This application of cell culture holds great promise in the areas of nutrition, pesticides, and pharmaceuticals, among others, especially when combined with genetic engineering.

TWO
CULTURE GUIDE TO SELECTED PLANTS

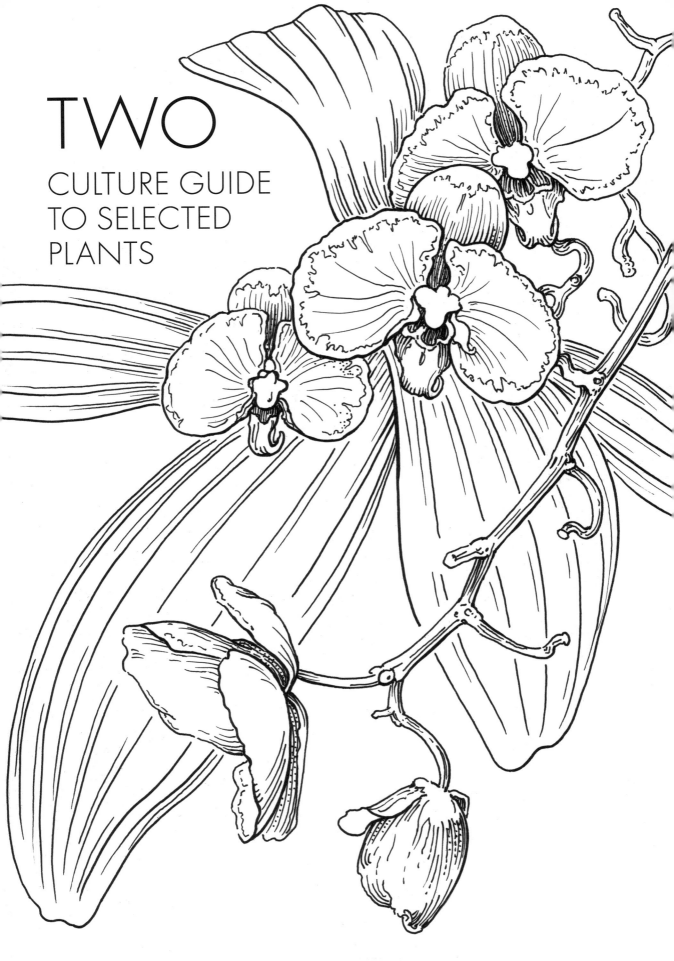

Introduction

Generally there are more ways than one to coax an explant or culture to grow *in vitro*. In reading the directions and formulas in this book, the novice might assume that the rules and protocols are binding and infallible; this is certainly not the case. In fact, these instructions are simply what has worked for someone, or close approximations thereof. Occasionally even the very authors of a formula have been unable to repeat their success under different circumstances. When dealing with living organisms, the variables are too numerous to calculate. The guidelines and the basic principles outlined are a place from which to start. The formulas given are derived from the references and other sources of information (see "Formula Comparison Table" at the back of the book). You will find more ideas and more references in other texts listed in the bibliography.

Use the procedures described in Section One (especially chapters 6, 7, and 8) as your guidelines for medium and explant preparation. If you do not have all of the usual additives for a particular formula, leave them out and see what happens. Many plants will grow on half-strength MS media plus sugar—probably not as well as on a more complete medium but well enough to get by. If a recommended cleaning process calls for 0.1% Tween 20 (which means 1 ml of Tween 20 per liter of water), try a drop or two of household dishwashing detergent instead. If the optimum temperature given is 73°F (23°C), most likely the cultures will do just fine in temperatures between 68°F and 79°F (20°C–26°C). You will soon learn which variables work best for you. Tissue culture is not an exact science. Whatever works is the best criteria. Experimentation and flexibility are vital components of successful tissue culture.

As stated previously in chapter 8, the 4 stages of culture growth beyond Stage 0—establishment (Stage I), multiplication (Stage II), rooting (Stage III), and acclimatization or hardening off (Stage IV)—are not necessarily distinct, and the same medium is often used for more than one stage and sometimes for at least 3 culture stages. Occasionally no culture medium is required for Stage III because rooting may be done *in vivo*, or perhaps a prerooting medium can be applied in the late-multiplication stage. You may find, again through experimentation and trial and error, that a medium that is specified for one stage may work just as well for the others.

In this newest edition of the book, the plants are now arranged alphabetically by genus. Both scientific and common names are given for all of the plants, as well as the family names. Within any family you will often find similar requirements or difficulties in tissue culture.

Murashige and Skoog (MS) Formula

Because the formula developed by Murashige and Skoog (MS) is used so frequently, it is not itemized in detail for each of the genera in Section Two. When the MS medium is required, it is indicated as MS salts, which includes MS major salts, MS minor salts, and iron. The list below gives the standard MS formula; the procedures for making stock solutions are discussed in chapter 6. If the formula for a particular species calls for 4628 mg/liter of MS salts, you will need 4530 mg of major salts, 33 mg minor salts, and 65 mg iron; if it calls for half-strength MS, or 2314 mg/liter, the major salts, minor salts, and iron should be halved accordingly.

Refer to chapter 5 for information on the other chemicals and substances mentioned throughout Section Two.

MS SALTS	MG/LITER
MAJOR SALTS	
Ammonium nitrate (NH_4NO_3)	1650
Calcium chloride ($CaCl_2 \cdot 2H_2O$)	440
Magnesium sulfate ($MgSO_4 \cdot 7H_2O$)	370
Potassium nitrate (KNO_3)	1900
Potassium phosphate (KH_2PO_4)	170
Subtotal	**4530**
MINOR SALTS	
Boric acid (H_3BO_3)	6.200
Cobalt chloride ($CoCl_2 \cdot 6H_2O$)	0.025
Cupric sulfate ($CuSO_4 \cdot 5H_2O$)	0.025
Manganese sulfate ($MnSO_4 \cdot H_2O$)	16.900
Potassium iodide (KI)	0.830
Sodium molybdate ($Na_2MoO_4 \cdot 2H_2O$)	0.250
Zinc sulfate ($ZnSO_4 \cdot 7H_2O$)	8.600
Subtotal	**32.830**
IRON	
Ferrous sulfate ($FeSO_4 \cdot 7H_2O$)	27.8
Na_2EDTA	37.3
Subtotal	**65.1**
Total mg/liter	**4627.93**

Actinidia deliciosa

KIWIFRUIT
Actinidiaceae, Chinese gooseberry family

The popularity of kiwifruit, *Actinidia deliciosa* (and the acreage devoted to its growth, particularly in Italy, New Zealand, France, China, Greece, and California), has increased dramatically because of the beneficial levels of Vitamin C, Vitamin E, and potassium in the fruit. Because *A. deliciosa* is dioecious (male and female are separate), only vegetative propagation is practical. Cuttings take 2–3 years to provide field plants, whereas field plants are produced in one year via tissue culture. Female cultivars such as 'Bruno', 'Chico', 'Saanichton 12', or 'Hayward' are commonly grown commercially with male pollinators such as 'Matua', 'Tomuri', or unselected males.

EXPLANT Meristems 0.2–0.5 mm long, or apical shoots 2–2.5 cm long.

TREATMENT OF DORMANT SHOOTS Take shoots in the winter and store at 1°C (34°F) for 2 months or at 4°C (39°F) for one month before stimulating bud break. A spray with fungicide might be necessary while the shoots are in storage.

Actinidia media

COMPOUND	STAGE I MG/LITER	STAGE II (LIQUID) MG/LITER	STAGE III (OPTIONAL) MG/LITER
MS salts	4,628	4,628	2,314
Sodium phosphate	170	170	–
Adenine sulfate	80	80	–
Inositol	100	100	100
Thiamine HCl	0.4	0.4	0.4
IBA	–	0.5–1.0	–
BA	0.5–1.0	0.5–1.0	–
Sucrose	30,000	30,000	20,000
Agar	7,000	–	7,000

pH 5.7

TREATMENT OF MERISTEMS Wash shoots in water and a drop of detergent for 20 minutes. Rinse in sterile distilled water. Dip in 70%–80% alcohol for 10–20 seconds. Rinse again in sterile water. Mix in 1:10 bleach with 0.1% Tween 20 for 15–20 minutes. Rinse 3 times in sterile distilled water. Again, dip in 70% alcohol for 10 seconds. Rinse 2 times in sterile distilled water. Dissect out meristems, sterilizing the instruments frequently.

TREATMENT OF APICAL SHOOTS Dormant shoots can have buds forced out by removing them from refrigerated storage, placing them in water, and growing them at 24°C (75°F). Shoots can also be taken in late spring when the plants are actively growing. Cut 3-cm shoot tips and remove expanded leaves. Stir in 6% bleach with 0.1% Tween 20 for 20 minutes. Rinse 3 times in sterile distilled water. Discard 1 cm from shoot base and place remaining tip on agar medium.

MEDIA An agitated liquid medium for Stage II has provided excellent results in some experiments, and therefore is recommended here. The addition of GA_3, along with BA has been shown to induce organogenesis from kiwifruit callus. Dispense medium into 125-ml flasks and cap with cotton and aluminum foil. During Stage III, include a small amount of IBA and increase the

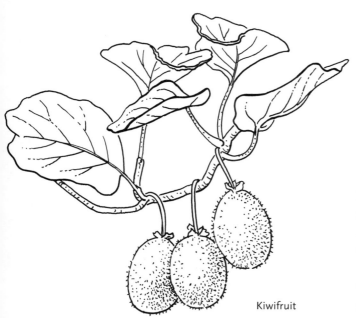

Kiwifruit

subculture period to 8 weeks. Employing other auxins in Stage III has proven to foster excessive callus. Stage III medium, which is half-strength MS, may well be omitted, and Stage II shoots can be rooted directly in soil mix. Using a preliminary dip of 2 mg/liter IBA may be advantageous for rooting.

Research has shown an increased rate of shoot proliferation with kiwifruit if the boron (B) concentration is raised to 2 mmol and if methionine is raised to 2 µM (Sotiropoulos et al. 2006). The practical limitations of this change need to be weighed against the increased number of shoots.

LIGHT 200–300 f.c. from fluorescent light for 16 hours light/8 hours dark.

TEMPERATURE 25°C (77°F).

REFERENCES Cristina Pedroso et al. 1992, Griffiths 1994, Ludvova and Ostrolucka 1998, Moncaleán et al. 2001, Moncaleán et al. 2009, Monette 1986, Murashige and Skoog 1962, Prado et al. 2005, Sotiropoulos et al. 2006.

Alstroemeria hybrids
INCA LILY, LILY-OF-THE-INCAS
Alstroemeriaceae, *Alstroemeria* family

Alstroemeria plants are distinctive herbaceous, monocotyledonous plants that are grown for their long-lived, colorful cut flowers. Native to Chile and Brazil in South America, they are the fifth most popular cut flower in the United States. Hybridization efforts by author Mark Bridgen has produced floriferous plants that bloom all summer in the cooler, northern climates until fall's first hard freeze and are winter hardy up to USDA zone 5 in the United States. *Alstroemeria* 'Mauve Majesty' and 'Tangerine Tango' are 2 of the most recent introductions into the market.

All *Alstroemeria* plants are vegetatively propagated by rhizome division or through tissue culture because plants are sterile and do not produce seeds. Commercially, tissue culture is the most viable way to produce these recalcitrant monocots. *Alstroemeria* cultures grow best in Stages II and III on a liquid medium; the plants do **not** need to be shaken as long as the explants are not completely submerged. The best medium to use is Murashige & Skoog medium with the addition of 2 mg/liter BA.

Alstroemeria media

COMPOUND	STAGE I MG/LITER	STAGE II MG/LITER	STAGE III MG/LITER
MS salts	4,628	4,628	4,628
Inositol	100	100	100
Glycine	2.0	2.0	2.0
Nicotinic acid	0.5	0.5	0.5
Pyridoxine HCl (B_2)	0.5	0.5	0.5
Thiamin-HCl (B_6)	0.1	0.1	0.1
BA	2.0	2.0	–
IBA	–	–	1.0
Sucrose	30,000	30,000	30,000
Gelrite	2,500	–	–

pH 5.7 to 5.8

EXPLANT 2.0- to 2.5-cm apical tips from actively growing plants. Disinfecting *Alstroemeria* is very difficult because the shoot tips are in close contact with growing media that contains numerous fungi and bacteria.

TREATMENT Remove actively growing shoot tips from underground rhizomes. Shoot tips should be vigorously washed under running water with a soft brush to remove any soil particles. Gently agitate tips in 2:10 bleach for 10–15 minutes. Rinse 3 times in sterile distilled water.

MEDIA MS salts and vitamins with BA (2 mg/liter). For Stage I, gellan gum should be used to solidify the medium; afterwards only liquid medium should be used.

LIGHT 300–400 f.c. from fluorescent light for 16 hours light/8 hours dark; continuous light with no dark period can also be used.

TEMPERATURE 18°C–21°C (64°F–70°F).

REFERENCES Bridgen, Kollman et al. 2009, Bridgen, Craig et al. 1990, Bridgen, King et al. 1991, Chiari and Bridgen 2000, Murashige and Skoog 1962.

Amelanchier alnifolia
SERVICEBERRY, SASKATOON SERVICEBERRY
Rosaceae, Rose family

Serviceberry (*Amelanchier alnifolia*) is a deciduous shrub or small tree that can grow to 7.6 m. This hardy ornamental is a native of the Pacific Northwest, and it is growing in popularity as new cultivars such as 'Smokey', 'Northline', 'Penbina', and 'Thiesson' appear in the trade. The sweet berries are antioxidants that supply vitamin B_2 and biotin. Although it is among the easier woody plants to tissue culture, it tends to suffer from vitrification (hyperhydricity) and problems of acclimatization.

Harris offers a formula different from the one outlined here. He used higher BA initially (3 mg/l), followed by an agitated liquid phase, without hormones, to condition shoots for rooting. Shoots were planted in perlite/vermiculite (3/1) soil mix and dampened with household fertilizer and 1 mg/liter IBA. The plants were covered with plastic and placed under cool-white fluorescent light in continuous light.

EXPLANT Apical 1–2 cm of young, actively growing shoots.

TREATMENT Remove all outer leaves from shoots and rinse in a stream of tap water for 30 minutes. Dip in 70% alcohol. Stir in 1:10 bleach for 10 minutes. Rinse 3 times in sterile distilled water. Trim to 5-mm shoot tips.

MEDIA Several simple modifications of MS have been used successfully for *Amelanchier* culture. The one presented here has low BA content to help avoid vitrification. Three-quarter-strength MS is used for Stage I. The addition of GA_3 helps to prevent bud dormancy from occurring.

LIGHT 200–300 f.c. from cool-white fluorescent light for all stages. For multiplication, 16 hours light/8 hours dark; for rooting and early acclimatization, continuous light.

TEMPERATURE 22°C–24°C (72°F–75°F).

REFERENCES Harris 1984, 1985, Murashige 1974, Murashige and Skoog 1962, Pruski et al. 1990.

Amelanchier media

COMPOUND	STAGE I MG/LITER	STAGE II MG/LITER	STAGE III MG/LITER
MS salts	3,471	4,628	4,628
Sodium phosphate	–	100	–
Inositol	100	100	100
Thiamine HCl	0.4	0.4	0.4
IAA/NAA	–	–	–
BA	2.0–2.5	2.0–2.5	–
GA_3	0.05	0.05	0.05
Sucrose	30,000	30,000	30,000
Agar	6,000	6,000	6,000
Charcoal	–	–	0.6

pH 5.7

Anthurium andraeanum
FLAMINGO FLOWER
Araceae, Arum family

Anthurium is one of the largest and most variable genera of the arum family. *Anthurium andraeanum* is an ornamental, monocotyledonous plant that is known for its unique cut flowers and attractive foliage. Cultivars like 'Arizona', 'Kaumana', 'Midori', 'Nitta', 'Ozaki', 'Paradise Pink', 'Sierra', and 'Tinora Red' are among the varieties of flamingo flowers cultivated by tissue culture methods. The disinfecting process described below has proven to be the most successful method of combating a persistent contamination problem.

Anthurium media

COMPOUND	STAGE I MG/LITER	STAGE II MG/LITER	STAGE III MG/LITER
MS salts	2,314	4,628	2,314
Inositol	100	100	100
Thiamin-HCl	0.5	0.5	0.5
Nicotinic acid	0.5	0.5	0.5
Pyridoxine HCl	0.5	0.5	0.5
Kinetin	–	0.1	0.2
NAA	–	–	0.1
IAA	0.2	0.2	–
BA	0.2	0.2	–
Sucrose	20,000	20,000	20,000
Agar	–	7,000	7,000

pH 5.5

Because *Anthurium* is a tropical plant and subject to contamination, it will benefit from indexing, as discussed in chapter 11.

EXPLANT Vegetative buds.

TREATMENT Cut nodal sections of approximately 1 × 1 × 0.5 cm size, each containing a bud, and then wash them 5 minutes in detergent (5% v/v). Soak for 15–20 minutes in 1:10 bleach with 0.1% Tween 20. Under a microscope, remove leaf coverings and excise buds no larger than 2 mm at their base. Soak buds in 1:10 bleach for 30–45 minutes. Rinse in sterile distilled water for 5 minutes.

MEDIA For Stage I, transfer 5- or 10-ml aliquots of the liquid medium into test tubes, cap and sterilize. After cooling, aseptically add one dissected bud to each test tube. Place the inoculated tubes on a rotator or wheel and agitate at approximately 2–5 rpm for 6 weeks. For Stage II, transfer 2-node shoot sections to the same medium, but with agar.

LIGHT Fluorescent light at 100–200 f.c. for 16 hours light/8 hours dark.

TEMPERATURE 25°C–28°C (77°F–82°F).

REFERENCES Chen et al. 2011, Kunisaki 1980, Martin et al. 2003, Murashige 1974, Murashige and Skoog 1962, Pierik et al. 1974.

Asparagus officinalis
ASPARAGUS
Asparagaceae, Asparagus family

Asparagus (*Asparagus officinalis*) is a perennial, dioecious monocot that is rarely propagated from seed because the resulting plants are too varied for optimum commercial production. Plants are not multiplied by division of the crown because it is too labor intensive. Male plants of *Asparagus* are preferred for production because they produce more and better stalks, and no seeds that would grow undesirable female volunteers. Some male plants are hermaphroditic and produce "super males," which are ideal subjects for tissue culture.

Tissue culture of *Asparagus* is not easy. *In vitro* multiplication tends to be slow, erratic, and very clone dependent. It is largely a 2-step process:

Flamingo flower

Asparagus media

COMPOUND	FIBROUS ROOTS/ CROWN MG/LITER	STORAGE ROOTS/ CROWN MG/LITER	MULTI- PLICATION/ DIVISION MG/LITER
MS salts	2,314	4,628	4,628
Adenine sulfate	40	40	40
Inositol	400	400	400
Thiamine HCl	4.0	4.0	4.0
Nicotinic acid	20	20	20
Pyridoxine HCl	20	20	20
IAA	–	0.1	0.4
NAA	0.1	–	–
BA	–	0.5	–
2iP	–	–	0.04–0.2
Kinetin	0.1	–	–
Ancymidol	1.3	1.3	1.3
Sucrose	20,000	30,000	30,000
Agar	7,000	7,000	7,000
Charcoal	–	–	0.0–1.5

pH 5.7

first to grow out the buds; then to induce nodal microcuttings to grow crowns with storage and nonstorage roots.

EXPLANT Lateral buds.

TREATMENT Fragments of green, edible spears that include a lateral bud should be dipped in 70% ethanol for 5 minutes. Then, spears about 15- to 20-cm long should be agitated in 1:10 bleach for 10 minutes followed by a wash in sterile distilled water for 5 minutes. Lateral buds should be excised from the stem and placed in medium for fibrous root development. The cladophylls (stems that function as leaves) that grow from the buds may be divided into 3- to 4-cm pieces and transferred every 3 weeks. Fibrous roots should develop in 4–6 weeks. Transfer plantlets that have developed fibrous roots to medium for storage root development. When these have developed a crown and both fibrous and storage roots, which take 6–9 months, transfer to multiplication/division medium. Each transferred segment must consist of a piece of crown with 2 or 3 shoots and both kinds of roots. In this stage, transfer every 6–8 weeks; alternate the medium with fibrous root medium

MEDIA Modified MS, with hormones NAA (0.1 mg/liter), kinetin (0.1 mg/liter), IAA (0.1 to 0.4 mg/liter), and 2iP (0.04 to 0.2 mg/liter). The high levels of inositol, thiamine HCl, nicotinic acid, and pyridoxine HCl are unique. Clonal differences make it difficult to select the optimum amounts of hormones and vitamins. Chin (1982) found that adding ancymidol (brand name A-Rest) at 5 mg/liter promoted stronger plantlets, reduced height, and helped prevent callus. Note that ancymidol is classified as a restricted use pesticide.

LIGHT 300 f.c. from cool-white fluorescent light for 16 hours light/8 hours dark. For final stage, increase intensity to 500–1000 f.c.

TEMPERATURE 25°C (77°F).

REFERENCES Cheetham et al. 1992, Chin 1982, Gorecka et al. 2003, Khalil et al. 2001, Kunachak et al. 1987, Munoz et al. 2006, Murashige et al. 1972, Yang and Clore 1973, 1974.

Begonia rex-cultorum
REX BEGONIA
Begoniaceae, Begonia family

Rex begonias are a rhizomatous type of *Begonia* that are grown for their outstanding, multicolored leaves that come in every color, texture, and pattern. They are mostly grown as potted houseplants, but they can also be placed in shade gardens. Wax begonias (*Begonia semperflorens-cultorum* hybrids) have small (3- to 3.5-cm) flowers that bloom profusely. Tuberous begonia (*Begonia ×tuberhybrida*) is noted for its huge flowers in

Begonia media

COMPOUND	STAGES I & II MG/LITER	STAGE III MG/LITER
MS salts	4,628	4,628
Inositol	100	100
Thiamine HCl	1.5	1.5
Nicotinic acid	0.5	0.5
Pyridoxine HCl	0.5	0.5
NAA	0.1	0.1
BA	0.1–0.5	–
Glycine	2.0	2.0
Sucrose	30,000	20,000
Agar	8,000	8,000

pH 5.7

rich colors. Rex begonias are not noted for their flowers, but their large, colorful leaves more than make up for the lack of flowers.

Rex begonias are one of a small group of plants that can be propagated by leaf section cuttings; this ability to produce clones adventitiously is very helpful for tissue culture.

EXPLANT Leaf petiole segments 5 mm long or leaf discs 2 cm in diameter and containing a piece of the main vein.

TREATMENT Dip in 70% ethanol for 30 seconds followed by 10–15 minutes in 1:5 bleach. Rinse 3 times in sterile water.

MEDIA MS salts, vitamins, and growth hormones. Reduce sugar to 20,000 mg/liter for rooting. Stage III may be omitted, rooting directly from Stage II to potting mix under high humidity.

LIGHT For Stages I and II, 100 f.c. from cool-white fluorescent light for 16 hours light/8 hours dark; for Stage III, increase intensity to 300 f.c.

TEMPERATURE 23°C–25°C (73°F–77°F).

REFERENCES Bowes and Curtis 1991, Cassells and Morrish 1985, 1987, Espino et al. 2004, Murashige 1974, Murashige and Skoog 1962.

Brassica oleracea var. *italica*
BROCCOLI
Brassica oleracea var. *botrytis*
CAULIFLOWER
Brassicaceae, Mustard family

Brassica oleracea is a diverse species that includes several vegetable crops, namely, broccoli, Brussels sprouts, cauliflower, kohlrabi, kale, and some cabbage. The production of broccoli (*B. oleracea* var. *italica*) and cauliflower (*B. oleracea* var. *botrytis*) clones that are multiplied by tissue culture can be useful for F_1 hybrid varieties and for maintaining clonal germplasm in breeding programs. Another important application of *Brassica* tissue culture is the multiplication of parents of a self-incompatible hybrid that may be particularly desirable. These tissue cultured parents can be planted in the field for open-pollinated seed production.

EXPLANT 2- to 5-cm-long flower buds, or 1-cm-long lateral shoot tips.

TREATMENT To decrease contamination, precondition stock plants by holding in a greenhouse at 16°C (61°F) for 3 weeks. Obtain young flower heads bearing 2- to 5-cm-long buds. Stir heads and lateral shoots in 1:10 bleach for 15 minutes.

Brassica media

COMPOUND	STAGES I & II MG/LITER	STAGE III MG/LITER
MS salts	4,628	4,628
Sodium phosphate	170	170
Adenine sulfate	80	80
Inositol	100	100
Thiamine HCl	0.4	0.4
Kinetin	2.0	–
IBA	1.0	1.0
Sucrose	30,000	30,000
Agar	8,000	8,000

pH 5.7

Excise 2- to 5-mm-long buds; rinse buds and shoots in 1:100 bleach for 15 minutes. Double wash with sterile distilled water.

Transplant rooted plantlets to a peat/perlite (1/1) soil mix in trays under intermittent mist for 3 days, at which time humidity can be reduced. Plants are ready to transplant to pots in 7 weeks.

MEDIA MS salts with sodium phosphate, inositol, thiamine, and adenine sulfate. For Stages I and II, add 2 mg/liter of kinetin and 1 mg/liter IBA. Omit kinetin for the rooting stage.

LIGHT For Stages I and II, 200–300 f.c. from cool-white fluorescent light, with 16 hours light/8 hours dark. For Stage III, increase intensity to 600 f.c.

TEMPERATURE 23°C (73°F).

REFERENCES Anderson and Carstens 1977, Anderson and Meagher 1977, Kubota and Kozai 1994, Murashige 1974, Murashige and Skoog 1962, Rihan et al. 2011, Sheng et al. 2011.

Broccoli

Catharanthus roseus
MADAGASCAR PERIWINKLE
Apocynaceae, Dogbane family

Madagascar periwinkle (*Catharanthus roseus*) is a valued herbaceous plant that continuously blooms with pink, purple, or white flowers. In addition to its ornamental value, it is popular in the pharmaceutical industry because of its ability to produce more than 120 types of secondary metabolites such as terpenoid indole alkaloids that may be used to treat cardiac diseases and certain tumors in mammals (Swanberg and Dai 2008). Physical and chemical extraction of products from *Catharanthus* cell cultures have been well detailed by Morris et al. (1985).

EXPLANT Stem pieces from 10- to 15-day-old seedlings, or seeds.

TREATMENT Seedlings should be washed in running tap water. Mix in 1:10 bleach with 0.1% Tween 20 for 10–15 minutes. Rinse in 3 changes of sterile water. Seeds can be placed in 70% alcohol for 2 minutes followed by 2:10 bleach for 15 minutes and a triple rinse in sterile water. Stem pieces should be placed under running tap water followed by an alcohol dip for 10 seconds. Stir in 1:10 bleach for 20 minutes. Rinse in 3 changes of

Catharanthus media

COMPOUND	STAGES I & II MG/LITER	STAGE III MG/LITER
MS salts	4,628	4,628
Inositol	100	100
Thiamine HCl	10	10
Nicotinic acid	1.0	1.0
Pyridoxine HCl	1.0	1.0
NAA	1.0	–
IBA	–	0.5
BA	0.5–1.5	–
Sucrose	30,000	30,000
Agar	7,000	7,000

pH 5.7

sterile distilled water. Cut into 1-cm pieces. Rinse again in sterile water. Place on MS medium with agar.

MEDIA The medium for growth is MS plus hormones. Lower BA concentrations (0.5 mg/l) are used for seed cultures.

LIGHT 300–400 f.c. from cool-white fluorescent light, with 16 hours light/8 hours dark for seedling pieces, total darkness for seeds.

TEMPERATURE 25°C–28°C (77°F–82°F).

REFERENCES Morris et al. 1985, Pietrosiuk et al. 2007, Rani et al. 2011, Swanberg and Dai 2008, Verma and Mathur 2011, Zenk et al. 1977.

Cattleya media

COMPOUND	STAGES I & II MG/LITER	STAGE III MG/LITER
Calcium nitrate	1,000	1,000
Magnesium sulfate	250	250
Potassium chloride	500	500
Potassium phosphate	250	250
MS minor salts	32.8	32.8
MS iron	5.1	5.1
Urea	500	500
Sucrose	20,000	20,000
Coconut milk	100 ml	10 ml
Agar	–	6,000

pH 5.5

Cattleya spp. and hybrids
CATTLEYA
Orchidaceae, Orchid family

Cattleya is one of the most widely known orchids in the world because of its large, showy flowers that have been used mostly as a long-lasting cut flower, but more recently as a potted plant. The flowers are available in all colors except a true blue and black. Conventional orchid propagation is achieved by the division and cultivation of backbulbs (old bulbous roots), which take many years to grow. Vegetative propagation of *Cattleya* by backbulbs would never be able to meet the demand for this popular monocotyledonous plant because it takes about 10 years to produce only 6–12 commercial-size plants. Breeding pure lines by conventional hybridizing is not an alternative because all the cultivated orchids are genetically complex hybrids (highly heterozygous).

Meristem culture makes it possible to rapidly multiply an unlimited number of grand prize winners. When orchid seeds germinate, they first produce protocorms, bulbous tissue with root hairs. Apical meristems in culture follow a similar development.

Cattleyas are sympodial, meaning that the growth of each shoot originates in the axils of the scale leaves. The growth of each shoot is terminated after 1 or 2 seasons.

In tissue culture, cattleyas are more demanding than some other orchids and are comparatively slow growing; it can sometimes take 2 months on a rotator before greening is observed.

EXPLANT Axillary buds or meristems from 2- to 5-cm-long shoots. These are taken from actively growing shoots from backbulbs, or preferably from a new "green" bulb to help avoid contamination and provide better growth. However, green bulbs are scarce since a plant usually produces only one new bulb per year. An explant should be taken before the new shoot is 5 cm long because the meristem ceases to grow at an early stage. If the eyes of a backbulb are poorly developed, place the backbulb in a bag of damp peat moss and allow it to swell before excising the bud.

TREATMENT Wash the shoot in water with detergent. Dip the shoot in alcohol for 10 seconds. Mix in 1:5 bleach for 15 minutes. Rinse briefly in sterile distilled water and dry on a sterile paper towel. Under a dissecting microscope, carefully remove the overlapping leaves with frequently sterilized tools. Look for a bud about 2 mm in size at the base of a leaf. Sever the bud just below the point of attachment. Use the bud or excise the meristem.

The cut surfaces of many orchid varieties brown quickly because of the oxidation of phenolic compounds. These phenolic compounds are toxic to the plantlets and leach into the medium, so the cultures should be transferred every 2 or 3 days until the browning and bleeding stops. To help prevent browning, dissect in fresh sterile antioxidant solution or coconut milk.

Place the bud or meristem explant in 5–10 ml of liquid medium in a test tube that will fit your rotator; the small amount of liquid provides optimum aeration for the explant. Slowly agitate (at 1–2 rpm) the liquid cultures on a rotator or wheel. The initial growth that is desired for multiplication is a small mass of protocorms, which are almost microscopic tuberous structures that may be multiplied or induced to form plantlets. They often show greening within one week and protocorm masses should be ready for division and transfer in 2 months, or when they have grown to 0.5 cm. As soon as the mass grows to 1 cm, it should be divided and put back into liquid or agar medium. When the cultures are grown on agar, frequent division will upset the orientation, or polarity, of the explants and will delay plantlet formation. Once polarity is established, the cultures put out shoots and roots and mature.

MEDIA Modified Knudson's C medium. Orchids may be started on agar medium, but *Cattleya* cultures are best started in a liquid medium and cultured initially in an agitated liquid medium. Using an agitated liquid medium will accomplish the following:

- Maximizes protocorm tissue by inhibiting expression of polarity.
- Dilutes the growth inhibitor that is produced in response to excision and sterilization.
- Improves aeration.
- Increases the contact of tissue surface with the nutrients.

Supplement media with coconut milk to produce protocorms and avoid callus. Coconut milk is a frequent addition to orchid media. It may be purchased from many commercial suppliers, or you may extract it yourself from a coconut, as described in chapter 5. Stage III plantlets may not need coconut milk, and agar is used during that stage.

Cattleyas cannot use nitrates as a nitrogen source because they lack the enzyme, nitrate reductase, that reduces nitrates. Using an ammonium ion, as in ammonium sulfate, is not an answer to the problem either because it renders the pH too low. Therefore, 500 mg of urea is substituted for ammonium sulfate in Knudson's formula for these orchids. Cattleyas also require potassium, so 500 mg of potassium chloride is added.

LIGHT 100–300 f.c. from cool-white fluorescent light. For Stage I, continuous light. For Stages II and III, either continuous light or cycles of 16 hours light/8 hours dark. Cool-white light has been shown to be beneficial for *Cattleya* growth.

TEMPERATURE 26°C (79°F).

REFERENCES Arditti 1977, Cybularz-Urban et al. 2007, Hartmann et al. 2010, Knudson 1946, Morel 1964, 1974, Murashige and Skoog 1962, Sagawa and Kunisaki 1982, Scully 1967, Vacin and Went 1949, van Overbeek et al. 1941, Wimber 1963.

Chrysanthemum ×*grandiflorum*
GARDEN MUM
Asteraceae, Aster family

Mums are one of the top 10 most popular cut flowers and garden plants. The plants are easily propagated vegetatively by stem cuttings or division. Other genera in the aster family have responded to these same tissue culture procedures, including *Anthemis*, *Argyranthemum*, *Calendula*, *Leucanthemum*, *Matricaria*, and *Tanacetum*.

EXPLANT Young shoot tips or unopened flower buds.

TREATMENT Wash stems in water and a few drops of detergent. Dip in 70% ethanol for 30–40 seconds followed by 2:10 bleach for 10–15 minutes and a triple rinse in sterile water.

MEDIA Chrysanthemums are an easy tissue culture subject, and media formulas can vary. Stage III can be avoided; 1- to 2-cm-long shoots can be removed from culture and stuck directly into a potting medium for rooting.

LIGHT 300 f.c. from cool-white fluorescent light, 16 hours light/8 hours dark. Mums are photoperiodic plants and flower buds will be initiated with short days.

TEMPERATURE 22°C–25°C (72°F–77°F).

REFERENCES Murashige and Skoog 1962, Rout and Das 1997, Song et al. 2011.

Citrus ×*limon*
LEMON
Rutaceae, Rue family

Citrus is one of the most important commercial fruit crops in the world; it is an important source of vitamin C and is grown in more than 100 countries. The lemon (*Citrus* ×*limon*) is a small evergreen tree that produces a popular sour-tasting fruit for juices and drinks such as lemonade. The sustainable development of the citrus industry is mainly dependent on a continuous supply of new and improved cultivars. Traditional propagation of this plant is by budding, cutting, or layering—all expensive and time-consuming techniques. Micropropagation overcomes some constraints to *Citrus* propagation and is important for the asexual propagation of clonal plants.

EXPLANT Shoots from rapidly expanding new branches.

Chrysanthemum media

COMPOUND	STAGES I & II MG/LITER	STAGE III MG/LITER
MS salts	4,628	2,314
Adenine sulfate	80	–
Inositol	100	–
Thiamine HCl	0.4	–
Nicotinic acid	0.5	–
Pyridoxine HCl	0.5	–
IAA	0.5	0.1
BAP	1.5	–
Sucrose	30,000	–
Agar	7,000	–

pH 5.8

Citrus media

COMPOUND	STAGES I & II MG/LITER	STAGE III MG/LITER
MS salts	4,628	4,628
Myo-inositol	100	100
Glycine	2.0	2.0
Thiamine HCl	0.1	0.1
Pyridoxine HCl	0.5	0.5
Nicotinic acid	0.5	0.5
IBA	–	3.0
BA	2.0	–
GA	2.0	–
Sucrose	30,000	30,000
Agar	8,000	8,000

pH 5.7

TREATMENT Remove growing shoot tips about 5–6 cm long, detach leaves, and wash stem sections with water and detergent. Shake for 5 minutes in 70% ethanol and then for 15–20 minutes in a 2:10 bleach solution with detergent. Rinse 3 times in sterile water. Remove 1-cm nodal sections and culture with the basal end of the node in the medium.

MEDIA Use modified MS medium with vitamins. Include the hormones BA and GA for multiplication and IBA for the rooting stage.

LIGHT 350–450 f.c. from cool-white fluorescent light for 16 hours light/8 hours dark.

TEMPERATURE 24°C–27°C (75°F–81°F).

REFERENCES Marutani-Hert et al. 2011, Murashige and Skoog 1962, Pérez-Tornero et al. 2010, Singh et al. 1994.

Corymbia/Eucalyptus media

COMPOUND	STAGES I & II MG/LITER	STAGE III MG/LITER
MS salts	2,314	1,157
D-Biotin	0.1	0.1
Calcium pantothenate	0.1	0.1
IBA	0.1	–
NAA	0.01	–
BA	0.2	–
Sucrose	20,000	20,000
Gelrite	3,500	3,500

pH 5.5

Corymbia ficifolia
RED-FLOWERING GUM, EUCALYPTUS
Myrtaceae, Myrtle family

Red-flowering gum (*Corymbia ficifolia*, synonym *Eucalyptus ficifolia*) is a common ornamental tree in the myrtle family. Its tidy, rounded shape, elegant foliage, and showy summer flowering make it a popular tree especially in hot areas with sandy, dry soils. Tissue culture of eucalypts, a large group of woody plants that includes the genera *Corymbia* and *Eucalyptus*, has particular significance for forest plantations as well as potential for horticulturally desirable species such as *Eucalyptus caesia* and *E. macrocarpa*. *Eucalyptus radiata* and *E. polybractea* are useful oil-producing species. Other eucalypts have been cultured with varied success on the same or different media, but *Corymbia ficifolia* (syn. *E. ficifolia*) appears to have the most success. Potential major problems include contamination, particularly with adult material, and bleeding, which usually occurs in fresh material. Bleeding can be minimized by soaking in ascorbic acid and by an initial incubation period of 3–5 days in the dark. Rooting and establishment also may face problems. Success rate of rooting can be very low, especially with certain cultivars.

EXPLANT Nodes or apical buds from seedlings, adult trees, or juvenile material growing from the base of adult trees.

TREATMENT Remove 5- to 7-cm-long stems from juvenile sections of the tree. Cut off leaves, leaving about 1 cm of petiole. Wash nodes in running tap water for one hour. Wash in distilled water with 0.1% detergent for 5 minutes. Soak in 0.2% ascorbic acid solution for 2 hours. Agitate in 2:10 bleach for 20 minutes. Rinse briefly in 3 changes of sterile distilled water. Trim to about 1 cm of main stem at node or apex. Place on agar medium and incubate for 24 hours in the dark, then transfer to normal lighting.

Multiple buds may appear in 2 months from the initial explant. During this period, there should be several transfers to fresh medium. Roots may appear in 3 weeks after being placed in rooting medium.

Establishment in sand/loam (1/1) soil takes a great deal of care. Humidity should be lowered very gradually. Cover plants with plastic wrap and place in deep shade for 24 hours. Then place in about ¾ shade. After growth has started, increase lighting very gradually up to full sun—leaves are subject to scorching with an abrupt change to brighter light intensity.

MEDIA Use half-strength MS with low sugar for Stages I and II. Decrease MS to quarter strength for Stage III. Include BA and NAA hormones for multiplication, and IBA for rooting stage.

LIGHT Place tubed explants in dark for 24 hours. Move to 100–300 f.c. from cool-white fluorescent light for 16 hours light/8 hours dark.

TEMPERATURE 24°C–27°C (75°F–81°F).

REFERENCES Barker et al. 1977, Chetty and Watt 2011, de Fossard 1976a, 1976b, 1977, de Fossard and Fossard 1988, Hartney 1982, Hartney and Eabay 1984, Murashige and Skoog 1962.

Cucumis sativus
GARDEN CUCUMBER
Cucurbitaceae, Gourd family

Cucumbers are one of the most economically important vegetable species in the world. Although most cucumbers are propagated by seeds, micropropagation is used to multiply rootstocks for grafting and gynoecious hybrids that produce all, or primarily, female flowers. Tissue culture offers a convenient method for multiplying these plants that is preferable to cuttings.

EXPLANT Axillary buds 1–3 mm long, taken from nodal sections of young plants, or shoot tips from seedlings.

TREATMENT Place leaf axils in 1:10 bleach for 5 minutes. Rinse in 3 changes of sterile distilled water for 2 minutes each. Excise 1- to 3-mm long axillary buds. If using seeds, sterilize the seeds in 95% ethanol for 2 minutes or 100% bleach for 10 minutes followed by a triple rinse in sterile water.

MEDIA MS salts plus vitamins, NAA, and BA.

Cucumis media

COMPOUND	STAGES I & II MG/LITER	STAGE III MG/LITER
MS salts	4,628	2,314
Inositol	100	100
Casein hydrolysate	200	–
Thiamine HCl	1.0	–
Nicotinic acid	0.5	0.5
Pyridoxine HCl	1.0	–
NAA	0.1	–
IBA	–	0.5–1.0
BA	0.2–0.5	–
Sucrose	30,000	30,000
Agar	7,000	7,000

pH 5.7

LIGHT 200–400 f.c. from cool-white fluorescent tubes and incandescent bulbs for 16 hours light/8 hours dark. If using sterile seeds, place them in dark for 48 hours to germinate.

TEMPERATURE 25°C–26°C (77°F–79°F).

REFERENCES Abrie and van Staden 2001, Ahmad and Anis 2005, Handley and Chambliss 1979, Konstas and Kintzios 2003, Lin et al. 2011, Murashige 1974, Murashige and Skoog 1962, Sarowar et al. 2003, Vasudevan et al. 2004, Wehner and Locy 1981.

Cymbidium spp. and hybrids
CYMBIDIUM, BOAT ORCHID
Orchidaceae, Orchid family

Cymbidiums have made a comeback in popularity recently because of the diverse variety of cultivars that are available, especially the miniature cymbidiums. Not only do these orchids made great cut flowers, but the plants are more cold tolerant than many other orchids and they produce large sprays of blooms in the winter. Although they are monocots, cymbidiums are relatively easy to micropropagate via tissue culture. They

Knudson's C *Cymbidium* media

COMPOUND	STAGES I & II MG/LITER	STAGE III MG/LITER
Ammonium sulfate	500	500
Calcium nitrate	1,000	1,000
Magnesium sulfate • H_2O	250	250
Potassium phosphate	250	250
MS minor salts	32.8	32.8
MS iron	65.1	65.1
BA	5.0	—
IAA	—	1.0
Sucrose	20,000	20,000
Coconut milk	100 ml	100 ml

pH 5.8

Vacin and Went *Cymbidium* media

COMPOUND	STAGES I & II MG/LITER	STAGE III MG/LITER
Ammonium sulfate	500	500
Calcium phosphate	200	200
Magnesium sulfate • H_2O	250	250
Magnesium sulfate • $4H_2O$	7.5	7.5
Potassium nitrate	525	525
Potassium phosphate	250	250
Ferric tartrate	28	28
Thiamine HCl	0.4	0.4
BA	5.0	—
IAA	—	1.0
Sucrose	20,000	20,000
Coconut milk	100 ml	100 ml

pH 5.8

Knudson's C medium is preferred, but modified Vacin and Went's *Cymbidium* orchid medium is satisfactory; liquid media may produce faster and more abundant growth. For Stages II and III, use either Knudson's C or Vacin and Went's medium. Vacin and Went's medium has also been used successfully for culture of *Dendrobium* and *Phalaenopsis* orchids. Paek and Kozai (1998) found that 5% sucrose in the medium is best for shoot induction from rhizomes.

LIGHT 100–300 f.c. from cool-white fluorescent light. For Stage I, continuous light. For Stages II and III, 16 hours light/8 hours dark.

TEMPERATURE 22°C–25°C (72°F–77°F).

REFERENCES Arditti 1977, Knudson 1946, Morel 1960, 1964, 1974, Paek and Kozai 1998, Paek and Yeung 1991, Sagawa and Kunisaki 1982, Shimasaki and Uemoto 1990, Vacin and Went 1949, Wimber 1963.

Dianthus caryophyllus
CARNATION, CLOVE PINK
Caryophyllaceae, Pink or carnation family

Carnations (*Dianthus* spp.) are common garden plants and cut flowers that are grown for their colorful and fragrant flowers. Various perennial species are grown in the garden including *D. chinensis*, *D. deltoides*, *D. gratianopolitanus*, *D. plumarius*, and others. Virus infections are a major problem with carnations, so virus-free stock plants are produced through meristem culture.

EXPLANT 1-mm-long shoot tip.

TREATMENT Remove newly developed shoot tips from vegetative stock plant stems with approximately 4 or 5 nodes. Remove all leaves and wash stems thoroughly in running water. Surface sterilize the plants with 95% ethanol for 5 seconds followed by a wash in 2:10 bleach with 0.1%

are sympodial in growth habit, have excellent regenerative powers, and can be cultured initially on a solid medium, from dormant or active axillary buds, even if the shoots are longer than 5 cm.

EXPLANT 2- to 5-mm buds from axillary shoots. These may be dormant buds from backbulbs.

TREATMENT See *Cattleya*.

MEDIA For Stage I, either agar or liquid modified

Dianthus media

COMPOUND	STAGE I MG/LITER	STAGE II MG/LITER	STAGE III MG/LITER
MS salts	4,628	4,628	4,628
D-Pantothenic acid calcium salt	5.0	5.0	5.0
Inositol	50	100	100
Thiamine HCl	0.1	0.1	0.1
Nicotinic acid	0.5	0.5	0.5
Pyridoxine HCl	0.5	0.5	0.5
Glycine	2.0	2.0	2.0
Casein hydrolysate	300	–	–
IBA	2.0	2.0	0.5
BAP	2.0	2.0	–
Sucrose	30,000	30,000	30,000
Agar	7,000	7,000	7,000

pH 5.8

Tween 20 for 10–15 minutes. Rinse 2 times in sterile distilled water for at least 3 minutes each. Aseptically remove 1-mm-long stem tips with 1 or 2 pairs of leaf primordia and place on agar medium.

MEDIA Hyperhydricity (vitrification) can be a problem with tissue cultures of *Dianthus*. If this occurs, thick, elongated, wrinkled, and light green leaves are produced. Hyperhydricity is less of a problem if nodes are used as explants and if there is good ventilation in the culture vessels; higher levels of phosphate or higher agar concentrations has also helped to prevent this problem. Standard MS salts are used for carnations, but with modifications. In Stage II, omit the casein hydrolysate, and in Stage III, reduce IBA to 0.5 mg/liter or root 2-cm-long shoots directly from Stage II into pellets that combine container and medium in one unit, such as Jiffy-7.

LIGHT Continuous light, 200–300 f.c. from cool-white fluorescent light.

TEMPERATURE 22°C–25°C (72°F–77°F)

REFERENCES Davis et al. 1977, Earle and Langhans 1975, Fraga, Alonso, Ellul et al. 2004, Gutiérrez-Miceli et al. 2010, Levy 1981, Mele et al. 1982, Mohamed 2011, Murashige and Skoog 1962.

Dieffenbachia seguine
DUMB CANE
Araceae, Arum family

Dieffenbachia seguine (syn. *D. maculata*) is an attractive tropical foliage plant that is commonly used as a houseplant because of its decorative leaves that have white, cream, or yellow variegation, marbling, and blotches. This monocotyledonous plant grows well with low light, warm temperature, and a humid atmosphere. The sap is toxic; take care that children and pets do not chew it because the sap can cause swelling and loss of speech—hence the common name of dumb cane. If the skin comes in contact with the sap, irritation and a rash can occur.

Dianthus

Dieffenbachia media

COMPOUND	STAGES I & II MG/LITER	STAGE III MG/LITER
MS salts	4,628	2,314
Sodium phosphate	170	–
Thiamine HCl	0.4	0.4
Myo-inositol	100	100
Pyridoxine HCl	0.5	0.5
Nicotinic acid	0.5	0.5
IAA	0.01	–
IBA	–	1.5
BAP	5.0	0.5
Glycine	2.0	–
Sucrose	30,000	30,000
Agar	7,000	7,000

pH 5.7 to 5.8

Because tropical foliage plants are highly susceptible to systemic fungal, bacterial, and viral infections, keep *Dieffenbachia* in a dry, clean, lighted atmosphere for 1–3 weeks before removing explant tissue. This practice reduces the number of microbial contaminants on the explants.

EXPLANT Fresh stems with 7–9 nodes, each stem about 12 cm long. Lateral buds or meristems.

TREATMENT OF LATERAL BUDS Remove lateral buds from the 0.5- to 1.0-cm-long stem segments, keeping approximately 3 mm of surrounding nodal stem tissue around each bud. Dip stem segments in 70% ethanol (v/v) for 30 seconds and rinse in sterile water. Rinse lateral buds in 2:10 bleach with 0.1% Tween 20 for 10 minutes. Rinse well in 3 changes of sterile distilled water about 5 minutes each. Aseptically trim each bud so that only one sheath remains over the tip. Leave approximately 1 mm of surrounding stem tissue. Aseptically place the explants upside down in test tubes containing MS initiation agar medium, with one explant per test tube.

TREATMENT OF MERISTEMS Cut apical shoot tip with 3 immature leaves surrounding the apical bud. Disinfect shoot tip cuttings as for lateral buds. Place on sterile paper toweling covering the platform of the dissecting microscope. Aseptically remove all but 1 or 2 of the primordial leaves. Using a sterile scalpel and tweezers, excise the meristem and adjoining 1–2 mm of stem tissue. Transfer the explant to the agar surface in a test tube containing MS initiation medium.

MEDIA Stages I and II are grown on an agar medium containing MS salts and Gamborg vitamins, sodium phosphate, thiamine HCl, glycine, BAP, and IAA. For Stage III, omit the sodium phosphate, glycine, and IAA, and add IBA.

LIGHT 200–300 f.c. from cool-white fluorescent light for 16 hours light/8 hours dark.

TEMPERATURE 25°C–26°C (77°F–79°F).

REFERENCES Azza and Khalafalla 2010, Gamborg et al. 1968, Knauss 1976, Litz and Conover 1977, Murashige and Skoog 1962, Shen et al. 2007.

Epiphyllum chrysocardium
HEART OF GOLD, ORCHID CACTUS
Cactaceae, Cactus family

The attractive tropical cactus *Epiphyllum chrysocardium* (now *Selenicerens chrysocardium*) is known as heart of gold because of the yellow filaments at the center of its large white flowers. Tissue culture of Cactaceae can play an important role in the rescue of endangered species.

EXPLANT Stem cuttings 1.5–2 cm long.

TREATMENT Dip in 30% alcohol. Rinse in sterile distilled water. Mix in 1:10 bleach for 15 minutes. Rinse 3 times in sterile distilled water.

The sunken shoot primordia should emerge in about 2 weeks after placing on agar medium. Allow shoots to grow 5 or more buds before transferring to rooting medium.

MEDIA Use half-strength MS salts with inositol, pyridoxine, thiamine, and sucrose. Recom-

Epiphyllum media

COMPOUND	STAGES I & II MG/LITER	STAGE III MG/LITER
MS salts	2,314	2,314
Inositol	500	500
Thiamine HCl	1.0	1.0
Pyridoxine HCl	5.0	5.0
IBA	–	0.01
NAA	0.1	–
BA	1.0	–
Sucrose	20,000	20,000
Agar	8,000	8,000

pH 5.7

mended hormones for Stages I and II are BA and NAA. For rooting, omit BA and NAA and add IBA.

LIGHT 16 hours light/8 hours dark at 100–300 f.c.

TEMPERATURE 26°C (79°F).

REFERENCES Lazarte et al. 1982, Mauseth 1979, Murashige 1974, Murashige and Skoog 1962.

Ficus carica
FIG
Moraceae, Mulberry family

The fig (*Ficus carica*) is one of the earliest plants to be domesticated and a favorite fruit for many people. The plant's origin is the Mediterranean region so it grows best in mild-temperate climates, with acceptable growth in tropical and subtropical areas. The deciduous fig plant grows as a small tree or large shrub with large leaves and produces numerous fruit throughout the growing season. Because fig seeds are not viable, the plants are propagated asexually, but cuttings have poor rooting. Tissue culture is used to produce virus-free plants.

EXPLANT Actively growing apical shoots. Milky latex will be produced from fresh cuttings, but the use of antioxidants will decrease loss from polyphenolics. Keeping the new explants in the dark for one week after initial culture has also shown some benefit.

TREATMENT Wash explants thoroughly under running tap water for 45 minutes followed by immersion in 2:10 bleach for 5 minutes. Rinse in sterile water 3 times. Immerse in 70% ethanol for 30 seconds and rinse with a sterile antioxidant solution of 0.01% L-ascorbic acid.

MEDIA MS medium with 2 mg/liter BAP and 0.2 mg/liter NAA for Stages I and II. For rooting, use half-strength, liquid MS medium with 2 mg/liter IBA and 0.2% activated charcoal.

Ficus media

COMPOUND	STAGES I & II MG/LITER	STAGE III MG/LITER
MS salts	4,628	2,314
NAA	0.2	–
BAP	2.0	–
IBA	–	2.0
Sucrose	30,000	30,000
Agar	8,000	–
Activated Charcoal	–	2,000

pH 5.7

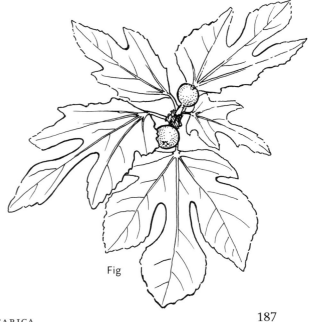

Fig

LIGHT 150–200 f.c. from cool-white fluorescent light for 16 hours light/8 hours dark.

TEMPERATURE 24°C–25°C (75°F–77°F).

REFERENCES Gunver and Ertan 1998, Kumar, Lakshmanan et al. 1998, Kumar, Radha et al. 1998, Murashige and Skoog 1962, Nobre and Romano 1998, Sharma 2010.

Fragaria ×ananassa
GARDEN STRAWBERRY
Rosaceae, Rose family

The common garden strawberry is a hybrid between *Fragaria chiloensis* and *F. virginiana*. Strawberry plants are traditionally propagated asexually by removal of runners, however commercially they are micropropagated to avoid virus and disease problems. More commercial strawberries are propagated by micropropagation than any other food crop. Tissue cultured plants tend to produce more runners, which is an advantage for multiplication in the field but may mean less fruit for the first year. Plantlets can be stored *in vitro* in the dark at 2°C–4°C (36°F–39°F) for several months before planting in the field.

Fragaria media

COMPOUND	STAGES I & II MG/LITER	STAGE III MG/LITER
MS salts	4,628	4,628
Inositol	100	100
Thiamine HCl	0.4	0.4
Nicotinic acid	0.5	0.5
Pyridoxine HCl	0.5	0.5
IBA	–	1.0
BA	1.0	–
TDZ	0.4	–
Sucrose	30,000	30,000
Agar	7,000	7,000

pH 5.7

EXPLANT Meristems from young (5-cm-long) runner tips that have not yet leafed out. If source plants are not certified as virus free, then the 5- to 8-mm-long meristem should be excised from apical and lateral buds as described in chapter 8.

TREATMENT Place about 12 runner tips in a beaker with 100 ml distilled water and 2 drops of detergent. Cover and stir on stirrer for 5 minutes (or shake by hand in a tightly covered jar). Remove the tips and place them in a clean beaker with 1:10 bleach. Stir for 10 minutes. Stir in 2 rinses of sterile distilled water for 10 seconds each. In the hood, rinse once more in distilled water and place on sterile toweling in a petri dish under a dissecting microscope. Excise meristems as described in chapter 8.

MEDIA MS salts and standard vitamins, with BA and TDZ for Stages I and II. Rooting can be directly from Stage II into a potting mix or in Stage III medium with IBA.

LIGHT 300 f.c. from cool-white fluorescent light for 16 hours light/8 hours dark.

TEMPERATURE 26°C (79°F).

REFERENCES Anderson and Haner 1978, Boxus et al. 1977, Damiano 1980, Haddadi et al. 2010, Kyte and Kyte 1981, Murashige and Skoog 1962, Sakila et al. 2007.

Freesia ×hybrida
FREESIA
Iridaceae, Iris family

Freesias are grown for their beautiful and fragrant cut flowers. They are also versatile when grown outdoors in the garden or as potted, indoor plants. Tissue culture of *Freesia* can improve propagation rates over the conventional division by corms.

EXPLANT 2-mm-diameter transverse slices of cormlets.

Freesia medium

COMPOUND	STAGES I, II & III MG/LITER
MS salts	4,628
Inositol	100
NAA	0.1
Kinetin	0.5
Sucrose	20,000
Agar	7,000

pH 5.8

TREATMENT If corms are dormant, they can be given a heat treatment—30°C (86°F) for 4 weeks—before culture. Remove the dry, outer leaf sheaths and wash in soapy water for 30 minutes. Agitate corms in 70% ethyl alcohol for 2 minutes. Mix in 1:10 bleach with 0.1% Tween 20 for 10 minutes. Rinse in 3 changes of sterile distilled water for 2 minutes each. Slice 2-mm cross sections. Place on agar medium in continuous dark. To obtain corms *in vitro*, expose shoots to 15°C (59°F) for 10 weeks and then place at 23°C (73°F).

Growth will occur in about 4 weeks in the dark-grown cultures. The subsequent light period needs about 2 weeks, after which the plantlets can be grown-on in moist peat under fluorescent lights and 20°C (68°F). Water with MS nutrient solution for one week, then with tap water. Plant in soil in greenhouse when plants are 8 weeks out of culture.

MEDIA MS modified with NAA at 0.1 mg/liter and kinetin at 0.5 mg/liter promotes both shoot and root development.

LIGHT Continuous darkness until formation of shoots and roots, then place in 100 f.c. from fluorescent light for 16 hours light/8 hours dark.

TEMPERATURE 23°C–25°C (73°F–77°F).

REFERENCES Anderson and Mielke 1985, Ascough et al. 2009, Doi et al. 1992, Gao et al. 2010, Meyer et al. 1975, Murashige 1974, Murashige and Skoog 1962.

Gerbera jamesonii
TRANSVAAL DAISY
Asteraceae, Aster family

Because of its colorful and long-lasting flowers, Transvaal daisy (*Gerbera jamesonii*) is one of the leading cut flowers in the world and also a popular potted and garden plant. Although *Gerbera* spp. grow easily from seed, most commercial cultivars are tissue cultured to produce large numbers of plants of specific colors.

EXPLANT 2 cm of emerging shoot tip. Whole, young capitula (flower heads) may also be used; although the process is less destructive of the plants and the flowers are more easily sterilized, these explants may be less responsive.

TREATMENT OF SHOOT TIPS Because contamination is a persistent problem, use greenhouse material that is as clean as possible. Obtain emerging shoot tips with 2–3 mm of base from the crown. Wash well in running water. Mix in 1:10 bleach with 0.1% Tween 20 for 30 minutes. Rinse in 3 mixes of sterile water for 15 minutes. Use whole or excise under microscope emerging buds with 1 mm of base, or use 3 mm of shoot tip.

Gerbera media

COMPOUND	STAGES I & II MG/LITER	STAGE III MG/LITER
MS salts	4,628*	2,314
Inositol	100	100
Adenine sulfate	80	–
Thiamine HCl	30	30
Nicotinic acid	10	10
Pyridoxine HCl	1.0	1.0
IAA	0.1	0.5–1.0
BA	1.0	–
Glycine	2.0	–
Sucrose	30,000	20,000
Agar	8,000	8,000

*Use 2,314 mg/liter of MS salts for flower buds

pH 5.7

Dip for one minute in 1:100 bleach and place on agar medium.

TREATMENT OF FLOWERS Wash young flower buds 0.8–1.0 cm diameter in soapy water and surface sterilize with 70% ethanol for 30 seconds followed by a 2:5 bleach treatment for 15 minutes. Rinse 3 times in sterile water. Remove external bracts of the buds before culturing.

MEDIA MS salts with 0.1 mg/liter IAA and 2.0 mg/liter BA for Stages I and II. For Stage III, increase IAA to 0.5 to 1.0 mg/liter. Reduce MS to half strength for rooting stage. Two weeks in Stage III medium may be sufficient before transplanting to potting mix.

LIGHT 100–300 f.c. from fluorescent lights for 16 hours light/8 hours dark.

TEMPERATURE 21°C–25°C (70°F–77°F).

REFERENCES Gantait et al. 2011, Kanwar and Kumar 2008, Murashige and Skoog 1962, Murashige et al. 1974, Pierik and Segers 1973, Posada et al. 1999, Preil and Schum 1985.

Hemerocallis spp. and hybrids
DAYLILY
Xanthorrhoeaceae, Grass tree family

Daylilies (*Hemerocallis* spp.) are among the top 10 herbaceous plants for the garden because of the wide selection of flower colors. They are known for their ability to survive extreme and harsh environmental conditions and still flower dependably each year. Meristems cannot be used as explants to micropropagate these plants because they are difficult to clean and the plant dies if the meristems are removed. Transplant plantlets to a well-drained soil-less mix.

EXPLANT 1-mm-long flower buds in unexpanded crown tissue.

TREATMENT Remove and discard bract tissue from flower buds. Wash inflorescence in water

Hemerocallis media

COMPOUND	CALLUS MG/LITER	PLANTLETS MG/LITER
Ammonium nitrate	1,650	1,650
Calcium chloride	440	440
Magnesium sulfate	370	370
Potassium nitrate	1,900	1,900
Potassium phosphate	300	300
MS minor salts	32.8	32.8
MS iron	65.1	65.1
Adenine sulfate	160	160
Inositol	100	100
Thiamine HCl	0.4	0.4
NAA	10	0.5
Kinetin	0.1	0.1
Casein hydrolysate	500	500
Malt extract	500	500
Sucrose	40,000	30,000
Agar	6,000	6,000

pH 5.5

with 0.1% Tween 20. Mix in 100% bleach with Tween 20 for 2 minutes. Rinse 3 times in sterile distilled water.

MEDIA Modified MS with increased potassium phosphate, plus casein hydrolysate, malt extract, and adenine sulfate. Added hormones are NAA at 10 mg/liter and kinetin at 0.1 mg/liter for callus formation; the NAA is lowered to 0.5 mg/liter for plantlet formation.

LIGHT 300–400 f.c. continuous light or 16 hour light/8 hour dark.

TEMPERATURE 23°C–26°C (73°F–79°F).

REFERENCES Adelberg et al. 2007, Adelberg et al. 2010, Fitter and Krikorian 1985, Griesbsach 1989, Heuser and Apps 1976, Heuser and Harker 1976, Meyer 1976.

Heuchera spp. and hybrids
ALUMROOT, CORAL BELLS
Saxifragaceae, Saxifrage family

Alumroot (*Heuchera americana*) and other heucheras are low-growing hardy, herbaceous perennials valued for attractive foliage and grown in the garden for color accent. The basic form has leaves of green with silver between the veins, but new hybrids have been developed with many different colors and leaf forms. Heucheras grow as neat, low mounds that form slender flower spikes rising 30 cm (1 ft.) or more above the foliage in late spring or early summer. The flowers of *H. sanguinea* (coral bells) are especially showy. Traditional propagation by leaf cuttings is very easy with *Heuchera*, so it is not surprising that they are easily tissue cultured from petioles and leaves.

EXPLANT Shoot tips or 1-cm cross sections of petiole.

TREATMENT Wash the shoot tips in water. Surface sterilize in 1:10 bleach with 0.1% Tween 20 for 10 minutes. Rinse twice with sterile distilled water. Excise the basal portion and outer leaves of the shoot and excise about 5-mm-long shoot tip explants.

MEDIA For shoot tips or leaf/petiole sections, use a modified MS with 0.1–0.5 mg/liter of NAA and 0.5–1.0 mg/liter of BA. For rooting, use IBA at 1.0 mg/liter and omit the BA and NAA.

Heuchera media

COMPOUND	STAGES I & II MG/LITER	STAGE III MG/LITER
MS salts	4,628	4,628
Glycine	2.0	2.0
NAA	0.1–0.5	–
IBA	–	1.0
BA	0.5–1.0	–
Sucrose	20,000	20,000
Agar	7,000	7,000

pH 5.6

LIGHT 300 f.c. fluorescent lighting for 16 hours light/8 hours dark.

TEMPERATURE 25°C (77°F).

REFERENCES Hosoki and Kajino 2003, Kitto et al. 1990, Murashige and Skoog 1962.

Hippeastrum spp.
AMARYLLIS
Amaryllidaceae, Amaryllis family

Hippeastrum has been commonly called amaryllis for so long that it is not surprising when it is confused with the genus *Amaryllis*, the belladonna lily, another attractive bulbous plant in the Amaryllidaceae. *Hippeastrum* is a valuable landscape plant and is commonly sold as a seasonal gift, potted and attractively packaged, ready for the recipient to water. It is one of a number of bulbous crops with a relatively slow rate of seasonal multiplication. Seed-propagated plants may take 2 or 3 years to flower.

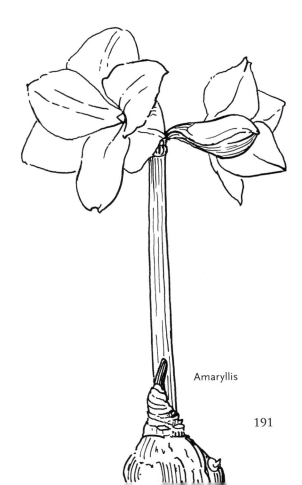

Amaryllis

Hippeastrum media

COMPOUND	STAGES I & II SCALES MG/LITER	STAGES I & II OTHER EXPLANTS MG/LITER	STAGE III MG/LITER
MS salts	4,628	–	4,628
Ammonium nitrate	–	1,650	–
Calcium chloride	–	440	–
Magnesium sulfate	–	370	–
Potassium nitrate	–	1,900	–
Potassium phosphate	–	300	–
MS minor salts	–	32.8	–
MS iron	–	65.1	–
Adenine sulfate	–	2.5	–
Inositol	100	100	–
Thiamine HCl	0.1	1.0	–
Nicotinic acid	0.5	1.0	–
Pyridoxine HCl	0.5	1.0	–
NAA	5.0	2.5	–
Kinetin	0.5	0.5	–
BA	1.0	–	–
Glycine	2.0	–	–
Casein hydrolysate	1,000	500	500
Malt extract	–	500	500
Sucrose	30,000	30,000	30,000
Gelrite	2.5	2.5	2.5

pH 5.5

EXPLANT Shoot tips, ovary tissue, scape slices, young flower buds (grown 10 cm above bulb neck), or 1- to 2-cm twin basal sections of bulb segments. Micropropagation from pieces of the bulb presents major challenges to Stage I disinfestation.

TREATMENT OF BULBS Trim bulbs to remove roots, leaves, and outer scales. Wash the bulbs in warm, soapy water followed by a rinse in warm, flowing water for 30 minutes. Rinse in 95% ethanol for one minute and 1:5 bleach for 20 minutes. Agitate in 3 sterile water rinses. Cut 1-cm squares from inner scales with a piece of the basal plate attached. Wash in 1:10 bleach for 2 minutes. Rinse well in sterile water.

TREATMENT OF SCAPES AND TIGHT FLOWER BUDS Wash in water with 0.1% Tween 20. Rinse by mixing in sterile water for 20 minutes. Dip in 90% alcohol for 10 seconds. Agitate for one minute in sterile water. Mix for 15 minutes in 1:10 bleach or 8% calcium hypochlorite. Trim buds. For the scapes, cut slices 3 mm thick and invert on medium.

MEDIA Modified MS with additives, depending on the reference used.

LIGHT 100–300 f.c. from fluorescent light. A 16- to 24-hour photoperiod is standard.

TEMPERATURE 15°C–26°C (59°F–79°F).

REFERENCES Christie 1985, Hussey 1975, Ilczuk et al. 2005, Mii et al. 1974, Murashige 1974, Murashige and Skoog 1962, Okubo et al. 1999, Olate and Bridgen 2004, Seabrook and Cumming 1977, R. H. Smith et al. 1999, Tombolato et al. 1994, van Aartrijk and van der Linde 1986.

Hosta spp. and hybrids
HOSTA, PLANTAIN LILY
Asparagaceae, Asparagus family

Hostas are popular herbaceous perennial landscape plants grown in temperate regions for their decorative foliage, attractive white to blue flowers, and ability to grow well in shade. Slow to propagate by traditional division, these plants can be micropropagated much faster. In addition, because of the recent attack of hosta virus X (HVX), many cultivars such as 'Gold Standard', 'Striptease', and 'Sum and Substance' are cleaned and then stock plants are micropropagated to maintain pathogen-free plants. Compared with many other species, hostas are slow to micropropagate but that is because they are monocotyledonous plants. There is room for experimentation with micropropagation of Hosta species and cultivars. The fact that some regenerate directly in light conditions and others require darkness for

Hosta media

COMPOUND	STAGE I MG/LITER	STAGE II MG/LITER	STAGE III MG/LITER
Ammonium nitrate	1,650	1,650	1,650
Calcium chloride	440	440	440
Copper sulfate	25	25	25
Magnesium sulfate	370	370	370
Potassium nitrate	1,900	1,900	1,900
Potassium phosphate	300	300	300
Sodium phosphate	170	170	170
MS minor salts	32.8	32.8	32.8
MS iron	65.1	65.1	65.1
Adenine sulfate	92	92	–
Inositol	100	100	100
Thiamine HCl	0.4	0.4	0.4
NAA	0.5	0.5	0.5
BA	0.2	0.2	0.1
Glycine	2.0	2.0	–
Casein hydrolysate	500	500	500
Sucrose	30,000	30,000	30,000
Agar	7,000	7,000	7,000

pH 5.5

callus production and/or shoot initiation suggests that the relationship of plant readiness to hormone balance needs further study.

EXPLANT Flower scapes about 30 cm long with immature flower buds 0.5–1 cm long.

TREATMENT Cut inflorescence, including about 6 cm of scape; remove flower buds. Wash buds in running water for one minute. Immerse buds and scape pieces in 70% ethyl alcohol for 30 seconds. Mix in 1:10 bleach with 0.1% Tween 20 for 10 minutes. Rinse well in 3 changes of sterile distilled water. Cut scape pieces into 3-mm sections.

MEDIA Modified MS with vitamins, NAA (0.5 mg/liter), and BA (2 mg/liter). Reduce BA to 0.1 mg/liter for Stages II and III. Meyer added adenine sulfate and casein hydrolysate for bud culture in standard light conditions. Maki et. al. (2005) recommended adding ancymidol at 3.2 mmol to increase shoot production.

LIGHT Some hostas initiate shoots after 4 weeks of darkness (including *Hosta lancifolia*, *H. plantaginea* 'Royal Standard', *H.* 'Honeybells', and *H.* 'Aoki'), followed by 200–300 f.c. from fluorescent light with 16 hours light/8 hours dark. Some species initiate shoots without a period of continuous darkness, such as *H. sieboldiana* from flower buds. Still others produce callus in the dark, followed by shoot growth in standard light conditions.

TEMPERATURE 22°C–26°C (72°F–79°F); 0°C–1°C (32°F–34°F) for cold storage.

REFERENCES Adelberg et al. 2000, Lubomski 1989, Maki et al. 2005, Meyer 1980, Murashige and Skoog 1962, Papachatzi et al. 1980a, 1980b, Saito and Nakano 2002, van Aartrijk and van der Linde 1986, Zillis and Zwagerman 1979, Zillis et. al. 1979.

Hypoxis hemerocallidea
AFRICAN POTATO, YELLOW STAR PLANT
Hypoxidaceae, Star lily family

Hypoxis hemerocallidea (syn. *H. rooperi*) is a tuberous perennial herb with long strap-shaped leaves and yellow star-shaped flowers. It comes from Africa and is related to the North American native *H. hirsuta*, a delightful alpine species. *Hypoxis hemerocallidea* has been used for centuries in Africa as food and as a medicinal. The medicinal qualities of *Hypoxis* species have been gaining recognition in the United States. The tissue culture techniques apply to other corms.

EXPLANT Upper half of corm, approximately 1.5- to 3-cm-long pieces.

TREATMENT Remove the leaves and roots from the corm and wash extensively with water containing a few drops of detergent. Sterilize by washing in 95% ethanol for 5 minutes and then for 30 minutes in 100% bleach with 3 or more rinses in sterile water. Remove the outer tissues

Hypoxis media

COMPOUND	STAGES I & II MG/LITER	STAGE III MG/LITER
MS salts	4,628	4,628
Inositol	100	100
Thiamine HCl	1.0	1.0
NAA	1.0	0.1
Kinetin	3.0	–
Casein hydrolysate	1,000	1,000
Sucrose	20,000	20,000
Agar	8,000	8,000

pH 5.7 to 5.8

(epidermal layers) of the corm and cut the inner tissues (pith) into large slices about 3 cm thick. Wash the slices with sterile water and further treat with 3:10 bleach for 30 minutes, with 3 or more rinses in sterile water. Cut the slices still further into smaller explants approximately 1–1.5 cm thick, making sure each piece contains some of the cambial (meristematic) layer.

Corm explants can rapidly turned black, which results in the browning of the medium. If this happens, subculture corm explants every week for 3–4 subcultures and then every 3 weeks.

MEDIA Standard MS salts with kinetin and NAA. Sucrose, inositol, thiamine HCl, and casein hydrolysate are other additives. The same medium without kinetin and lower concentration of NAA is used for rooting.

LIGHT 100–300 f.c. from cool-white fluorescent light for 16 hours light/8 hours dark.

TEMPERATURE 25°C–27°C (77°F–81°F).

REFERENCES Appleton and van Staden 1995, Murashige and Skoog 1962, Ndong et al. 2006, Page and van Staden 1984, van Aartrijk and van der Linde 1986.

Ipomoea batatas
SWEET POTATO VINE
Convolvulaceae, Morning glory family

Sweet potato (*Ipomoea batatas*) is an important vegetable known for its large, starchy, and sweet-tasting tuberous roots. In recent years, it has also become popular as an ornamental plant that is grown for its black, variegated, or lime green leaves and is sometimes called the "tuberous morning glory." The sweet potato vine is propagated by dividing the tuberous root or by stem cuttings since the vines rarely set seed. The sweet potato is only distantly related to the potato (*Solanum tuberosum*).

EXPLANT Shoot tips from apical and axillary buds of sprouted roots.

TREATMENT Dip the shoots in 3:10 bleach plus 0.2% Tween 20 for 20 minutes. Rinse 2 or 3 times in sterile distilled water for a few minutes each. Dip the shoots in 70% ethanol for one minute and again rinse 3 times with sterile water. Aseptically remove leaves from around the shoot tip and culture the growing point.

MEDIA MS salts with myo-inositol, thiamine, l-arginine, pantothenic acid, and gibberellic acid.

Ipomoea medium

COMPOUND	STAGES I, II & III MG/LITER
MS salts	4,628
Myo-inositol	100
Thiamine HCl	0.4
L-arginine	100
Ascorbic acid	200
Pantothenic acid	2
Gibberellic acid	20
Sucrose	30,000
Kinetin	0.5
IAA	0.2
Agar	8,000

pH 5.7 to 5.8

LIGHT Keep explants in the dark for the first week of culture, followed by 200–300 f.c. from cool-white fluorescent light with 16 hours light/8 hours dark for shoot production, and with 8 hours light/16 hours dark for microtuber production.

TEMPERATURE 25°C (77°F).

REFERENCES Dagnino et al. 1991, Gosukonda et al. 1995, Murashige and Skoog 1962.

Iris germanica
BEARDED IRIS
(rhizomatous)
Iridaceae, Iris family

Bearded iris (*Iris germanica*) is an herbaceous perennial garden plant with colorful and fragrant flowers that arise from rhizomes. In tissue culture, it is slow and difficult to grow, but persistence and patience are duly rewarded.

EXPLANT 2-mm cross sections of peduncle (flower cluster stalks) of young inflorescences that are still tightly closed.

TREATMENT With a sterilized knife, cut young inflorescences into 15-cm-long pieces. Remove and discard bract tissue. Quickly dip in 70% ethanol, and then stir in 1:10 bleach with 0.1% Tween 20 for 20 minutes. In the hood, rinse in 3 changes of sterile water for 2 minutes each. Vertically slice 2-mm sections of peduncles. Place with cut surface down on agar medium.

Callus will develop from the explant in 6–12 weeks. Cut callus into several pieces and place in the light. Transplant the shoots and roots that develop from the edges of the callus pieces (in 2–5 months) to a peat/perlite (1/1) soil mix. Keep under mist for 2 weeks, and then gradually reduce mist and humidity.

MEDIA Modified MS with added casein hydrolysate and adenine sulfate. Add 2.5 mg/liter NAA and 0.5 mg/liter kinetin for callus formation.

LIGHT Continuous dark for 6–12 weeks, followed by 100–300 f.c. from fluorescent light for 16 hours light/8 hours dark.

TEMPERATURE 26°C (79°F).

REFERENCES Hussey 1975, 1976, Kawase et al. 1995, Meyer et al. 1975, Mielke and Anderson 1989, Murashige 1974, Murashige and Skoog 1962.

Iris medium
(rhizomatous)

COMPOUND	STAGES I, II & III MG/LITER
MS salts	4.628
Adenine sulfate	160
Inositol	100
Thiamine HCl	0.4
NAA	2.5
Kinetin	0.5
Casein hydrolysate	500
Sucrose	30,000
Agar	7,000

pH 5.5

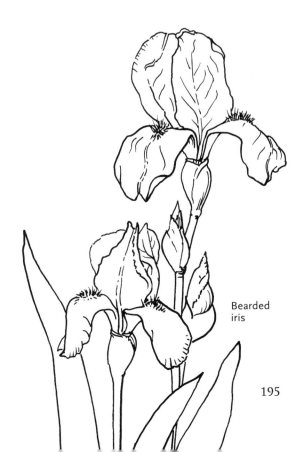

Bearded iris

Iris ×hollandica

DUTCH IRIS
(bulbous)
Iridaceae, Iris family

Dutch irises (*Iris* ×*hollandica* and cultivars) have true bulbs and are commonly used as cut flowers. The bulbs are small but produce large blue, white, yellow, and purple flowers. Several other *Iris* species with true bulbs are winter-hardy and popular as garden plants.

EXPLANT 0.5- to 1.5-mm-long shoot tips.

TREATMENT Bulbs should be stored in the dark at 5°C (41°F) for 6 weeks. Peel off and discard roots, dry tunics, and outer scales. Wash thoroughly under running tap water for 30 minutes and dip in 95% ethanol for 3 minutes. Surface sterilize the bulbs in 8:10 bleach with continuous stirring for 20 minutes and rinse 3 times in sterile water. Bulbs containing basal discs can be cut into 4 sections.

Another approach is to wait until the central shoot tip begins to elongate. Once the shoots have elongated to 0.5–1.0 cm, trim away all but the basal plate with the tiny sheath leaves covering the new shoot tip (0.5–1.5 mm). Cut away and discard the basal plate, saving only the shoot tip. Stir in 1:10 bleach for 15 minutes followed by a thorough rinse in sterile water. Excise shoot tips about 0.5–0.8 mm long.

The first transfer into multiplication occurs at about 6 weeks. Two shorter transfer periods should see the formation of 1- to 2-mm buds. Transfer the individual 1- to 2-mm buds to multiplication medium. Following 6 months in the bulbing medium, plant in a cool (10°C [50°F]) greenhouse for rooting.

MEDIA Three stages of medium are used: multiplication, intermediate, and bulbing. The first medium is for shoot multiplication, the intermediate medium is to dilute hormones before bulblet development, and the third medium is for bulblet development.

Iris media
(bulbous)

COMPOUND	MULTI-PLICATION MG/LITER	INTER-MEDIATE MG/LITER	BULBING MG/LITER
MS salts	4,628	4,628	4,628
Adenine sulfate	80	–	–
Inositol	100	100	100
Thiamine HCl	0.4	0.4	0.4
IAA	1.0	–	–
NAA	–	0.1	–
BA	1.5	–	–
Kinetin	–	0.01	–
Sucrose	30,000	60,000	60,000
Agar	7,000	6,000	6,000

pH 5.7

LIGHT For the multiplication and intermediate stage, 100–300 f.c. from fluorescent light for 16 hours light/8 hours dark. For the bulbing stage, continuous dark for 4 weeks up to 6 months.

TEMPERATURE Multiplication and intermediate stage, 20°C (68°F); bulbing stage, 4 weeks at 5°C (41°F) followed by 25°C–28°C (77°F–82°F) for 5–8 weeks, followed by 5°C (41°F) in the dark.

REFERENCES Anderson et al. 1990, Meyer et al. 1975, Mielke 1984, Mielke and Anderson 1989, Murashige 1974, Murashige and Skoog 1962, Nasircilar et al. 2011, van der Linde et al. 1988.

Juglans regia

ENGLISH WALNUT, PERSIAN WALNUT
Juglandaceae, Walnut family

Walnuts (*Juglans* spp.) are very valuable trees for nut and wood production, but their propagation is difficult and time-consuming. Micropropagation of walnuts is also difficult but produces true-to-type results, greater consistency, and plants that grow on their own roots without grafting. Success is slow because of problems stemming from contamination, exudates, and lack of response. A few walnut cultivars are now being commercially micropropagated.

English walnut (*Juglans regia*), also known as Persian walnut, is usually grafted on a rootstock, commonly *J.* 'Paradox', a hybrid between *J. hindsii* and *J. regia*. Eastern black walnut (*J. nigra*) is prized for its wood and is commonly grown from seed, but can be micropropagated by tissue culture. Juvenile nodal segments of walnut and embryonic axes from seeds can be used to remove the initial propagules.

EXPLANT Pretreated 25-cm-long stems from mature trees or 3- to 4-cm-long nodal segments from seedlings.

TREATMENT OF MATURE PLANTS Juvenile material that is 2–3 months old is best to use for explants if possible. Wash stems in running tap water. Cut off leaves, leaving 2-cm petiole bases (for orientation). Cut stems, under water, into 5-cm-long nodal pieces. Wash for 3 hours under running tap water. Agitate in 1:5 bleach with 2 drops of detergent for 15 minutes, more or less depending on the strength of the shoot. Rinse individual stem pieces in sterile distilled water. To decrease explant exudation, transfer onto fresh medium 1, 3, 5, and 8 days after culture.

TREATMENT OF SEEDLINGS Nuts should be imbibed for 24 hours in water and disinfected with a 30- to 60-second dip in 95% ethanol followed by a wash in 1:5 bleach for 5–10 minutes. Rinse in 3 sterile water washes. Allow the seeds to germinate aseptically on a quarter-strength MS medium with no growth regulators and 6 g/liter agar. Allow the germinated seedlings to grow for 8 weeks before excising nodal segments.

MEDIA Double-phase medium has been shown to double the success rate. This is accomplished by pouring sterile liquid medium over explants after their transfer onto the solidified medium. The liquid phase covers the surface of the solid medium. Gellan gum as a solidifying agent has

Juglans media

COMPOUND	STAGES I & II MG/LITER	STAGE III MG/LITER
MS salts	4,628	4,628
Adenine sulfate	20.0	20.0
Inositol	100	100
Thiamine HCl	2.0	2.0
Nicotinic acid	1.0	1.0
IBA	0.01	0.15
BA	1.0	–
Glycine	2.0	2.0
Sucrose	30,000	30,000
Gelrite	2,300	2,300
Activated charcoal	–	10,000

pH 5.4 to 5.7

been shown to be better than agar. Hyperhydricity (vitrification) has been frequently seen with walnut, but this MS medium and level of BA should avoid it.

LIGHT 300 f.c. from a combination of cool-white and warm-white fluorescent lights for 16 hours light/8 hours dark.

TEMPERATURE 22°C–25°C (72°F–77°F).

REFERENCES Bosela and Michler 2008, Cornu and Jay-Allemand 1989, Driver 1985, McGranahan et al. 1987, McGranahan and Leslie 1988, Revilla et al. 1989, Rodriguez 1982, Scaltsoyiannes et al. 1997.

Kalanchoe blossfeldiana
KALANCHOE
Crassulaceae, Orpine family

Kalanchoe (*Kalanchoe blossfeldiana*) is a succulent potted plant with dark green leaves and large umbels of flower clusters of red, pink, or yellow that last for weeks as a houseplant. Requiring only 4–6 weeks from culture initiation to potted plantlets, this species is ideal for classroom tissue culture demonstration.

Kalanchoe media

COMPOUND	STAGES I & II MG/LITER	STAGE III MG/LITER
MS salts	4,628	4,628
Sodium phosphate	120	120
Adenine sulfate	80	–
Inositol	100	100
Thiamine HCl	0.4	0.4
IAA	1.0	–
IBA	–	1.0
Kinetin	1.0	–
GA_3	–	0.2
Sucrose	30,000	30,000
Agar	7,000	7,000

pH 5.7

EXPLANT Leaf blade sections (1.5 cm long and wide), stem sections (2 cm long), or shoot tips (2 mm long).

TREATMENT Wash 3- to 5-cm-long cuttings in warm water with detergent. Rinse briefly in water. Mix in 2:10 bleach with 0.1% Tween 20 for 10–15 minutes. Rinse in 3 changes of sterile water for one minute each. Trim to size.

MEDIA Modified MS with kinetin and IAA. For Stage III rooting, omit kinetin and IAA, and add IBA and GA_3.

LIGHT 100–300 f.c. from cool-white fluorescent light for 8 hours light/16 hours dark.

TEMPERATURE 22°C–26°C (72°F–79°F).

REFERENCES Griffiths 1994, Murashige and Skoog 1962, Sanikhani et al. 2006, Smith 1992, Smith and Nightingale 1979.

Kalmia latifolia
MOUNTAIN LAUREL
Ericaceae, Heath family

Mountain laurel (*Kalmia latifolia*) is a noteworthy shrub native to North America and a pleasant companion for *Rhododendron*, especially in moderate, moist climates. It is not, however, easy to propagate by cuttings; tissue culture has allowed hybrids to be more efficiently produced. Lloyd and McCown used mountain laurel in developing and publishing their woody plant medium (WPM), which has proved very useful in tissue culture. Stage III is optional, and plantlets may be rooted directly in 100% peat, at 30°C (86°F), high humidity, and continuous light. *Kalmia* can also be cultured by following the directions and media for *Rhododendron*.

EXPLANT 2- to 3-cm-long softwood shoot tips in the spring or early summer.

TREATMENT Carefully remove all leaves more than 1 cm long. Dip shoot tips in 70% ethyl

Kalmia woody plant media

COMPOUND	STAGES I & II MG/LITER	STAGE III (OPTIONAL) MG/LITER
Ammonium nitrate	400	400
Boric acid	6.2	6.2
Calcium chloride	96	96
Calcium nitrate	556	556
Cupric sulfate	0.025	0.025
Magnesium sulfate	370	370
Manganese sulfate	22.3	22.3
Potassium phosphate	170	170
Potassium sulfate	990	990
Sodium molybdate	0.025	0.025
Zinc sulfate	8.6	8.6
MS iron	65.1	65.1
Inositol	100	100
Thiamine HCl	1.0	0.1
Nicotinic acid	0.5	0.5
Pyridoxine HCl	0.5	0.5
2iP	2.0	–
Glycine	2.0	2.0
Sucrose	20,000	20,000
Agar	6,000	6,000

pH 5.2

alcohol. Mix in 1:10 bleach with 0.1% Tween 20 for 15 minutes. Rinse in 3 changes of sterile distilled water for one minute each. Trim away any damaged material. If brown, phenols are "bleeding" into the medium; transfer the cuttings to fresh media.

MEDIA WPM is a significant change from MS. It has low nitrate, high potassium, and low sugar. The minor elements, vitamins, and hormones are standard, with 1–2 mg/liter 2iP used for multiplication.

LIGHT Continuous light at 100–300 f.c. from cool-white fluorescent light.

TEMPERATURE 28°C–30°C (82°F–86°F).

REFERENCES Anderson 1975, 1978, Lloyd and McCown 1980, McCown and Lloyd 1983, Murashige and Skoog 1962, Nishimura and Fujimori 2000, Norton and Norton 1985.

Lilium spp.
LILY
Liliaceae, Lily family

The genus *Lilium* includes many different species, varieties, and hybrids valued for their cut flowers and as long-lived garden plants. The Easter lily (*L. longiflorum*) is the most common lily on the market. Most commercially micropropagated lilies belong to 3 groups: Asiatic hybrids, the LA (Longiflorum) hybrids, and oriental lilies. The high demand for lily bulbs require that these plants be micropropagated to supply the large numbers, keep the plants disease free, and to allow for international shipping.

Lilium cultures can be chilled *in vitro* before planting, but most growers chill *in vivo*. For cold storage of bulblets for spring planting, wash the bulblets, and then dip in fungicide. Pack in barely dry vermiculite in mesh bags in trays and store in darkness at 0°C–1°C (32°F–34°F).

Lilium media

COMPOUND	STAGES I & II MG/LITER	STAGE III (OPTIONAL) MG/LITER
MS salts	4,628	4,628
Adenine sulfate	80	–
Inositol	100	100
Thiamine HCl	0.4	0.4
BAP	2.0	–
NAA	0.2	0.2
Activated charcoal	–	2.0
Sucrose	30,000	30,000
Agar	7,000	7,000

pH 5.75

EXPLANT 4- to 5-mm-long sections from bulb scale bases.

TREATMENT Separate inner bulb scales such that they retain a portion of the basal plate. Wash in tap water with a few drops of detergent. Rinse in running tap water for 15–30 minutes. Mix in 1:10 bleach for 10–20 minutes. Rinse in 3 changes of sterile distilled water for one minute each. Cut 4- to 5-mm cubes from the base of scales, including some basal plate tissue.

MEDIA Lily bulbs usually respond well to MS media with additional adenine sulfate, thiamine, and inositol plus growth regulators. Activated charcoal is sometimes added during Stage III rooting.

LIGHT Both intermittent light and continuous darkness are used, depending on the propagule. Continuous darkness is preferred during Stage III as it leads to the formation of more bulblets and fewer leaves, which is better for transplanting. Cultures of bulb sections are better under 200–300 f.c. from fluorescent light, 16 hour light/8 hour dark.

TEMPERATURE 25°C–30°C (77°F–86°F), with the warmer end of the spectrum preferable to the cooler. Chilling temperatures at 10°C (50°F) can be used to store bulbs before culture for up to 10 weeks.

REFERENCES Anderson 1977, Gayed et al. 2006, Han et al. 2005, Hussey 1975, Murashige and Skoog 1962, Nesi et al. 2009, Nhut 1998, Stimart and Ascher 1981a, 1981b, Takayama and Misawa 1980, Takayama et al. 1982, van Aartrijk and van der Linde 1986.

Malus domestica
APPLE
Rosaceae, Rose family

Apple (*Malus* spp.) is the third most important fruit crop in the world after banana and grape. Due to their nature, commercial varieties need to be propagated vegetatively and are usually grafted. Tissue culture requirements for apples can vary in small detail from species to species, cultivar to cultivar, and even clone to clone. Some experimentation is usually necessary to determine the precise factors for optimum response.

EXPLANT Stem pieces with 1 or 2 nodes, apical buds from actively growing shoots with lateral buds showing, or semi-dormant buds. Buds from 0.2 to 0.6 cm long show less browning and higher survival.

TREATMENT OF NODAL SECTIONS Collect 7- to 10-cm-long shoots from the first growth in spring or semi-dormant buds. Remove and discard leaves, leaving 1 cm of petiole attached to the stem. Wash with vigorous agitation in tap water and 1% Tween 20 for 10 minutes. Rinse in running tap water for 2 minutes. Stir in 1.5:10 to 1:5 bleach with 0.1% Tween 20 for 10–20 minutes— the harder the shoot, the stronger the bleach and/ or the longer the time in bleach solution. In the transfer hood, rinse 3 times for a few seconds in sterile distilled water. Cut 1- or 2-node pieces (less than 1 cm long), with a short piece of stem above and below the node. Be sure to sterilize your instruments frequently between cuts. Place lower stem stub into agar so that explant is upright and the bud is above the agar level. Explant buds should break and grow in 3 weeks; promptly transfer to Stage II medium. Transfer cultures at least every 3 weeks.

TREATMENT OF VEGETATIVE BUDS Harvest buds when the plants are dormant and stored under refrigeration until required. Transfer to Stage II medium as soon as explant starts growing. Transfer at least every 3 weeks, more often if there is bleeding or no response. In Stage II, in preparing for rooting, etiolate the cultures by placing them in continuous dark (in cardboard boxes) in the growing room. After 2–3 weeks, place the cultures in the light for 1–2 weeks to green up before rooting.

MEDIA For Stage I, use half-strength MS salts and full-strength iron, plus inositol and thiamine HCl, with no hormones. Use agar because gellan gum may cause vitrification. Stage II usually requires full-strength MS, but BA and IBA requirements vary. Try BA at 2.0 mg/liter and IBA at 0.1 mg/liter. In Stage III, return to Stage I medium. Vitrification (hyperhydration) can be common on leaves *in vitro*, but loosening the lids of the culture vessels will enable sufficient gas exchange to prevent this.

Malus media

COMPOUND	STAGE I MG/LITER	STAGE II MG/LITER	STAGE III MG/LITER
MS major salts	2,265	4,530	2,265
MS minor salts	16.4	32.8	16.4
MS iron	65.1	65.1	65.1
Inositol	208	208	208
Thiamine HCl	2.5	2.5	2.5
IBA	–	0.1	–
BA	–	2.0	–
Sucrose	30,000	30,000	30,000
Agar	6,000	6,000	6,000

pH 5.7

LIGHT 300–400 f.c. from cool-white fluorescent light for 16 hours light/8 hours dark.

TEMPERATURE 24°C–26°C (75°F–79°F).

REFERENCES Dobránszki and Teixeira da Silva 2010, Dunstan 1981, Hazell 1998, Lane 1982, Magyar-Tábori et al. 2010, Murashige and Skoog 1962, Suttle 1983, Zimmerman and Broome 1980a.

Mammillaria elongata
GOLDEN STAR CACTUS
Cactaceae, Cactus family

Mammillaria, the largest genus in the Cactaceae and a favorite of collectors, has a large variety of stem forms, flower colors, sizes, and spine variations. These cacti are valued as potted houseplants because of their interesting shapes and easy care. Due to careless and unethical pillaging, however, some cacti are classified as endangered species. Tissue culture has helped to save many of them.

EXPLANT Tubercles, the round nodules or warty outgrowths, with part of the apical branch at their base. Tubercles taken from the basal part of the branch will produce callus, not shoots.

TREATMENT Mix excised branches for one hour in a fungicide with 0.5% Tween 20 added. Trim off spines. Disinfect with 70% ethanol for 60 seconds followed by 1:10 bleach for 15–30 minutes. Remove the tubercles and give them a second disinfection in 2:10 bleach for 20 minutes. Rinse in 3 changes of sterile distilled water for 2 minutes each. Successful disinfection for *Mammillaria* can be difficult to obtain.

These plants are easy to root and Stage III is not necessary. For *ex vitro* root initiation, transfer to the greenhouse in pots containing a peat/vermiculite/perlite (2/1/1) soil mix. Cover with plastic. To harden off, over 2 weeks gradually enlarge a pinhole-sized opening in the plastic cover.

Mammillaria medium

COMPOUND	STAGES I & II MG/LITER
MS salts	4,628
Sodium phosphate	85
Adenine sulfate	80
Inositol	100
Thiamine HCl	30
Nicotinic acid	10
Pyridoxine HCl	1.0
BA	0.1
NAA	0.2
Sucrose	30,000
Agar	8,000

pH 5.7 to 5.8

MEDIA MS salts with organics, which includes inositol, adenine sulfate, and vitamins. Growth regulators are BA and NAA.

LIGHT 200–300 f.c. from cool-white fluorescent light for 16 hours light/8 hours dark.

TEMPERATURE 25°C–27°C (77°F–81°F).

REFERENCES Johnson and Emino 1979, Mauseth 1979, Murashige and Skoog 1962, Papafotiou et al. 2001, Ramirez-Malagon et al. 2007.

Miltonia spp. and hybrids
MILTONIA
Odontoglossum spp. and hybrids
ODONTOGLOSSUM
Oncidium spp. and hybrids
ONCIDIUM
Orchidaceae, Orchid family

General information on tissue culture of orchid genera is given under *Cattleya*. Like that genus, the 3 genera discussed here—*Miltonia*, *Odontoglossum*, and *Oncidium*—are epiphytic orchids that grow from sympodial pseudobulbs, but they require a slightly different medium than that of *Cattleya*.

Miltonia, *Odontoglossum*, and *Oncidium* media

COMPOUND	STAGES I & II MG/LITER	STAGE III MG/LITER
Ammonium sulfate	500	500
Calcium nitrate	1,000	1,000
Magnesium sulfate	250	250
Potassium phosphate	250	250
MS minor salts	32.8	32.8
MS iron	65.1	65.1
BA	1–2	–
NAA	0.1	0.1
Sucrose	20,000	20,000
Coconut milk	100 ml	100 ml
Agar	–	6,000

pH 5.5

Miltonias are very easy to grow and have a lovely fragrance compared to other orchid species; flower colors range from dark purple to white. Odontoglossums produce arching inflorescences of white, red, purple, brown, yellow, or blotched with many colors. Oncidiums have flowers that are mostly golden yellow or brown with or without reddish-brown barring, white and pink blooms, or deep red colors.

EXPLANT Axillary buds or meristems from 2- to 5-cm-long shoots from backbulbs or new green bulbs. An explant should be taken before the new shoot is 5 cm long because the meristem ceases to grow at an early stage. If the eyes of a backbulb are poorly developed, place the backbulb in a bag of damp peat moss and allow it to swell before excising the bud.

TREATMENT See *Cattleya*.

MEDIA Modified Knudson's C medium. These orchids may be started on agar medium, but they are usually cultured initially in agitated liquid medium, which is the preferred method (see *Cattleya*). Coconut milk may not be needed for Stage III plantlets.

LIGHT 100–300 f.c. from cool-white fluorescent light. For Stage I, continuous light. For Stages II and III, either continuous light or cycles of 16 hours light/8 hours dark.

TEMPERATURE For initiation, 26°C (79°F); for multiplication, shoot-tip formation, and rooting, 22°C–25°C (72°F–77°F).

REFERENCES Arditti 1977, Hartmann and Kester 2010, Knudson 1946, Mata-Rosas et al. 2011, Morel 1960, 1964, 1974, Murashige and Skoog 1962, Nuraini and Shaib 1992, Sagawa and Kunisaki 1982, Scully 1967, Vacin and Went 1949, van Overbeek et al. 1941, Wimber 1963.

Miscanthus sinensis
CHINESE SILVER GRASS
Poaceae, Grass family

Chinese silver grass (*Miscanthus sinensis*) is widely cultivated in temperate regions around the world and is also a living source of energy in the form of phytomass for combustion and an inexpensive source for large-scale ethanol. It is an herbaceous perennial plant that grows from 80 cm to 2 m (32 in. to 6.5 ft.) tall, forming dense clumps from an underground rhizome. The miscanthus grasses also have ornamental value because of their upright, arching growth habit and plumes of flowers that bloom above the foliage.

EXPLANT Axillary buds from rhizomes.

TREATMENT Remove nodal buds from actively growing stem segments, remove all foliage, and sterilize by dipping in 95% ethanol for one minute and 1:10 bleach with 0.1% Tween for 20 minutes. Rinse 3 times in sterile water.

MEDIA Stages I and II are grown on an MS medium with 1 mg/liter BAP and 0.4 mg/liter NAA solidified with gellan gum. Some success with 8 g/liter agar has also been reported. Stage III uses 0.1 mg/liter NAA in a liquid medium.

Miscanthus media

COMPOUND	STAGES I & II MG/LITER	STAGE III MG/LITER
MS salts	4,628	4,628
Cysteine-HCl	50.0	50.0
NAA	0.4	0.1
BAP	1.0	–
Sucrose	30,000	30,000
Gelrite	2,500	–

pH 5.7 to 5.8

If explant browning occurs, pre-soaking in 150 mg/liter citric acid or ascorbic acid can be used or 3 g/liter activated charcoal can be added to the medium.

LIGHT 300–400 f.c. from cool-white fluorescent light for either 16 hours light/8 hours dark or continuous light.

TEMPERATURE 22°C–26°C (72°F–79°F).

REFERENCES Glowacka et al. 2010a, 2010b, Gubisova et al. 2012, Murashige and Skoog 1962.

Musa acuminata × *M. balbisiana*
BANANA
Musaceae, Banana family

The popular food crop banana was once referred to as its own species, *Musa sapientum*, but it is actually a hybrid cultivar of the wild seeded bananas *M. balbisiana* and *M. acuminata*. It is one of the earliest plants to be domesticated by humans for agriculture around 8000 b.c. and was used for food, fiber, and construction materials. The banana is also an attractive tropical plant that is commonly used in the garden because of its decorative leaves.

Because tropical foliage plants are highly susceptible to systemic fungal and bacterial infections, keep stock plants in a dry, clean, lighted atmosphere for 1–3 weeks before removing explant tissue. This practice reduces the number of microbial contaminants on the explants.

EXPLANT Growing points from rhizomes measuring at least 20 cm (8 in.) long.

TREATMENT Remove sword suckers, approximately 40 cm (16 in.) long, that contain the shoot tip. Treat for 10 minutes in a 2:10 bleach solution with 0.1% Tween 20. Rinse twice in sterile distilled water about 5 minutes each. Aseptically remove the shoot tips from the ensheathing leaf bases and include 2–3 mm of basal stem tissue. Remove meristems by systematically dissecting overlapping leaf bases until the apical meristem is exposed. When the *in vitro* plants have 2 or 3 leaves, they can be transferred directly to a potting medium

MEDIA Stages I and II are grown on an agar medium containing MS salts and additional calcium chloride. Stage III uses half-strength MS plus 0.5 mg/liter NAA. Martin et al. showed that the addition of 50–100 mg/liter calcium chloride after the 6th transfer decreases or eliminates shoot tip necrosis and dieback.

Musa media

COMPOUND	STAGES I & II MG/LITER	STAGE III MG/LITER
MS salts	4,628	2,314
Calcium chloride	100	100
Adenine sulfate	2.0	2.0
Thiamine HCl	0.4	0.4
Myo-inositol	100	100
IAA	2.0	–
NAA	–	0.5
BAP	1.5	–
Coconut milk		
Sucrose	20,000	20,000
Agar	7,000	7,000

pH 5.8

LIGHT 300–400 f.c. from cool-white fluorescent light for either 16 hours light/8 hours dark or continuous light.

TEMPERATURE 25°C–27°C (77°F–81°F).

REFERENCES Drew et al. 1989, Martin et al. 2007, Murashige and Skoog 1962, Sahijram et al. 2003.

Nandina domestica
HEAVENLY BAMBOO
Berberidaceae, Barberry family

Heavenly bamboo (*Nandina domestica*) is not a bamboo, as its name suggests. It is an ornamental, evergreen shrub that is especially noted for red berries and leaves in the fall in cooler areas. Cultivars 'Harbour Dwarf' and 'Compacta' (dwarf) have been successfully tissue cultured. The yellow exudate in culture media is due to the alkaloid berberine, as opposed to many other cultures where the exudates are presumed to be phenolics.

EXPLANT Lateral or terminal buds.

TREATMENT Remove outer leaves from 5-cm-long shoot tips. Wash shoots in warm water with liquid detergent for 10 minutes. Rinse briefly with water. Agitate in 1:10 bleach with 0.1% Tween 20 for 20 minutes. On sterile paper toweling or filter paper, excise buds under a dissecting microscope.

MEDIA Gamborg's B5 medium is used with MS iron for Stages I and II. MS salts are used at one-third strength for Stage III rooting. MS may be used instead of Gamborg's B5 for all stages, but the latter is preferred. Unique to this formula, charcoal is used to help initiation.

LIGHT 600 f.c. from wide spectrum lamps, 16 hours light/8 hours dark. Cool-white fluorescent lamps are generally satisfactory for tissue culture, but because 2 different studies of *Nandina* used wide spectrum lamps, which provide more far-red rays, they are suggested here.

Nandina media

COMPOUND	STAGE I MG/LITER	STAGE II MG/LITER	STAGE III MG/LITER
MS salts	–	–	1,543
Ammonium sulfate	134	134	–
Boric acid	3.0	3.0	–
Calcium chloride	150	150	–
Cobalt chloride	0.025	0.025	–
Cupric sulfate	0.025	0.025	–
Magnesium sulfate • H$_2$O	250	250	–
Manganese sulfate • 4H$_2$O	10	10	–
Potassium iodide	0.75	0.75	–
Potassium nitrate	2,500	2,500	–
Sodium molybdate	0.25	0.25	–
Sodium phosphate	150	150	–
Zinc sulfate	2.0	2.0	–
MS iron	65.1	65.1	–
Inositol	100	100	100
Thiamine HCl	0.4	0.4	0.4
NAA	0.1	0.1	3.0
BA	1.0	1.0	–
Sucrose	30,000	30,000	30,000
Charcoal	1,500	–	1,500

pH 5.7

TEMPERATURE 26°C (79°F).

REFERENCES Akira and Itokawa 1988, Gould and Murashige 1985, Keever and Findley 2002, Matsuyama 1978, Murashige and Skoog 1962, R. H. Smith 1983.

Nephrolepis exaltata
BOSTON FERN
Nephrolepidaceae, *Nephrolepis* family

Ferns are one of the most popular groups of ornamental houseplants grown for their beautiful foliage. Most ferns are commercially propagated through tissue culture from runners or spores. Spore propagation is seasonal and can be more difficult because the spores must go through a gametophyte (sexual) stage. Most Boston ferns (*Nephrolepis exaltata*) are propagated asexually from runners because this technique allows clones to be produced.

Probably more Boston ferns are propagated by tissue culture than any other ornamental crop. Not only can vast numbers of this houseplant be achieved by tissue culture, but also tissue cultured Boston ferns appear healthier, more uniform, and generally more pleasing than those propagated by conventional methods did. With a possible production period of less than 4 months, a million plants per year are within reach of even a modest facility. Two popular cultivars are 'Bostoniensis' and 'Fluffy Ruffles'.

EXPLANT 2.5-cm-long segments of runner tips.

TREATMENT Remove 10 cm of actively growing fern runner tip. Cut into 2.5-cm-long pieces. Stir pieces in 1:10 to 1.5:10 bleach for 15 minutes. In the transfer hood, dip in 70% alcohol. Rinse in 3 changes of sterile distilled water. Cut into 5- to 10-mm-long pieces. Place on agar or in liquid medium. Transfer runner pieces that grow in Stage I to the same medium for Stage II, or chop them up (after removing the larger fronds) and spread on agar medium of the same composition. The latter procedure is a time-saving process for mass production.

Nephrolepis media

COMPOUND	STAGES I & II MG/LITER	STAGE III MG/LITER
MS salts	2,314	2,314
Sodium phosphate	125	12.5
Inositol	500	500
Thiamine HCl	5.0	5.0
BA	1.0	–
NAA	0.1	0.1
Sucrose	20,000	20,000
Agar	8,000	8,000

pH 5.7

MEDIA Use half-strength MS salts with the hormones BA and NAA for Stages I and II. BA is omitted in Stage III.

LIGHT 100–300 f.c. from cool-white fluorescent light for 16 hours light/8 hours dark.

TEMPERATURE 25°C–27°C (77°F–81°F).

REFERENCES Cooke 1977b, 1979, Fernández and Revilla 2003, Hagiabad et al. 2007, Hvoslefeide 1991, Lazar et al. 2010, Loescher and Albrecht 1979, Murashige and Skoog 1962, Oki 1981, R. H. Smith 1992.

Panax ginseng
ASIAN GINSENG
Panax quinquefolius
AMERICAN GINSENG
Araliaceae, Aralia family

Ginsengs are medicinally important plants with many uses in the natural health industry. They are perennial herbs that have palmate leaves, grow to 1 m (3 ft.) in height, and have tuberous roots, which are valued as charms and are the source of a widely used tonic, especially in Asia. Asian ginseng (*Panax ginseng*) and American ginseng (*P. quinquefolius*) are cultivated in the United States, Canada, Korea, China, and Russia. Seed propagation is very slow, taking up to 7 years from seed to harvest, but tissue culture accelerates plant production.

EXPLANT Pith (soft tissue in the center of a root or stem) from 2-cm-long root slices.

TREATMENT Wash the root in running tap water. Agitate in 70% alcohol for one minute followed by a 15- to 20-minute agitated soak in 1:10 bleach with 2 drops of detergent. Rinse in 3 changes of sterile distilled water. Using a cylindrical cork borer, extract 1-cm-wide pieces of pith from the slices and place on medium.

Panax media

COMPOUND	STAGES I & II MG/LITER	STAGE III MG/LITER
MS salts	4,628	4,628
Inositol	100	100
Thiamine HCl	10	10
Nicotinic acid	1.0	1.0
Pyridoxine HCl	1.0	1.0
NAA	0.5	0.5
Kinetin	2.5	–
Casein hydrolysate	100	100
Sucrose	30,000	30,000
Activated charcoal	50	50
Agar	8.0	8.0

pH 5.7 to 5.8

MEDIA Schenk and Hildebrandt's (1972) medium is solid and includes 2.5 mg/liter kinetin and 0.5 mg/liter NAA.

LIGHT 300 f.c. from cool-white fluorescent light for 16 hours light/8 hours dark.

TEMPERATURE 22°C–24°C (72°F–75°F).

REFERENCES Choi et al. 1982, Furuya and Ushiyama 1994, Gamborg et al. 1976, Nitsch and Nitsch 1969, Obae et al. 2011, Uchendu et al. 2011, Wang 1990.

Pelargonium graveolens
SCENTED GERANIUM
Geraniaceae, Geranium family

The scented geranium (*Pelargonium graveolens*), sometimes called rose geranium, is a tender perennial plant treated as an annual. This ornamental makes an excellent house and garden plant because of its pleasing fragrance; the leaves are used to flavor tea and jelly and can be added to apple jelly to create a scent. There are approximately 200 species of *Pelargonium* and different scents such as apple, lemon, coconut, rose, and others can be found in some of the other species. Plants in the geranium family produce oil of geranium that is used in soaps, perfumes, insect repellents, and cosmetics.

Although geraniums can be propagated easily by stem cuttings, this method limits their ability to be produced in large-scale cultivation. Tissue culture is an effective method for rapid multiplication of desirable clones that have high oil content.

EXPLANT Nodal stem explants from stem sections.

TREATMENT Wash the stems in running water for one hour after removing leaves and stipules. Dip explants in 95% ethanol for 30 seconds followed by 1:10 bleach with 0.1% Tween 20 for 10 minutes. Rinse 3 times with sterile water.

MEDIA For Stages I and II, use full-strength MS with 1 mg/liter of BA plus 0.5 mg/liter GA. For rooting in Stage III, use half-strength MS with IAA at 1.0 mg/l; omit the BA and GA.

LIGHT 300–400 f.c. fluorescent lights for 16 hours light/8 hours dark.

TEMPERATURE 22°C–25°C (72°F–77°F).

REFERENCES Arshad et al. 2012, Cassells and Minas 1983, Minas 2007, Murashige and Skoog 1962, Satyakala et al. 1995.

Pelargonium media

COMPOUND	STAGES I & II MG/LITER	STAGE III MG/LITER
MS salts	4,628	2,314
Myo-inositol	100	100
IAA	–	1.0
BA	1.0	–
GA	0.5	–
Sucrose	30,000	30,000
Gellan gum (Phytagel)	2,500	2,500

pH 5.7 to 5.8

Phalaenopsis spp. and hybrids
MOTH ORCHID
Orchidaceae, Orchid family

In recent years, the moth orchid (*Phalaenopsis*) has become one of the most common potted orchids. Its ease of production, long period of bloom, and ability to be forced into flower year-round make it a popular plant. It produces graceful inflorescences that are loaded with large or miniature long-lasting flowers that are available in white, pink, purple, or yellow. *Phalaenopsis* orchids are monopodial (the main stem growth continues indefinitely) and seldom produce side shoots. As a result, they are usually tissue cultured from flower scape nodes in an agar medium.

EXPLANT Flower scape nodal sections or meristems.

TREATMENT OF NODAL SECTIONS With a clean, single-edged razor blade, cut a flower scape into single nodes, leaving 5–10 mm of stem on either side. Using tweezers or fine forceps, carefully remove the scale that covers the node. Agitate for 5 minutes in distilled water with several drops of detergent. In the transfer hood, soak nodal sections for 5 minutes in 1:1 solution of bleach and water. Rinse well in 3 changes of sterile water. Remove the bracts covering the undeveloped buds and make fresh cuts on both ends of the explants. Make a second sterilization in 1:10 bleach solution for 10 minutes followed by another triple rinse in sterile water. Place each section in a culture tube containing 10 ml agar medium, making sure of good contact with the medium.

TREATMENT OF MERISTEMS Remove outer leaves from 3-cm-long shoots. Dip in ethyl alcohol for 2 seconds, and then mix in 1:10 bleach for 15 minutes. Rinse in sterile water. Remove remaining leaves from shoots. Excise meristem consisting of apical dome, 2 leaf primordia, and a cube of tissue, all less than 0.5 mm.

Knudson's C *Phalaenopsis* media

COMPOUND	STAGES I & II MG/LITER	STAGE III MG/LITER
Ammonium sulfate	500	500
Calcium nitrate	1,000	1,000
Magnesium sulfate • H_2O	250	250
Potassium phosphate	250	250
MS minor salts	32.8	32.8
MS iron	65.1	65.1
Sucrose	20,000	20,000
Coconut milk	100 ml	–
Agar	–	6,000

pH 5.8

Vacin and Went *Phalaenopsis* media

COMPOUND	STAGES I & II MG/LITER	STAGE III MG/LITER
Ammonium sulfate	500	500
Calcium nitrate	–	1,000
Calcium phosphate	200	–
Magnesium sulfate • H_2O	250	250
Manganese sulfate • $4H_2O$	7.5	–
Potassium nitrate	525	–
Potassium phosphate	250	250
MS minor salts	–	32.8
MS iron	–	65.1
Ferric tartrate	28	–
Thiamine HCl	0.4	–
Sucrose	20,000	20,000
Coconut milk	100 ml	–
Agar	–	6,000

pH 5.8

MEDIA Grow and multiply buds or meristems in agar or liquid medium, using either modified Knudson's C or Vacin and Went's formula, with coconut milk. The cultures may be started in agitated liquid, but solid agar medium is usually satisfactory and is required for Stage III.

LIGHT 150–200 f.c. from cool-white fluorescent light for 16 hours light/8 hours dark.

TEMPERATURE 24°C–26°C (75°F–79°F).

REFERENCES Arditti et al. 1977, Nuraini and Shaib 1992.

Phlox subulata
MOSS PINK, CREEPING PHLOX
Polemoniaceae, Phlox family

Moss pink (*Phlox subulata*), a popular herbaceous garden perennial that forms an evergreen mat about 15 cm high, provides early color from small, 5-petaled flowers that bloom in rose, pinks, purples, or white. Tissue culture of phlox can help eliminate diseases that are transmitted by the usual crown-division or stem cuttings methods of multiplication.

EXPLANT 0.5- to 1-cm-long shoot tips excised from actively growing shoots.

TREATMENT Remove leaves that are over 1 cm long from 2-cm-long shoots. Wash shoots in running water, then surface-sterilize by immersion in 70% ethanol for 30 seconds followed by 1:10 bleach for 25 minutes. Rinse in sterile distilled water by dipping briefly in 3 separate rinses.

MEDIA MS salts with Nitsch's additives (biotin, glycine, pyridoxine HCl, thiamine HCl, nicotinic acid, and folic acid), and BA. Rooting medium is the same except BA is omitted and NAA is added.

LIGHT 100–200 f.c. from wide-spectrum fluorescent lights for 16 hours light/8 hours dark.

TEMPERATURE 22°C–26°C (72°F–79°F).

REFERENCES Fraga, Alonso, and Borja 2004, Murashige 1974, Murashige and Skoog 1962, Schnabelrauch and Sink 1979, Zhang et al. 2008.

Phyllostachys aurea
GOLDEN BAMBOO, FISH POLE BAMBOO
Poaceae, Grass family

Golden bamboo (*Phyllostachys aurea*) is one of the most popular and common bamboos grown. The plants reach 4.5- to 6-m (15- to 20-ft.) tall and are useful for screening. The culms are strong, straight, and green, eventually turning an attractive yellow if exposed to direct sunlight. The internodes are tightly compressed at the base, giving the culms a tortoise shell–type pattern. Like most *Phyllostachys* species, this one spreads rapidly and can be an aggressive spreader in warmer climates.

EXPLANT Nodal portions of the culm.

TREATMENT Remove nodal portions of the culm about 25 mm (1 in.) in length with a lateral bud. Soak in tap water with several drops of a detergent for 10–30 minutes. Surface sterilize with 70% ethanol for 10 minutes followed by 4:10 bleach with 0.1% Tween 20 for 20–60 minutes. Rinse 3 times with sterile water.

MEDIA Full strength MS medium is used for Stages I and II. MS salts are used at half strength with 3 mg/liter IBA for Stage III rooting. Liquid

Phlox media

COMPOUND	STAGES I & II MG/LITER	STAGE III MG/LITER
MS salts	4,628	4,628
Adenine sulfate	40	–
D-Biotin	–	0.2
Folic acid	0.4	–
Inositol	100	100
Thiamine HCl	0.5	0.5
Nicotinic acid	0.5	0.5
Pyridoxine HCl	0.5	0.5
NAA	–	0.5
BA	5.0	–
Glycine	2.0	2.0
Sucrose	30,000	30,000
Agar	7,000	7,000

pH 5.7

Phyllostachys media

COMPOUND	STAGES I & II MG/LITER	STAGE III MG/LITER
MS salts	4,628	2,314
Myo-inositol	100	100
Thiamine HCl	0.4	0.4
Nicotinic acid	0.5	0.5
Pyridoxine HCl	0.5	0.5
BA	2.5	–
IBA	0.2	3.0
Sucrose	30,000	30,000

pH 5.7

medium has been shown to be most beneficial for bamboo tissue culture. If cultures are static, the explants should not be submerged under the medium.

LIGHT 300–400 f.c. cool-white fluorescent lights for 16 hours light/8 hours dark.

TEMPERATURE 25°C–26°C (77°F–79°F).

REFERENCES Devi and Sharma 2009, Gielis 1995, Murashige and Skoog 1962, Ogita 2005, Ogita et al. 2008

Pinus spp.
PINE
Pinaceae, Pine family

The micropropagation of adult conifer trees can be challenging. The major problem with the process is related to the juvenile versus adult forms of the plant. Media formulations have been developed for several species of pine (*Pinus*). Many people use MS in dilute or other modified forms. Rooting and hardening off are especially difficult with pine cultures. As with other genera, requirements may vary among the species, within species, and from clone to clone. However, micropropagation is making an impact on this economically important genus.

EXPLANT Apical buds or embryos from viable seeds.

TREATMENT OF APICAL BUDS Remove dormant apical buds and sterilize them in 20% bleach for 15 minutes in sterile distilled water. Dissect out any firm white embryos and place on Stage I medium.

TREATMENT OF SEEDS Agitate seeds in 50% bleach for 15 minutes. Rinse 3 times in sterile distilled water. Next, soak the seeds for 40 hours in sterile distilled water, rinse them in 33% bleach for 5 seconds and then rinse again.

MEDIA Low salts and low sucrose prevail in pine media. Some roots will form in Stage II; if not, try rooting in Stage III medium, which is a modified

Pinus media

COMPOUND	STAGE I MG/LITER	STAGE II MG/LITER	STAGE III MG/LITER
MS major salts	–	–	2,314
Ammonium sulfate	200	–	–
Calcium chloride	150	–	–
Calcium nitrate	300	–	–
Magnesium sulfate	250	740	–
Potassium chloride	300	65	–
Potassium nitrate	1,000	80	–
Sodium phosphate	170	170	–
Sodium sulfate	–	200	–
MS minor salts	32.8	32.8	16.4
MS iron	65.1	65.1	32.6
Inositol	10	10	–
Thiamine HCl	1.0	1.0	–
Nicotinic acid	0.1	0.1	–
Pyridoxine HCl	0.1	0.1	–
IBA	–	–	1.0
NAA	2.0	–	1.0
BA	10.0	10.0	–
Sucrose	30,000	30,000	30,000
Agar	7,000	7,000	7,000

pH 5.7 to 5.8

half-strength MS with NAA and IBA. Alternatively, try adding 10 mg/liter of IBA to Stage II medium for 4 weeks, then return to Stage II without the IBA to induce root formation. Activated charcoal at 2 grams/liter has been shown with some *Pinus* species to be beneficial for *in vitro* growth.

LIGHT 300–500 f.c. from cool-white fluorescent light for 16 hours light/8 hours dark.

TEMPERATURE 22°C–25°C (72°F–77°F).

REFERENCES Abdullah et al. 1985, Beaty et al. 1985, de Diego et al. 2010, Pullman et al. 2005, Sommer and Caldas 1981, Thompson and Zaerr 1981, von Arnold and Eriksson 1981.

Platycerium stemaria
STAGHORN FERN, ELKHORN FERN
Polypodiaceae, Polypod family

Staghorn ferns (*Platycerium stemaria* and other species in the genus) are very popular foliage plants because of their uniquely shaped leaves or fronds and their ability to grow well in a variety of locations. They are not parasitic plants; rather, they are epiphytic, growing on another plant but deriving moisture and nutrients from debris on the host plant, from the air, and from rain. Staghorn ferns can be mounted on a piece of bark or wood, on the stems of tree ferns, in moss, or in coarse soil mix or ground bark; their roots penetrate the wood for support. These ferns are often commercially micropropagated by using the spores that are produced on the underside of the fertile fronds. However, because some of the more desirable species in the genus, such as queen elkhorn fern (*Platycerium wandae*), produce very few offshoots (basal shoots) for vegetative propagation, they are good candidates for tissue culture.

EXPLANT 3-mm-long shoot apex from offshoot.

Platycerium media

COMPOUND	STAGES I & II MG/LITER	STAGE III MG/LITER
MS salts	4,628	4,628
Sodium phosphate	170	170
Adenine sulfate	80	–
Thiamine HCl	0.4	0.4
Nicotinic acid	1.0	1.0
Pyridoxine HCl	1.0	1.0
BA	1.0	–
IAA	15	–
Sucrose	30,000	30,000
Agar	8,000	8,000

pH 5.7

TREATMENT Obtain offshoots with fronds less than 5 cm wide. Wash in running tap water for 5 minutes. Remove and discard larger fronds and root mass. Excise 1-cm-long shoot tips. Mix in 1:10 bleach for 10 minutes, and then follow with 3 rinses of sterile water. Remove hairs. Mix in 1:20 bleach for 5 minutes. Excise 3-mm apical dome. Rinse in sterile water. Place on Stage I medium. The explants may turn black overnight, but growth can be expected in 6 weeks. Subdivide and transfer after 2 months. Make further transfers at 3-week intervals.

A time-saving practice for commercial production involves homogenizing Stage III cultures. Place about 40 Stage III plantlets in a sterile blender with 50 ml of sterile water. Blend at low speed for 5 seconds. Aseptically pipet 10-ml portions from the blender onto sterile media in culture dishes, jars, or flasks. In 2 months, transfer the rooted plantlets (possibly as many as 200) to potting mix, tree fern, or sphagnum. Harden off under mist or plastic cover.

MEDIA Modified MS, with IAA included for Stages I and II but omitted in Stage III. Lily (*Lilium*) medium also gives excellent results for tissue culturing staghorn ferns.

LIGHT 100–300 f.c. from cool-white fluorescent light for 16 hours light/10 hours dark.

TEMPERATURE 23°C–25°C (73°F–77°F).

REFERENCES Ambrožič-Dolinšek et al. 2002, Cooke 1979, Hennen and Sheehan 1978, Murashige and Skoog 1962.

Primula vulgaris
PRIMROSE, ENGLISH PRIMROSE
Primulaceae, Primrose family

Primrose (*Primula vulgaris*, syn. *P. acaulis*), a low-growing, herbaceous perennial, is popular as a spring-blooming garden plant and indoors as a potted plant because of its colorful flowers of purple, yellow, red, pink, or white. Other *Primula* species have become popular, including *P. denticulata*, *P. elatior*, and *P. obconica*, with attractive clusters of flowers in a head or umbel at the top of the flower stem. Wilbur Anderson, the pioneer of *Rhododendron* tissue culture, developed the formula for primrose to increase clonal lines for F_1 hybrid seed production. Most primroses are difficult to clean because the shoot tip area is in such close contact with the soil.

EXPLANT Shoot buds or seedlings.

TREATMENT Remove small division from parent plant. Wash well in tap water with a few drops of liquid detergent. Cut away and discard outer leaves. Carefully excise shoot bud with a few millimeters of base. Agitate excised shoot bud or seedling in 1:10 bleach with 0.1% Tween 20 for 20 minutes. Rinse in 1:100 bleach, followed by 3 washes in sterile distilled water.

MEDIA Anderson's inorganics and organics (see *Rhododendron*) specially formulated for double *Primula acaulis*. Half-strength salts are used for Stage III.

LIGHT 100–300 f.c. from cool-white fluorescent light for 16 hours light/8 hours dark.

Primula media

COMPOUND	STAGES I & II MG/LITER	STAGE III MG/LITER
Ammonium nitrate	400	200
Boric acid	6.2	3.1
Calcium chloride	440	220
Cobalt chloride	0.025	0.013
Cupric sulfate	0.025	0.013
Magnesium sulfate • H_2O	370	185
Manganese sulfate • $4H_2O$	16.9	8.5
Potassium iodide	0.30	0.15
Potassium nitrate	480	240
Sodium molybdate	0.25	0.13
Sodium phosphate	380	190
Zinc sulfate	8.6	4.3
Ferrous sulfate	55.7	27.8
Na_2EDTA	74.5	37.3
Adenine sulfate	80	–
Inositol	100	100
Thiamine HCl	0.4	0.4
IAA	1.5	0.5
BA	2.5	–
Sucrose	30,000	30,000
Agar	7,000	7,000

pH 5.7

TEMPERATURE 22°C–25°C (72°F–77°F).

REFERENCES Anderson 1984, Okrslar et al. 2007, Schween and Schwenkel 2002, 2003, Yamamoto et al. 1998.

Prunus spp.
PLUM, PEACH, APRICOT, AND CHERRY
Rosaceae, Rose family

The genus *Prunus* includes plum (*P. domestica*), peach (*P. persica*), apricot (*P. armeniaca*), almond (*P. dulcis*), sweet cherry (*P. avium*), and sour cherry (*P. cerasus*), among other species. More than 400 species in the northern temperate regions of the world are grown for their fruit and as ornamental

Prunus media

COMPOUND	STAGES I & II MG/LITER	STAGE III MG/LITER
MS salts	2,314	2,314
Ascorbic acid	2.0	2.0
Glycine	2.0	–
Inositol	100	–
Thiamine HCl	1.0	–
Nicotinic acid	0.5	–
Pyridoxine HCl	0.5	–
Ferulic acid	1.0	–
IBA	0.5	3.0
BA	0	–
GA	2.0	–
Agar	6,000	6,000
Sucrose	30,000	30,000

pH 5.7 to 5.8

plants. Although the micropropagation procedures for the different *Prunus* species vary, there are similarities that carry across the different species.

EXPLANT Actively growing shoot tips and axillary buds, 1–2 cm long.

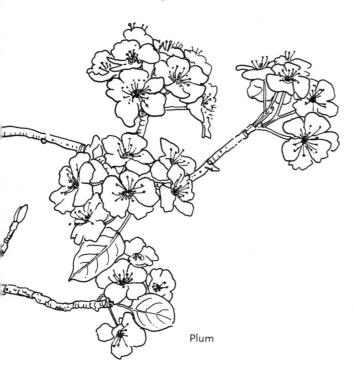
Plum

TREATMENT Obtain 1- to 2-cm-long, actively growing shoot tips. Surface sterilize with 70% ethanol for 30 seconds followed by continuous agitation in 15% bleach and 0.02% Tween-20 solution for 15–30 minutes. Rinse 3 times in sterile distilled water. Trim to 0.5 cm.

MEDIA MS salts are used throughout with BA at 2.0 mg/liter in Stages I and II, but not at Stage III. Rooting can be directly into soil from Stage II, or use Stage III medium, omitting BA and increasing IBA to 3.0 mg/liter.

LIGHT For Stages I and II, 200–300 f.c. from cool-white fluorescent for 16 hours light/8 hours dark.

TEMPERATURE 25°C–26°C (77°F–79°F).

REFERENCES Almehdi and Parfitt 1986, Feliciano and de Assis 1983, Kalinina and Brown 2007, Koubouris and Vasilakakis 2006, Miller et al. 1982, Murashige and Skoog 1962, Parfitt and Almehdi 1986.

Rhododendron spp. and hybrids
RHODODENDRON AND AZALEA
Ericaceae, Heath family

Rhododendron species, which include both rhododendrons and azaleas, are popular garden ornamentals that grow well in acid soil and moderate temperatures. Plants come in a wide range of flower colors and leaf sizes. Mostly evergreen, the more than 1000 species in the genus range from low shrubs to trees. They respond well to tissue culture, a good means of securing large numbers of plants. In the field, tissue cultured plants tend to branch more than plants propagated traditionally by cuttings.

EXPLANT 5- to 7-cm shoot tips of actively growing plants, with lateral buds showing.

TREATMENT Wash 5-cm-long shoot tips in soapy water for 5–30 minutes. Remove leaves and

Rhododendron woody plant media

COMPOUND	STAGES I & II MG/LITER	STAGE III MG/LITER
Ammonium nitrate	400	133
Boric acid	6.2	2.0
Calcium chloride	440	147
Cobalt chloride	0.025	0.008
Cupric sulfate	0.025	0.008
Magnesium sulfate · H_2O	370	123
Manganese sulfate · $4H_2O$	16.9	5.6
Potassium nitrate	480	160
Sodium molybdate	0.25	0.08
Sodium phosphate	380	127
Zinc sulfate	8.6	2.9
Ferrous sulfate	55.7	18.6
Na_2EDTA	74.5	24.8
Adenine sulfate	80	–
Myo-inositol	100	100
Thiamine HCl	0.4	0.4
IAA	1.0	–
IBA	–	5.0
2iP	2.0–5.0	–
Sucrose	30,000	30,000
Agar	7,000	7,000
Charcoal	–	800

pH 5.0 to 5.5

terminal bud. Dip in 70% alcohol for 10 seconds, then rinse briefly in distilled water. Mix in 1:10 bleach with 0.1% Tween 20 for 10–25 minutes, then rinse for one minute in 3 changes of sterile distilled water. Trim away 1 cm from base. Lay tips firmly on agar slant, making sure of good contact between shoot and agar.

Transfer every 2 weeks until growth appears, then every 4–6 weeks. As the lateral buds break, the explant stem should be gradually divided in 2 or 3 transfers. By the time the leaves are completely miniaturized, 3- to 4-fold multiplication can be expected every 6 weeks.

MEDIA Use WPM (see *Kalmia*) or Anderson's *Rhododendron* formula (above), which is a modified MS with lower nitrate and added iron, sodium phosphate, adenine sulfate, 2iP, and IAA. Rooting can be achieved directly from Stage II into peat/perlite (1/1) soil under mist, or use Stage III for a brief conditioning. For Stage III, use one-third-strength-salts, omit 2iP and IAA, and add IBA and charcoal.

LIGHT 100–300 f.c. from cool-white fluorescent light for 16 hours light/8 hours dark.

TEMPERATURE 21°C–25°C (70°F–77°F).

REFERENCES Anderson 1975, 1978, Briggs et al. 1994, Hebert et al. 2010, Kyte and Briggs 1979, Lloyd and McCown 1980, McCown and Lloyd 1983, McCullouch and Briggs 1982, Meyer 1982, Norton and Norton 1985, W. A. Smith 1981, Wong 1981.

Rhododendron

Rosa spp. and hybrids
ROSE
Rosaceae, Rose family

Roses (*Rosa* spp.) are among the most important cut flower crops in the world and are garden favorites in most temperate climates. There are more than 100 species of rose and many more hybrids that are continually being introduced. Some are erect shrubs while others are climbing plants. The large, showy flowers range from red to yellow, white, and pink. Tissue culture is a viable alternative to grafting roses. Unfortunately, many roses are somewhat uncooperative when subjected to tissue culture.

EXPLANT 4- to 5-cm-long actively growing shoot tips from disease-free plants.

TREATMENT Remove all leaves except tiny terminal ones. Wash shoots in water with a few drops of detergent and dip in 70% ethanol for 20 seconds. Sterilize in 1.5:10 bleach for 20 minutes. Follow by 3 rinses in sterile distilled water for 2 minutes each. Trim base to make 1- to 3-cm-long explants. It is possible to omit Stage III, rooting shoots directly from Stage II into a sand/peat (1/1) potting mix. Place rooted plantlets in potting mix under high humidity. Root initials may grow best if plantlets *in vivo* are grown in the dark for a few days.

MEDIA Modified MS agar medium with BA at 2 mg/liter for Stages I and II; omitted for Stage III. The addition of NAA or IBA at 0.1 mg/liter at Stage III is helpful for root initiation.

LIGHT 100–300 f.c. from cool-white fluorescent light for 16 hours light/8 hours dark.

TEMPERATURE 25°C (77°F).

REFERENCES Canli and Skirvin 2008, Carelli and Echeverrigaray 2002, Davies 1980, Hasagawa 1979, 1980, Jabbarzadeh and Khosh-Khui 2005.

Rosa media

COMPOUND	STAGES I & II MG/LITER	STAGE III MG/LITER
MS salts	4,628	4,628
Inositol	100	100
Thiamine HCl	1.0	1.0
Nicotinic acid	0.5	0.5
Pyridoxine HCl	0.5	0.5
BA	2.0	–
NAA	0.1	0.1
Glycine	2.0	–
Sucrose	30,000	–
Agar	7,000	–

pH 5.7

Rubus fruticosus
BLACKBERRY
Rosaceae, Rose family

Blackberry is a bramble with a tasty, edible fruit that can be produced by any of several *Rubus* species, including *R. fruticosus*. The plants typically have biennial canes (stems) and perennial roots. Blackberries are not as popular as raspberries, but are just as delicious and can yield more fruit per bush than raspberries. Some blackberry cultivars even bear fruit after summer raspberries are gone. Blackberries respond well to culture, multiplying 3–6 times every 3–4 weeks.

EXPLANT 5-cm-long shoot tips.

TREATMENT Excise actively growing shoots; remove leaves, except for the youngest ones enclosing the terminal bud. Agitate shoot tips or nodal sections in water with a few drops of detergent for one minute. Dip in 70% ethanol for 30 seconds followed by 20 minutes in 3:10 bleach with 0.1% Tween 20. Rinse briefly in sterile water, followed by 5 minutes each in 3 changes of sterile water. Trim to 2–4 cm. As a new shoot is produced by an explant, subculture the shoot with an attached node of the original explant.

Rubus fruticosus media

COMPOUND	STAGES I & II MG/LITER	STAGE III MG/LITER
MS salts	4,628	4,628
Inositol	100	100
Thiamine HCl	0.4	0.4
IBA	–	2.0
BA	2.0	–
GA$_3$	0.5	–
Sucrose	30000	30,000
Gelrite	3,000	3,000
Charcoal	–	3000

pH 5.7

MEDIA Modified MS with IBA, BA, and GA$_3$. Stage I may be in agitated liquid or on 5 g/liter agar. Use Stage III medium, or root directly from Stage II in fine soil-less potting mix.

LIGHT 200–400 f.c. from warm-white fluorescent lamps for 16 hours light/8 hours dark.

TEMPERATURE 24°C–25°C (75°F–77°F).

REFERENCES Borgman and Mudge 1986, Donnelly et al. 1985, Fernandez and Clark 1991, Jafari Najaf-Abadi and Hamidoghli 2009, Kyte and Kyte 1981, McNicol and Graham 1990, Murashige and Skoog 1962, Wu et al. 2009, Zimmerman and Broome 1980c.

Rubus occidentalis
AMERICAN RASPBERRY
Rosaceae, Rose family

American raspberry (*Rubus occidentalis*), a shrub native to the eastern United States, is valued for its delicious-tasting black or red berries that are high in anthocyanins (antioxidants). Most raspberries (*Rubus* spp.) are difficult to micropropagate. The larger the explant, the sooner it will start growing but the more apt it is to be contaminated. Excised buds may be cleaner than shoots, but they will take longer to grow. In any case, explants take several weeks to react. They should be transferred every 2 weeks or more frequently. Experimenting and fine-tuning with hormone strength, gel strength, and salt concentrations are important to success. Tissue culture is the fastest way to introduce new raspberry cultivars to the trade. Effective management of the stock plants is essential for maximum success of *in vitro* propagation of *Rubus*. The stock plants should be chilled at 4°C (39°F) to stimulate vigorous growth of new sprouts; doing this results in the highest initiation rate.

EXPLANT Stem sections 3–5 cm long or 2- to 3-cm-long shoot tips from active new growth.

TREATMENT Wash shoots in water with 0.1% Tween 20 for 15 minutes. Immerse in 70% ethanol for 5 seconds. Stir in 1:10 bleach with 2 drops of detergent for 10–25 minutes. Treatment time should correspond to tenderness of the shoot. Rinse 3 times in sterile distilled water. Trim to

Rubus occidentalis media

COMPOUND	STAGES I & II MG/LITER	STAGE III MG/LITER
MS salts	2,341	–
Ammonium nitrate	–	400
Calcium chloride	–	440
Magnesium sulfate	–	370
Potassium nitrate	–	480
Sodium phosphate	–	380
MS minor salts	–	32.8
MS iron	–	65.1
Inositol	200	200
Thiamine HCl	1.0	1.0
IBA	0.1	1.0
BA	1.0	–
Sucrose	30,000	15,000
Agar	7,500	7,500
Charcoal	–	2,000

pH 5.7 to 5.8

single-node sections or excise buds. Shade and humidity are critical in growing-on these tissue cultured plantlets. Light seedling mixes will help the transition period, during which the plantlets must grow new leaves before they are able to flourish in a typical greenhouse environment.

MEDIA The initiation medium is half-strength MS. For *in vitro* rooting in Stage III, a modified form of Anderson's medium is recommended.

LIGHT A reduction in light intensity in the culture rooms improves shoot formation, elongation and root development. 100–200 f.c. from cool-white fluorescent light for 16 hours light/8 hours dark.

TEMPERATURE 22°C–25°C (72°F–77°F).

REFERENCES Anderson 1980, Borgman and Mudge 1986, Donnelly et al. 1985, Gebhardt 1985, Kyte and Kyte 1981, McNicol and Graham 1990, Murashige and Skoog 1962, Wu et al. 2009, Zimmerman and Broome 1980c.

American raspberry

Saintpaulia ionantha
AFRICAN VIOLET
Gesneriaceae, Gesneria family

Saintpaulia ionantha is the species most responsible for the hybrid African violets that are available today. Traditional propagation by leaf cuttings is very easy, so it is not surprising that African violets are easily tissue cultured. Large, commercial growers either micropropagate their own plants or buy in micropropagated African violets.

EXPLANT 1 × 1 cm squares of leaf, or 1-cm cross sections of petiole.

TREATMENT Wash the whole leaf and petiole in water with 1% detergent. Agitate in 1:10 bleach with 0.1% Tween 20 for 10 minutes. Rinse in 3 washes of sterile distilled water. Cut leaves into 1 × 1 cm squares and petioles into 2-mm cross sections. Adventitious plants arise quickly in 6–8 weeks.

MEDIA For leaf or petiole sections, use a modified MS with 1.0 mg/liter of NAA and 1.0 mg/liter of BA. Even with no growth regulators in the medium, adventitious shoots and roots will form.

LIGHT For Stages I and II, 300 f.c. from cool-white

Saintpaulia medium

COMPOUND	STAGES I, II & III MG/LITER
MS salts	4,628
Sodium phosphate	170
Inositol	100
Thiamine HCl	0.4
Nicotinic acid	0.4
Pyridoxine HCl	0.4
NAA	1.0
BA	1.0
Sucrose	30,000
Agar	6,000

pH 5.7 to 5.8

fluorescent light for 16 hours light/8 hours dark. For Stage III, reduce light intensity to 50–100 f.c.

TEMPERATURE 22°C–25°C (72°F–77°F).

REFERENCES Bilkey et al. 1978, Cassells and Plunkett 1984, Cooke 1977a, Murashige and Skoog 1962.

Sequoia sempervirens
COASTAL REDWOOD
Cupressaceae, Cypress family

Coastal redwood (*Sequoia sempervirens*) is an extremely long-lived tree mainly distributed in the coastal areas of northern California and southwestern Oregon. It is a valuable big timber species due to its rapid growth and a promising ornamental plant because of its beautiful tree shape and red bark. For a conifer, this tree lends itself to tissue culture with relatively few problems. *In vitro* cultures from stems, embryos, needles, cotyledons, and hypocotyls have been reported. Cleaning and growing-on are the most critical steps.

EXPLANT 1- to 3-cm-long shoot tips and stem segments.

TREATMENT Wash shoot tips in water with a few drops of liquid detergent. Rinse in water, then rinse in 70% alcohol for 1–5 minutes. Mix in 1:10 bleach with 0.1% Tween 20 for 10–20 minutes. Rinse in 1:100 bleach for 10 seconds followed by 3 rinses in sterile water. Boulay suggested removing 4-cm-long shoots from the rooting medium, cutting them just above the medium, soaking them in commercial rooting hormone for 24 hours, and then planting them in perlite/vermiculite (4/1) soil mix.

MEDIA MS salts with supplements and a trace of IAA suffice for Stages I and II. Rooting will occur in Stage III when cultures are transferred to half-strength MS salts, half vitamins, lower sugar, no kinetin, both IAA and IBA as auxins, and charcoal. Gellan gum is satisfactory for Stages II and III.

Sequoia media

COMPOUND	STAGES I & II MG/LITER	STAGE III MG/LITER
MS salts	4,628	2,314
Sodium phosphate	160	–
Adenine sulfate	80	–
Inositol	100	50
Thiamine HCl	0.4	0.2
IAA	0.5	2.0
IBA	–	3.0
Kinetin	2.0	–
Sucrose	30,000	20,000
Gelrite	2,000	2,000
Charcoal	–	600

pH 5.6

LIGHT 100–300 f.c. from cool-white fluorescent light for 16 hours light/8 hours dark. After 4 weeks in rooting, a 2-week period in darkness may help.

TEMPERATURE 22°C–25°C (72°F–77°F).

REFERENCES Ball 1978, Boulay 1979, Liu et al. 2006, Murashige and Skoog 1962, Sul and Korban 1994, 2005.

Solanum tuberosum
POTATO
Solanaceae, Nightshade family

The potato (*Solanum tuberosum*) is a crop of worldwide importance and an integral part of the diet of a large part of the world. Potato production ranks 4th among major food crops, and plants are vegetatively propagated. Because potatoes are very subject to disease, tissue culture is an ideal way to multiply disease-free stock.

The best method of micropropagation is by nodal propagation. Commercially, the propagation of potatoes through meristem tip culture is a routine procedure for the recovery of virus free plants and the production of certified virus free

Solanum medium

COMPOUND	STAGES I, II & III MG/LITER
MS salts	4,628
Inositol	100
Thiamine HCl	100
Nicotinic acid	50
Pyridoxine HCl	50
Glycine	200
Sucrose	30,000
Agar	7,000

pH 5.7

seeds. The establishment of the primary explant requires approximately 8 weeks. Nodal multiplication rates vary from variety to variety, from line to line (a line is all the plants derived from a single tuber), and even from different locations of the node on the shoot (basal nodes vs. apical nodes). With the particular medium given here, roots will develop along with shoots, and therefore a separate rooting medium is not needed. Transfer to soil is readily accomplished in a mist bench or a humidity tent.

A second micropropagation technique for potatoes involves the production of miniaturized tubers (microtubers) on potato shoots cultured *in vitro*. Microtuber production is a very useful method to propagate and store valuable potato stocks. Depending on the cultivar, potato microtuber development can be promoted under short-day photoperiods and a higher level of sucrose to the medium. Field-planted microtubers perform as well as regular seed tubers.

EXPLANT Single nodal segments from apical and axillary buds of sprouted tubers.

TREATMENT Take a clean potato, wash it with soapy water, and scrub the eyes (buds) to remove dirt. Place the tuber in a paper bag and let it sprout in the dark at 21°C (70°F); this may take a couple months if the potato is freshly dug. Remove the sprouts from the tuber and surface sterilize them in 1:10 bleach plus a few drops of detergent for 10–15 minutes. Rinse 2 or 3 times in sterile distilled water for a few minutes each.

MEDIA MS salts with inositol, thiamine, glycine, nicotinic acid, and pyridoxine HCl. For microtuber production, use higher levels of sucrose—8 grams/liter.

LIGHT 300 f.c. from cool-white fluorescent light in cycles of 16 hours light/8 hours dark for shoot production and 8 hours light/16 hours dark for microtuber production.

TEMPERATURE 21°C–23°C (70°F–73°F).

REFERENCES Goodwin and Adisarwanto 1980, Khuri and Moorby 1996, Mohamed and Alsadon 2010, Murashige and Skoog 1962.

Syngonium podophyllum
ARROWHEAD VINE
Araceae, Arum family

Arrowhead vine (*Syngonium podophyllum*) is native to the tropical regions of Central and South America. It is one of the most popular houseplants because of its attractive foliage variegation and tolerance to low light environments. Commercially, arrowhead vine is almost exclusively propagated by tissue culture because of the threat of spreading diseases through traditional cuttings. Interesting variations of arrowhead vines have the potential for economical multiplication through tissue culture.

EXPLANT 1- to 5-mm-long shoot tips.

TREATMENT Obtain 8- to 10-cm-long shoot tips from young plants including the youngest leaf and stem. Defoliate stems to expose the nodal buds and immerse in 70% ethanol for 45–60 seconds. Soak in 2:10 bleach with 0.1% Tween 20 for 25–40 minutes. Rinse well in 3 changes of sterile distilled water. Excise 1–5 mm of shoot apex.

Syngonium media

COMPOUND	STAGE I MG/LITER	STAGE II MG/LITER	STAGE III MG/LITER
MS salts	4,628	4,628	4,628
Sodium phosphate	340	340	–
Inositol	100	100	–
Thiamine HCl	0.4	0.4	–
IAA	1.0	1.0	–
2iP	3.0	2.0	–
Sucrose	30,000	30,000	–
Agar	–	–	7,000

pH 5.8

MEDIA Stages I and II are carried out in stationary or agitated liquid medium with MS salts, sodium phosphate, vitamins, 2iP, and IAA. For Stage III, use MS salts and agar.

LIGHT For Stages I and II, 300 f.c. from cool-white fluorescent light for 16 hours light/8 hours dark; for Stage III, increase light intensity to 800 f.c.

TEMPERATURE 24°C–25°C (75°F–77°F).

REFERENCES Cui et al. 2008, Makino and Makino 1978, Murashige 1974, Murashige and Skoog 1962, Watad et al. 1997, Zhang et al. 2006.

Syringa spp. and hybrids
LILAC
Oleaceae, Olive family

Lilacs (Syringa spp.) are a common and popular ornamental shrub in the temperate regions of the world. This popular plant is not easy to propagate by stem cuttings, but it is easy to micropropagate. The French hybrid lilac (Syringa vulgaris) is among the easiest of the woody plants to establish in culture and micropropagate. Other species, generally identified as shrub lilacs (for example, S. microphylla and S. julianae) are slightly more recalcitrant. The tree lilacs (S. chinensis, S. reticulata, and S. pekinensis) can be successfully micropropagated, but are more difficult to establish and acclimatize to the culture system. However, once established, they perform much like the shrub lilacs, but with slightly lower multiplication rates. Lilacs have excellent axillary shoot proliferation potential in culture.

EXPLANT Actively growing lateral buds and shoot tips.

TREATMENT Most shrub forms of lilac are easy to establish in culture and can multiply rapidly (3 times) with a 4-week subculture cycle. Surface sterilize by dipping in 70% ethanol followed by 20 minutes in 2:10 bleach with a few drops of detergent. Rinse 3 times in sterile water.

Syringa woody plant media

COMPOUND	STAGES I & II MG/LITER	STAGE III MG/LITER
Ammonium nitrate	400	400
Boric acid	6.2	6.2
Calcium chloride	96	96
Calcium nitrate	556	556
Cupric sulfate	0.025	0.025
Magnesium sulfate · H_2O	370	370
Manganese sulfate · $4H_2O$	22.3	22.3
Potassium phosphate	170	170
Potassium sulfate	990	990
Sodium molybdate	0.025	0.025
Zinc sulfate	8.6	8.6
MS iron	65.1	65.1
Myo-inositol	100	100
Thiamine HCl	1.0	0.1
Nicotinic acid	0.5	0.5
Pyridoxine HCl	0.5	0.5
Glycine	2.0	2.0
2iP	1.0	–
NAA	–	0.01–0.1
Sucrose	20,000	20,000
Agar	4,000	4,000
Gellan gum (Phytagel)	1,400	1,400

pH 5.6 to 5.7

MEDIA WPM with 2% sucrose and 0.4 g/liter of ammonium nitrate (NH_4NO_3) (5 mmol) that includes 1 mg/liter 2iP and an agar and a gellan gum (in this case, Phytagel) mix as the gelling agents. Do not use BAP as the cytokinin source for lilac micropropagation. Lilac plants are easy to root *ex vitro* using a mix of peat moss and perlite.

LIGHT 100 f.c. from cool-white fluorescent light for 16 hours light/8 hours dark.

TEMPERATURE 21°C–24°C (70°F–75°F).

REFERENCES McCown 2001, McCown and Lloyd 1983, Murashige and Skoog 1962, Pierik et al. 1988.

Vaccinium corymbosum
HIGHBUSH BLUEBERRY
Ericaceae, Heath family

The fruit of many of the *Vaccinium* species are considered health-promoting foods because of their antioxidant and anti-inflammatory characteristics. Of these, highbush blueberry (*V. corymbosum*) is the most important commercial crop and is very much in demand.

EXPLANT Single-node cuttings during early summer from slightly hardened new growth.

TREATMENT Take softwood cuttings and trim off the leaves, making sure not to damage the axillary buds. Wash the cuttings in tap water and a few drops of detergent. Rinse in running water for 5–10 minutes. Agitate in 70% alcohol for 30–60 seconds. Mix in 3:5 bleach with 0.1% Tween 20 for 20–30 minutes. Rinse in 3 changes of sterile distilled water. Cut about 1- to 2-cm-long single-node stem pieces to have 5 mm of adjoining stem on either side of the axillary bud. Lay explants on agar and press lightly into the agar so that good contact is made between the total length of the stem and the agar. For Stage II, proliferating shoots should be subcultured every 6–8 weeks. Rooting of blueberries is usually easy and Stage III uses the same medium with no cytokinins. If you want to use an auxin in Stage III, use 0.5 mg/liter IBA. Remove from the culture all shoots that are at least 2–3 cm long. Root those shoots on peat/perlite/vermiculite (1/1/1) or milled sphagnum.

MEDIA Half-strength MS salts with vitamins, 2iP, reduced sugar, and agar. Omit 2iP in last transfer before rooting microcuttings directly from Stage II to Stage IV. Anderson's *Rhododendron* formula may be used as an alternative.

LIGHT 200–400 f.c. from warm-white and cool-white fluorescent lights for 16 hours light/8 hours dark.

TEMPERATURE 24°C–26°C (75°F–79°F).

REFERENCES Anderson 1975, 1978, Billings et al. 1988, Eccher and Noè 1989, Kyte and Briggs 1979, Liu et al. 2010, Lloyd and McCown 1980, McCullouch and Briggs 1982, Meiners et al. 2007, Nickerson 1978, Preece and Sutter 1991, Reed and Abdelnour-Esquivel 1991, Zimmerman and Broome 1980b.

Vaccinium media

COMPOUND	STAGES I & II MG/LITER	STAGE III MG/LITER
MS salts	2,314	–
MS vitamins	–	FULL STRENGTH
WPM	–	2,684
Inositol	100	–
Thiamine HCl	0.4	–
2iP	5.0	–
Zeatin	–	2.0
Sucrose	20,000	20,000
Agar	6,000	6,000

pH 5.2

Vitis vinifera
WINE GRAPE
Vitaceae, Grape or vine family

Grapes are a major fruit crop used for fresh fruit, jelly, juice, and wine production. The urgent need for virus-free and crown gall–free grapevines is beginning to be met by tissue culture propagation of *Vitis vinifera*. Grape plants can be rendered disease-free for vineyards, and tiny *in vitro* stock plants can be held in culture. For plants that are already disease-free, larger explants are advised for quicker establishment. Conventional methods of eradicating disease from grapevines is extremely time consuming and costly. Several years are required to propagate a few hundred plants from a "clean" plant, or 2 years to produce field-ready plants by grafting. In contrast, hundreds of tissue cultured plants can be field ready within a year, either on their own roots or by grafting.

EXPLANT 3-node shoot tips.

TREATMENT Cut 5-cm-long actively growing shoot tips then remove and discard the expanded leaves. Surface sterilize in 5% calcium hypochlorite solution plus 0.1% Tween for 15–20 minutes. Rinse 3 times in sterile distilled water. In the transfer hood, cut 3-node tips and place on medium. Tall jars or tubes are preferable to baby-food jars to allow enough space for micro-nodal cuttings for transfer. Cuttings may go directly from Stage II to soil for rooting, but they may be quicker to establish if put through Stage III. For Stage IV, use a peat/vermiculite (1/4) soil-less mix or coarse perlite. Place in high humidity for 3 weeks, and then continue with intermittent mist, gradually reducing humidity.

MEDIA MS medium with additives. Half-strength MS has been shown to be beneficial at Stage III.

LIGHT 300–400 f.c. from cool-white fluorescent light for 16 hours light/8 hours dark.

TEMPERATURE 24°C–28°C (75°F–82°F).

REFERENCES Barlass and Skene 1978, Chee et al. 1984, Harris and Stevenson 1982, Heloir et al. 1997, Krul and Myerson 1980, Lewandowski 1991, Mhatre et al. 2000, Monette 1983, Murashige and Skoog 1962, E. F. Smith et al. 1992.

Vitis media

COMPOUND	STAGE I MG/LITER	STAGE II MG/LITER	STAGE III MG/LITER
MS salts	4,628	4,628	2,314
Inositol	100	100	100
Thiamine HCl	0.4	0.4	0.4
IBA	–	–	0.5
BA	12.0	2.0	–
Sucrose	30,000	30,000	10,000
Agar	8,000	8,000	8,000

pH 5.8

THREE
RESOURCES

Formula Comparison Table

FORMULAS
(AMOUNTS IN MILLIGRAMS PER LITER)

COMPOUNDS	Anderson (1978)	Gamborg et al. (1976)	Gautheret (1942)	Heller (1953)	Hildebrandt, Riker & Duggar (1946)	Hoagland (1950)
Ammonium nitrate (NH_4NO_3)	400	–	–	–	–	–
Ammonium phosphate ($NH_4H_2PO_4$)	–	–	–	–	–	115
Ammonium sulfate ($[NH_4]_2SO_4$)	–	134	–	125	–	–
Boric acid (H_3BO_3)	6.2	3.0	0.05	1.0	3.0	2.86
Calcium chloride ($CaCl_2 \cdot 2H_2O$)	440	150	–	75	–	–
Calcium nitrate ($Ca[NO_3]_2 \cdot 4H_2O$)	–	–	500	–	800	945
Calcium phosphate, tribasic ($Ca_{10}[PO_4]_6[OH]_2$)	–	–	–	–	–	–
Cobalt chloride ($CoCl_2 \cdot 6H_2O$)	0.025	0.025	0.05	–	–	–
Cupric sulfate ($CuSO_4 \cdot 5H_2O$)	0.025	0.025	0.05	0.03	–	–
Magnesium sulfate ($MgSO_4 \cdot 7H_2O$)	370	250	125	250	720	250
Manganese chloride ($MnCl_2 \cdot 4H_2O$)	–	–	–	–	–	1.81
Manganese sulfate ($MnSO_4 \cdot 4H_2O$)	–	–	–	0.1	4.5	–
Manganese sulfate, monohydrate ($MnSO_4 \cdot H_2O$)	16.9	10	3.0	–	–	–
Potassium chloride (KCl)	–	–	–	750	130	–
Potassium iodide (KI)	–	0.75	0.5	0.1	0.375	–
Potassium nitrate (KNO_3)	480	2,500	125	–	160	607
Potassium phosphate (KH_2PO_4)	–	–	125	–	–	–
Potassium sulfate (K_2SO_4)	–	–	–	–	–	–
Sodium molybdate ($Na_2MoO_4 \cdot 2H_2O$)	0.25	0.25	–	–	–	–
Sodium nitrate ($NaNO_3$)	–	–	600	600	–	–
Sodium phosphate ($NaH_2PO_4 \cdot H_2O$)	380	150	125	125	132	–
Sodium sulfate (Na_2SO_4)	–	–	–	–	100	–
Zinc sulfate ($ZnSO_4 \cdot 7H_2O$)	8.6	2.0	0.18	1.0	3.0	–
Ferric sulfate ($Fe_2[SO_4]_3$)	–	–	–	50	–	–
Ferric tartrate ($Fe_2[C_4H_4O_6]_3$)	–	–	–	–	5.0	–
Ferrous sulfate ($FeSO_4 \cdot 7H_2O$)	55.7	27.8	–	–	–	5.0
Ethylenediaminetetraacetic acid, disodium salt (Na2EDTA)	74.5	37.3	–	–	–	–
Adenine sulfate	80	–	–	–	–	–
d-Biotin	–	–	–	–	–	–
Inositol	100	100	–	–	–	–
Nicotinic acid	–	1.0	–	–	–	–
Pyridoxine HCl	–	1.0	–	–	–	–
Thiamine HCl	0.4	10	–	–	–	–
6-(γ,γ-Dimethylallyamino)purine (2iP)	5.0	–	–	–	–	–
IAA	1.0	–	–	–	–	–
Kinetin	–	–	–	–	–	–
Glycine	–	–	–	–	–	–
Agar	6,000	–	–	–	–	–
Sucrose	30,000	30,000	–	–	–	–

Knop (1865)	Knudson (1946)	Linsmaier & Skoog (1965)	Lloyd & McCown (WPM) (1980)	Morel & Muller (1964)	Murashige & Skoog (1962)	Nitsch & Nitsch (1969)	Schenk & Hildebrandt (1972)	Vacin & Went (1949)	White (1963)
–	–	1,650	400	–	1,650	720	–	–	–
–	–	–	–	–	–	–	300	–	–
–	500	–	–	1,000	–	–	–	500	–
–	–	6.2	6.2	–	6.2	10	5.0	–	1.5
–	–	400	96	–	440	166	200	–	–
500–800	1,000	–	556	500	–	–	–	–	300
–	–	–	–	–	–	–	–	200	–
–	–	0.025	–	–	0.025	–	0.1	–	–
–	–	0.025	0.025	–	0.025	0.025	0.2	–	–
125–200	250	370	370	125	370	185	400	250	720
–	–	–	–	–	–	–	–	–	7.0
–	7.5	22.3	–	–	–	25	–	7.5	–
–	–	–	22.3	–	16.9	–	10	–	–
–	–	–	–	1,000	–	–	–	–	65
–	–	0.33	–	–	0.83	–	1.0	–	0.75
125–200	–	1,900	–	–	1,900	950	2,500	525	80
125–200	250	170	170	125	170	68	–	250	–
–	–	–	990	–	–	–	–	–	–
–	–	0.25	0.025	–	0.25	0.25	0.1	–	–
–	–	–	–	–	–	–	–	–	–
–	–	–	–	–	–	–	–	–	16.5
–	–	–	–	–	–	–	–	–	200
–	–	8.6	8.6	–	8.6	10	1.0	–	3
–	–	–	–	–	–	–	–	–	2.5
–	–	–	–	–	–	–	–	28	–
–	25	2.8	27.8	–	27.8	27.8	15	–	–
–	–	37.3	37.3	–	37.3	37.3	20	–	–
–	–	–	–	–	–	–	–	–	–
–	–	–	–	–	–	0.05	–	–	–
–	–	100	100	–	100	100	1,000	–	–
–	–	–	0.5	–	0.5	5.0	5.0	–	0.5
–	–	–	0.5	–	0.5	0.5	0.5	–	0.1
–	–	–	1.0	–	0.4	0.5	5.0	–	0.1
–	–	–	1.0	–	–	–	–	–	–
–	–	1.3	–	–	1.3	0.1	–	–	–
–	–	0.001–10	–	–	0.04–10	–	–	–	–
–	–	–	2.0	–	–	2.0	–	–	3.0
–	–	10,000	6,000	–	8,000	8,000	–	–	–
–	–	30,000	20,000	–	30,000	20,000	–	–	20,000

Notes: These are well-known basic formulas to which are added micronutrients, vitamins, growth regulators, and other compounds, as desired. Omitted are compounds of beryllium, titanium, nickel, and folic acid, all of which are specified in some original formulas but are very rarely used.

Conversion Formulas and Tables

Because some equipment specifications and all media formulas are expressed in metric terms, it is important to be able to read and understand the metric system. Familiarity with metric measurements comes quickly and easily with practice and makes media calculations relatively simple. Length, volume, weight, and temperature are the 4 basic properties for which measurements are taken. In the English system, these are measured in units that are based on different multiples—12 inches in a foot, 4 quarts per gallon, 16 ounces in a pound, and so on. In the metric system, however, all are measured in units based on multiples of 10. The prefixes of metric measurements indicate the multiple. The most common prefixes for metric measurements are as follows:

kilo (k)   = 10^3 = 1,000
centi (c) = 10^{-2} = 0.01 (one-hundredth)
milli (m) = 10^{-3} = 0.001 (one-thousandth)
micro (μ) = 10^{-6} = 0.000001 (one-millionth)
nano (n) = 10^{-9} = 0.000000001 (one-billionth)

In the metric system, length is measured in meters (m), volume in liters (l), weight in grams (g), and temperature in degrees Celsius (°C).

Commonly used units of length, volume, and weight

Length

UNIT	SYMBOL	METRIC	ENGLISH EQUIVALENT
kilometer	km	1000 m	0.62 miles
meter	m	100 cm	3.3 feet
centimeter	cm	10^{-2} m	
2.5 cm			1 inch
millimeter	mm	10^{-3} m	
micrometer [formerly micron]	μm [μ]	10^{-6} m	
nanometer	nm [formerly mμ]	10^{-9} m	
angstrom	Å	10^{-10} m	

Volume

UNIT	SYMBOL	METRIC	ENGLISH EQUIVALENT
liter	l	1000 ml	1.1 quarts
3.8 liters			1 gallon
milliliter	ml	10^{-3} liter	
microliter	μl	10^{-6} liter	
cubic centimeter	cm³	1 ml	

Weight

UNIT	SYMBOL	METRIC	ENGLISH EQUIVALENT
kilogram	kg	1000 g	2.2 pounds
gram	g	1000 mg	0.035 ounces
454 g			1 pound
milligram	mg	10^{-3} g	
microgram	μg	10^{-6} g	
nanogram	ng	10^{-9} g	
picogram	pg	10^{-12} g	

Temperature

Celsius   °C   °C = 5/9 × (°F−32)
Fahrenheit °F   °F = (9/5 × °C) + 32

- 0°C = 32°F, water freezes
- 22°C = 72°F, room temperature
- 37°C = 98.6°F, body temperature
- 100°C = 212°F, water boils

Concentration

Converting concentrations of selected auxins and cytokinins from µM (micromoles) to mg/L (milligrams per liter).

					MG/L VALUES				
Auxin/Cytokinin:	BAP	TDZ	2iP	KIN	ZEATIN	NAA	IAA	IBA	2,4D
Molecular weight:	225.2	220.25	203.3	215.2	219.3	186.2	175.2	203.2	221
µM									
0.01	0.0023	0.0020	0.0020	0.0022	0.0022	0.0018	0.0018	0.002	0.0022
0.05	0.0113	0.011	0.0102	0.0108	0.011	0.009	0.0088	0.0101	0.011
0.50	0.1126	0.11	0.1017	0.1076	0.1097	0.093	0.0876	0.1016	0.1105
1.00	0.2252	0.22	0.2033	0.2152	0.2193	0.1862	0.1752	0.2032	0.221
2.50	0.563	0.551	0.5082	0.538	0.5483	0.4655	0.438	0.508	0.5525
5.00	1.126	1.10	1.0165	1.076	1.0965	0.931	0.876	1.016	1.105
10.00	2.252	2.2	2.033	2.152	2.193	1.862	1.752	2.032	221
25.00	5.63	5.5	5.0825	5.38	5.4825	4.655	4.38	5.08	5.525
50.00	11.26	11	10.165	10.76	10.965	9.31	8.76	10.16	11.05
250.00	56.3	55.06	50.825	53.8	54.825	46.55	43.8	50.8	55.25
500.00	112.60	110.125	101.65	107.6	109.65	93.1	87.6	101.6	110.5

BAP = 6-benzylaminopurine; $C_{12}H_{11}N_5$
TDZ = Thidiazuron; 1-Phenyl-3-(1,2,3-thidiazol-5-yl)urea; $C_9H_8N_4OS$
2iP = IPA; 6-[γ, γ-dimethylallylamino] purine; $C_{10}H_{13}N_5$
KIN = Kinetin; 6-furfurylaminopurine; $C_{10}H_9N_5O$
ZEATIN = 6-[4-hydroxy-3-methylbut-2-enlyamino] purine; $C_{10}H_{13}N_5O$
NAA = 1-naphthaleneacetic acid; $C_{12}H_{10}O_2$
IAA = indole-3-acetic acid; $C_{10}H_9NO_2$
IBA = indole-3-butyric acid; $C_{12}H_{13}NO_2$
2,4D = 2,4-dichlorophenoxyacetic acid; $C_8H_6Cl_2O_3$

The Microscope: Introduction, Use, and Care

A microscope is often necessary to allow you to visualize particles too small to be seen with the naked eye. This is particularly important when dealing with microbial contaminants because most bacteria, yeast, and fungal spores range in size from 2 to 10 mm. The limit of visibility with the naked eye is about 100 mm (0.1 mm).

When properly illuminated, a microscope is able to provide both magnification and resolution.

The value of a microscope depends not so much on the amount of magnification, but more on the ability of the instrument to allow the viewer to clearly separate and distinguish (that is, resolve) 2 objects in close proximity. By definition, *resolution* is the minimum distance that can exist between 2 points for them to be observed as separate particles. Due to its better resolution, the simple microscope invented by Antonie van Leeuwenhoek was

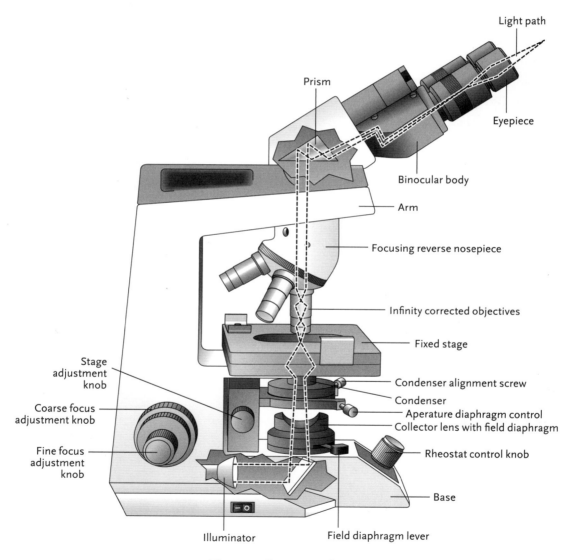

The parts of a compound microscope.

superior to Robert Hooke's compound microscope even though Hooke's provided greater magnification. The lenses employed by Hooke were optically flawed, which caused the images to blur. We now know that objective lenses with chromatic (color) and spherical distortions will diminish resolution. Thus, the best image is not the one that is largest, but the one that is clearest.

The optical and mechanical features of a microscope are shown in the accompanying illustration. The fixed stage of a microscope has a clamp for holding a glass specimen slide on the surface of the stage and which allows movement of the slide around the surface of the stage. The movement is actuated by 2 adjustment knobs located below the stage, one for horizontal movement and the other for vertical movement. Coarse and fine focus adjustment knobs bring the slide specimen into focus. The 2 main operating systems of a compound microscope are optics and illumination.

Optical System

The table following compares the major characteristics of the 3 most commonly used objectives. From the table you can see that the magnification of the oil immersion objective is almost 10 times greater than the magnification of the low-power objective. The total magnification is the product of the magnification of the ocular lens (usually 10×) and the magnification of the objective lens. For example, a 45× objective used with a 10× ocular magnifies the object on the slide 450 diameters—a diameter is the unit of measure used to describe the magnifying power of a lens.

The working distance is the distance, in millimeters, from the tip of the objective lens, when in focus, to the principal point of focus. You can see that the working distance of the oil immersion lens is dangerously close to the coverslip surface.

The iris diaphragm on a microscope is similar to the iris of the eye, which opens in diminished light and closes with excess light. With respect to the microscope, the objective lens being used

Comparison of three common microscope objectives

OBJECTIVE	LOW	HIGH DRY	OIL
Focal length	16 mm	4 mm	1.8 mm
Magnification*	10×	45×	96×
Working distance (mm)	7.0	0.6	0.15
Iris diaphragm opening	Least	Some	Greatest
Numerical aperture (N.A.)*	0.25	0.66	1.25

*These figures are inscribed on the outside surface of each objective.

determines how much to open or close the iris diaphragm. The oil immersion objective requires the largest iris diaphragm opening because the bottom tip of the lens is very small and admits little light. The light is also conserved by using a drop of lens immersion oil on the surface of the coverslip. The lens immersion oil deflects into the lens light rays that otherwise would be diffracted (bent outwards from the lens opening) in the absence of the oil.

The numerical aperture relates directly to the resolving power of the objective, but not to magnification. The oil immersion objective has a resolving power 5 times greater than that of the low-power objective.

Illumination System

The best resolution and magnification on a microscope can only be attained with optimal illumination. Centering the light on the specimen is especially important. On less-expensive microscopes, the centering is provided. In microscopes used for research purposes, the proper centering of light becomes critical. This is achieved by using the condenser alignment screw (shown in the diagram).

The field diaphragm lever adjusts the amount of light entering the condenser system. Move the condenser knob (not shown) up and down to achieve the best focus. The amount of light entering the condenser system can be further adjusted with the condenser aperture diaphragm control. The amount of light emitted by the condenser should be tailored to the size of the microscope objective lens. Be cautioned: too much light results in glare from stray light, which reduces resolution.

Microscope Operating Procedure

1. Plug in the microscope light source or illuminator. If you have a voltage transformer, adjust it to 3 volts for the low-power objective and increase to 6 volts for the higher power objectives.
2. Cover a stained slide with a coverslip. Place the slide, *specimen side up*, in the stage slide holder.
3. Using the stage adjustment knobs, position the specimen over the center of the light coming from the condenser.
4. Raise the objective nosepiece using the coarse focusing knob and rotate the low-power objective (10×) into place with the nosepiece.
5. While observing from the side, use the coarse focus adjustment knob to lower the low-power objective as far as it will go.
6. Using the condenser knob, lower the condenser about 0.1 mm (approximately paper thickness) below the highest position.
7. While looking through the ocular, lower the objective using the coarse focus adjustment knob until what appears as a blur comes in the microscope field. Complete the focusing process using the fine focus adjustment knob. If you are using a binocular microscope, focus first with one eye only, then close that eye and focus the specimen with the other eye using the ocular adjustment located on the eyepiece. While looking through the oculars with both eyes, turn the knob located between the oculars to adjust the distance between the 2 oculars so that they agree with the distance between your eyes. High-eyepoint oculars are available for people who wear glasses.
8. For final light adjustment, remove one of the oculars and look through the ocular tube. Adjust the condenser aperture diaphragm control until the edges of the diaphragm just disappear from view. This adjustment, often neglected, offers the best resolution.
9. Look around for an area you wish to study more intensively. Switch to the next higher objective (high dry, 45×). If your microscope is parfocal (the objectives can be changed without changing the focus), the specimen will either already be in focus or almost in focus, requiring only fine focus adjustment. If using a microscope that is not parfocal, you will need to repeat steps 4, 5, and 7 for the higher objective. For optimum resolution, step 8 should be repeated for each objective.
10. The highest magnification available for viewing a specimen is the oil immersion objective (100×). First, rotate the high dry objective to one side so that a drop of lens immersion oil can be placed on the coverslip.
11. If your objectives are parfocal, rotate the oil immersion objective lens into the oil. If done carefully, the objective lens will not come into contact with the coverslip surface. Use only the fine focus adjustment knob, taking care not to hit the slide. If your objectives are not parfocal, it will be necessary to look from the side—not through the ocular—while carefully lowering the oil immersion objective into the oil using the coarse focus adjustment knob. Now looking through the ocular, use the fine focus adjustment knob for final focusing. Newer microscopes usually have spring-loaded objectives and/or nosepieces that help prevent damage to

the lens when the objective touches the slide. Nevertheless, great care is required.

12. When you are finished observing the specimen, wipe excess oil from the oil immersion objective lens with lens paper and, if necessary, clean the ocular with lens paper moistened with distilled water. This is especially important in hot weather when sweat from your eyes, acidic in nature, contacts the ocular lens. The presence of other foreign particles on the ocular can be determined by rotating the ocular lens manually while looking through the microscope. Any dirt on the ocular lens also will rotate as you rotate the lens, thus indicating the presence of dirt. If there is dirt on the ocular, clean both the upper and lower surfaces of the ocular lens. If any dirt remains after cleaning, that is usually evidence either of a scratched lens surface or of dirt particles on the inside surface of the lens.

Microscope care

- Always remove excess oil from the oil immersion lens before storing.
- Before storing the microscope, rotate the nosepiece to the lowest objective and raise it to the highest position with the coarse adjustment knob.
- Use a dust cover and a box for storage.
- Always transport the microscope using both hands, one holding the arm of the microscope and the other under the base.
- Do not force microscope instrument knobs. Consult a specialist about any problems.
- Do not attempt to clean the *inside* of objective lenses. This also requires a specialist.

Professional Organizations

American Society for Horticultural Science (ASHS)
Web: ashs.org
Publications: *HortScience, HortTechnology,* and *Journal of ASHS* (ISSN 003-1062)

International Association for Plant Biotechnology (IAPB)
Web: iapbhome.com
Publication (with SIVB): *In vitro Cellular and Developmental Biology—Plant*

The International Plant Propagator's Society
Web: ipps.org
Publication: *Proceedings of the International Plant Propagator's Society*

Society for In Vitro Biology (SIVB)
Web: sivb.org
Publication (with IAPB): *In vitro Cellular and Developmental Biology—Plant*

Suppliers

Agdia
30380 County Road 6
Elkhart, Indiana 46514, U.S.A.
800-622-4342
www.agdia.com
ELISA and ImmunoStrips test kits for plant pathogens

American Laboratory
395 Oyster Point Blvd., #321
S. San Francisco, California 94080, U.S.A.
650-243-5600
www.americanlaboratory.com
Print and on-line magazine featuring equipment, instrumentation, and applications for chemistry and life sciences

Beaver Plastics
7-26318-TWP Road 531A
Acheson, Alberta, Canada T7X5A3
888-453-5961
www.spencer-lemaire.com
Planter containers for growing on, subirrigation systems

Caisson Labs
1740 N. Research Park Way
N. Logan, Utah 84341, U.S.A.
877-840-0500
www.caissonlabs.com
Plant research and tissue culture reagents

Carolina Biological Supply Company
2700 York Road
Burlington, North Carolina 27215, U.S.A.
800-334-5551
www.carolina.com
Lab supplies and equipment, educational kits, teacher resources

Cleanroom Filters and Supplies
177 N. Commerce Way
Bethlehem, Pennsylvania 18017, U.S.A.
800-334-9142
www.hepafilters.com
HEPA filters, laminar flow benches

Consolidated Plastics
4700 Prosper Drive
Stow, Ohio 44224, U.S.A.
800-362-1000
www.consolidatedplastics.com
Autoclavable containers and jars

Dent-Eq
5124 Roxborough Dr.
Hermitage, Tennessee 37076, U.S.A.
615-513-8842
www.dent-eq.com
Glass bead sterilizers

Edmund Scientifics
532 Main Street
Tonawanda, New York 14150, U.S.A.
800-728-6999
www.scientificsonline.com
Science and lab products with a focus on educators

EMD Millipore
290 Concord Road
Billerica, Massachusetts 01821, U.S.A.
781-533-6000
www.millipore.com
Water filtration and many other life science products

Eppendorf North America
102 Motor Parkway
Hauppauge, New York 11788, U.S.A.
800-645-3050
www.eppendorf.com
Wide range of products for the life sciences; micropipette specialists

Fungi Perfecti
P.O. Box 7634
Olympia, Washington 98507, U.S.A.
800-780-9126
www.fungi.com
Sterile culture and growing room equipment and supplies, desktop laminar flow hoods

Grainger
100 Grainger Parkway
Lake Forest, Illinois 60045, U.S.A.
www.grainger.com
847-535-1000
Motors

Growers Supply
1440 Field of Dreams Way
Dyersville, Iowa 52040, U.S.A.
800-245-9881
www.growerssupply.com
Greenhouse structures, environmental controls, growing supplies

Hemco Corporation
711 S. Powell Road
Independence, Missouri 64056, U.S.A.
800-779-4362
www.hemcocorp.com
Clean rooms, HEPA filters, lab furniture

International Greenhouse Company
1644 Georgetown Road
Danville, Illinois 61832, U.S.A.
888-281-9337
www.greenhousemegastore.com
Greenhouse structures, environmental controls, growing supplies

Labconco Corporation
8811 Prospect Avenue
Kansas City, Missouri 64132, U.S.A.
800-821-5525
www.labconco.com
Water purification systems

Laboratory Equipment
P.O. Box 3574
Northbrook, Illinois, 60065, U.S.A.
847-559-7560
www.laboratoryequipment.com
Print and on-line magazine featuring equipment information, vendors; free subscription

Leica Microsystems
1700 Leider Lane
Buffalo Grove, Illinois 60089, U.S.A.
800-248-0123
www.leica-microsystems.com
Microscopes

Liberty Industries
133 Commerce Street
East Berlin, Connecticut 06023, U.S.A.
800-828-5656
www.liberty-ind.com
Laminar airflow hoods, cleanrooms, HEPA filters

Li-Cor
4647 Superior Street
Lincoln, Nebraska 68504, U.S.A.
800-447-3576
www.licor.com
Light meters, other measurement instruments

Life Technologies
3175 Staley Road
Grand Island, New York 14072, U.S.A.
800-955-6288
www.lifetechnologies.com
Media and more life science products

LumiGrow
33 Commercial Blvd., Suite A
Novato, California 94949, U.S.A.
800-514-0487
www.lumigrow.com
LED grow lights for grow rooms and greenhouses

Magenta Corporation
3800 N. Milwaukee Avenue
Chicago, Illinois 60641, U.S.A.
773-777-5050
www.magentallc.com
Culture containers (sold through third-party distributors)

New Brunswick Scientific Company
44 Talmadge Road
Edison, New Jersey 08818, U.S.A.
800-448-2298
www.nbsc.com
Rotators, shakers, automatic preparation equipment

Northern Steel
600 Andover Park E.
Seattle, Washington 98188, U.S.A.
800-562-1086
Slotted angles for shelving supports

***Phyto*Technology Laboratories**
P.O. Box 12205
Shawnee Mission, Kansas 66282, U.S.A.
888-749-8682
www.phytotechlab.com
Products for tissue culture and plant biotechnology; extensive technical information

Plantmedia
Bio-World
4150 Tuller Road, Suite 228
Dublin, Ohio 43017, U.S.A.
800-860-9729
www.plantmedia.com
Products for tissue culture and plant biotechnology

Research Organics
4353 E. 49th Street
Cleveland, Ohio 44125, U.S.A.
800-321-0570
www.resorg.com
Biochemicals, Gellan (Gelrite), agar

Scientific Supply and Equipment
619 S. Snoqualmie
Seattle, Washington 98108, U.S.A.
800-441-0088
www.scientificsupplyandequipment.com
Chemicals, supplies, and equipment

Sigma-Aldrich
3050 Spruce Street
St. Louis, Missouri 63103, U.S.A.
800-521-8956
www.sigmaaldrich.com
Chemicals, tissue culture media and supplies, cell culture, biotech and indexing supplies, books

SKC
863 Valley View Road
Eighty Four, Pennsylvania 15330, U.S.A.
800-752-8472
www.skcinc.com
Air sampling pumps and accessories

Terra Universal
800 S. Raymond Avenue
Fullerton, California 92831, U.S.A.
714-578-6000
www.terrauniversal.com
Clean room handling and products, laboratory equipment, manuals and books

Thermo Fisher Scientific
75 Panorama Creek Drive
Rochester, New York 14625, U.S.A.
800-625-4327
www.thermoscientific.com
Wide range of laboratory equipment and supplies

Voigt Global Distribution
P.O. Box 1130
Lawrence, Kansas 66044, U.S.A.
877-484-3552
www.vgdllc.com
Lab equipment and supplies; distributor for DIFCO microbiological supplies

VWR
100 Matsonford Road
Building 1, Suite 200
Radnor, Pennsylvania 19087, U.S.A.
www.vwr.com
800-932-5000
Supplies and equipment

Ward's Natural Science Establishment
5100 W. Henrietta Road
Rochester, New York 14692, U.S.A.
800-962-2660
www.wardsci.com
Biological supplies, tissue culture supplies

Glossary

ABSCISSION LAYER a thin-walled layer of cells that forms for the normal separation of leaves or fruit from plants

ADVENTITIOUS growing from unusual locations, such as aerial roots from stems; or the formation of buds at locations other than nodes or stem tips

AERATION the process of mixing, shaking, or venting a culture to provide it with air

AGAR a polysaccharide gel derived from certain red algae. Agar is used for making semi-solid or gelatinous media for growing tissue cultures and microbes.

ALKALOID any of a number of colorless, crystalline, bitter organic substances with alkaline properties. Many are found in plants; they can have toxic effects on humans. Caffeine, quinine, and strychnine are examples of alkaloid substances.

AMINO ACID a nitrogenous organic compound that serves as a structural unit of proteins.

ANTHER CULTURE the *in vitro* culture of anthers to obtain haploid plants

ANTIOXIDANT a substance used to prevent phenolic oxidation, the browning of cultures because of bleeding phenolic exudates

APEX, APICES (pl.), apical (adj.) tip

APICAL DOMINANCE the inhibition of lateral growth because of auxins produced by the apical bud

APICAL MERISTEM the new, undifferentiated tissue at the microscopic tip of a shoot or root. The apical meristematic dome of stem tips, which is usually about 0.5 mm or less, along with the accompanying leaf primordia, are commonly used as explants.

ARTIFICIAL SEED a somatic embryo that has been coated or encapsulated and grown as true seed

ASEPTIC free from microorganisms. The terms *sterile technique* and *aseptic technique* are used somewhat interchangeably.

ASEXUAL without sex, vegetative

ATOM the smaller particle of an element that retains the chemical characteristics of that element

ATOMIC WEIGHT a system of relative weights of atoms

AUTOCLAVE a vessel for sterilizing using steam under pressure. It is usually distinguished from a pressure cooker by its size and design. Also the verb form, to sterilize in such a vessel.

AUTOTROPHIC able to produce one's own food from inorganic substances

AUXINS a group of growth hormones associated with root initiation, lateral bud inhibition, and cell enlargement

AXIL the point of attachment of lateral growth with the stem

BALLAST a type of transformer wired into fluorescent lamps

BIOSYNTHESIS the formation of chemicals by living cells

BLEEDING the purplish black or brown coloration occasionally found in media due to phenolic products given off by explants or transfers

BRIDGE a piece of filter paper or other device placed within a test tube of liquid medium to hold the culture out of the liquid. Also known as a raft or float.

CALLUS a proliferating mass of disorganized, mostly undifferentiated or undeveloped cells

CALLUS CULTURE the multiplication of callus cells in sterile culture

CAMBIUM the thin layer of tissue between the wood and bark of a stem

CARBOHYDRATE any of certain organic compounds composed of the elements carbon, hydrogen, and oxygen, including sugars, starches, and cellulose

CARCINOGEN any substance that causes cancer

CATALYST a substance that promotes the rate of a chemical reaction but which is not itself changed chemically

CELL the basic physical unit of living organisms

CELL CULTURE the multiplication *in vitro* of single cells or clumps of cells not organized as tissues; often included in the broad term *tissue culture*

CELL SUSPENSION CULTURE the culture of one or more cells suspended in a liquid medium

CENTRIFUGE a device for spinning test tubes of liquid cultures for separating particles, by density or size, via centrifugal force. Also the verb form, to separate particles via centrifugal force.

CHIMERA a plant that contains tissues of more than one genotype, such as certain plants with variegated foliage

CHLOROSIS the absence of green pigments in plants due to lack of light, nutrient deficiency, or disease. Chlorotic plants often have a yellowed appearance.

CHROMOSOME the structure in a cell nucleus that carries the genetic material contained within genes.

CLONE a plant produced asexually from a single plant, and thus of genotype identical to one another and to the parent plant

COCONUT MILK, COCONUT WATER the liquid endosperm of coconuts

COENZYME an organic substance that activates an enzyme or accelerates its action

COLEOPTILE a protective membrane surrounding the primary leaf of grass seedlings. It is noted for its auxin content, as demonstrated by Fritz Went's classic experiment.

COMPOUND a substance made up of 2 or more elements chemically combined in fixed proportions

CONDUCTIVITY a measure of the ability of water to conduct electricity. The conductivity level indicates the concentration of ions dissolved in the water, and thus the purity of the water. It is often measured in units called siemens (S).

CONTAMINANT any unwanted microorganisms that appear in tissue culture media and overgrow, inhibit, poison, or otherwise pollute the desired culture

COTYLEDON the first leaf or seed leaves. Monocots (monocotyledons) have one cotyledon and dicots (dicotyledons) have 2.

CROWN the often-fleshy base of a plant where the stem joins the root

CRYOPRESERVATION the storage of cells or whole tissue at subzero temperatures

CULTIVAR a named plant variety in cultivation

CULTURE a plant or plant part growing *in vitro*; to grow a plant *in vitro*.

CYTOKININS a group of growth hormones that induce bud formation and shoot multiplication and program cell division and enlargement

DECIDUOUS leaf fall at a particular season or stage of growth, as in trees and shrubs that lose their leaves for the winter, as opposed to evergreen

DEDIFFERENTIATE to revert to a nonspecialized or undifferentiated state, as in cells or tissues

DEIONIZE to remove soluble minerals and certain organic salts from water by passing the water through an ion-exchange unit

DICOTYLEDON (dicot) a flowering plant that contains 2 cotyledons (seed leaves). Dicotyledons are members of the class Dictyledoneae and include trees, shrubs, and many herbaceous plants.

DIFFERENTIATE to initiate the growth of new and varied tissues or organs for specialized functions. The process is known as differentiation.

DIPLOID having 2 sets ($2n$) of chromosomes, which is typical of vegetative (somatic) cells

DISINFECT to destroy pathogenic microorganisms using chemical agents (disinfectants)

DISTILL to remove dissolved minerals, particulates, and organic matter from water by boiling the water and then passing the steam and vapor through a condensing coil, from where the recondensed, distilled water is collected in a container

DMSO (dimethyl sulfoxide) a toxic solvent that is sometimes used in minute quantities to dissolve certain organic compounds for plant tissue culture media. It readily passes through skin.

DORMANCY a period in which there is no growth or little physiological activity; a resting stage. Dormancy can sometimes be broken by cold treatment or by applying gibberellic acid (GA3).

ELEMENT a substance composed of atoms that are of the same type and that cannot be separated into simpler substances by any usual chemical means

EMBRYO CULTURE the *in vitro* culture of embryos excised from seeds, or embryos induced to form from somatic cells (*see* somatic embryogenesis)

EMBRYOGENESIS the formation of an embryo, either in the seed (sexual) or in culture (asexual)

EMBRYOID an embryo-like somatic structure that develops in some cell and callus cultures and is capable of developing into an embryo and, subsequently, a whole plant

ENDOSPERM the nutritive tissue or liquid surrounding a developing embryo in a seed

ENZYME a specialized protein that acts as an organic catalyst in specific chemical reactions

EPINASTY an abnormal downward bending of a leaf due to the more rapid growth of the upper surface. It may be an indication of a nutritional deficiency or imbalance.

ETIOLATE to develop long, spindly growth with low chlorophyll. Etiolation results from insufficient light or lack of light

EXCISE to remove by cutting

EXPLANT the part of a plant used to start a culture

EXUDATE a substance that bleeds (exudes) into a medium from a culture, often brown or purplish black lethal phenolics

FLAMING the procedure of sterile technique whereby an instrument is dipped in alcohol and then set on fire to burn off the alcohol and further sterilize the instrument

FOOT-CANDLE (f.c.) a unit of measure of light intensity, based on the illumination on one square foot of surface that is one foot away from a standard candle. One foot-candle is roughly equivalent to 10 lux (0.093 f.c. = 1 lux). This unit is obsolete in academia, having been replaced by photosynthetically active radiation (PAR) units such as micromoles per square meter per second (mmol m-2 sec-1), microeinsteins per square meter per second (mE m-2 sec-1), or watts per square meter (wm-2). The unit foot-candle is retained in this work for its simplicity and because foot-candle–reading meters are inexpensive and readily available.

FORMULA (1) a chemical compound expressed in letters, numbers, or symbols to indicate its composition; (2) the ingredients, or recipe, for tissue culture media

GAMETE the male or female reproductive cell; also known as germ cells and as microspore or pollen (male) and megaspore or ovule (female). Gametes have a haploid supply of chromosomes (n) and are the product of meiosis.

GENES a segment of base pairs in a DNA molecule that codes for particular protein or proteins determining hereditary characteristics of the organism

GENETIC ENGINEERING the manipulation of genes and chromosomes to vary genetic characteristics

GENETIC INSTABILITY the tendency of cells to mutate

GENETIC VARIATION any variation in characteristics as a result of genetic inheritance or mutation

GENOTYPE the genetic make-up of an organism

GERM CELL *see* gamete

GIBBERELLINS a group of growth regulators that influence cell enlargement

GROWTH REGULATOR any organic compound that influences growth and multiplication, such as cytokinins and auxins. *See also* hormones

HABITUATED a state wherein a culture no longer responds to a substance, such as a hormone or nutrient, because the culture no longer requires that hormone or nutrient or is unable to utilize it

HAPLOID the number of chromosomes (n) in reproductive or germ cells, which is half the normal number of chromosomes ($2n$) in vegetative cells

HARDENING OFF the process of gradually acclimating plants to tougher environmental conditions, most notably the less-humid conditions of the field or greenhouse than is experienced in culture or during rooting

HEPA high efficiency particulate air filter, an essential component of a laminar airflow hood to help ensure sterile transfer work

HERBACEOUS PLANT a seed plant that lacks woody tissue

HETEROTROPHIC requiring external organic substances as sources of food and nutrients

HOOD *see* transfer chamber

HORMONE any naturally occurring or synthetic chemical, such as cytokinins and auxins, which affects growth and development. *See also* growth regulator.

HOT PLATE/STIRRER an electrical device that serves as both a hot plate and a magnetic stirrer; the heating and stirring features can be used simultaneously or independently of one another

HYBRID a plant created from a cross between 2 genetically different plants by means of sexual reproduction, or artificially by protoplast fusion and other methods of gene transfer in genetic engineering

HYDRATE a compound with chemically bound water. *See* water of hydration.

HYPOCOTYL the portion of a seedling stem below the cotyledons and above the roots

INCUBATE to grow under favorable conditions, usually in an incubator or other cabinet providing controlled environmental conditions

INDEXING the testing of plants for the presence of pathogens or contaminants

INDICATOR PLANT a plant that readily expresses the symptoms of a particular disease, onto which a part of another plant is grafted to determine if the latter plant has the suspect disease

INORGANIC CHEMICAL not organic, usually lacking carbon

INTERGENERIC said of a cross between 2 or more different genera

INTERNODE the section of stem between 2 nodes

INTERSPECIFIC said of a cross between 2 or more different species.

IN VITRO literally, *in glass* (Latin). Often used interchangeably with the terms "tissue culture" and "micropropagation," it refers to growing under artificial conditions.

IN VIVO literally, *in life* (Latin). Occurring naturally or under natural conditions; not *in vitro*.

ION any electrically charged atom or group of atoms, whether positively or negatively charged.

IONIZATION the process by which a compound chemically separates into negative and positive ions

JUVENILITY a state of growth associated with the young or immature stage in a plant's life cycle prior to its ability to flower

LAMINAR AIRFLOW the even flow of air in a specific, layered direction and without turbulence. *See* transfer chamber.

LAYERING a form of vegetative reproduction in which a shoot or twig comes in contact with the soil, forms roots, and grows on as a new plant

MACERATE to separate cells by cutting, grinding, soaking, or other physical or chemical action

MAGNETIC STIRRER a stirring device in which a whirling magnet attracts a magnetized stir bar placed inside a glass container on the plate. It is often combined with a hot plate to allow for heat while stirring.

MEDIUM, MEDIA (pl.) (1) the substrate for plant growth, specifically the mixture of certain chemical compounds to form a nutrient-rich gel or liquid in or on which tissue cultures are grown; (2) a soil mix

MEIOSIS the reduction division process by which haploid (*n*) sex cells (gametes) are produced. It is a reduction division because the number of chromosomes is reduced from 2*n* to *n*.

MENISCUS the curved upper surface of a column of liquid

MERISTEM the undifferentiated cells at the tips of roots and shoots, in stems as cambium, and in other locations on a plant. Meristematic tissue is the basis of plant growth and development. The term is used loosely in this book to denote a microscopic shoot tip, usually less than 1.5 mm long and consisting of 1 or 2 leaf primordia, used as an explant. Also, the verb form, to excise such small shoot or root tips.

MERISTEM CULTURE the *in vitro* culture of meristematic tissue. Also misused more broadly to denote micropropagation

MESOPHYLL the thin, soft tissue between the upper and lower epidermis of the leaf

METABOLISM the physiological activities of an organism

METABOLITE a product of metabolism. *See also* secondary product

MICRON an obsolete term for one one-thousandth of a millimeter (0.001 mm), now called a micrometer (mm). The symbol for a micron is m.

MICROPROPAGATION literally, propagation on a very small scale; vegetative multiplication *in vitro*.

MITES any nearly microscopic animals of the order Acarina, class Arachnida. They can be a devastating contaminant of plant tissue cultures.

MITOSIS the process of vegetative cell division in which chromosomes duplicate and then separate

MOLAR SOLUTION one mole (one gram molecular weight) of a substance in one liter of solution

MOLE the molecular weight of a compound expressed in grams

MOLECULAR WEIGHT the sum of atomic weights in a molecule

MOLECULE the smallest quantity into which a chemical compound can be divided and still retain the characteristic properties of that compound

MONOCOTYLEDON (MONOCOT) a flowering plant that contains a single cotyledon (seed leaf). Monocotyledons are members of the class Monocotyledoneae and include grasses, bulbs, and orchids.

MORPHOGENESIS the anatomical and physiological development of an organism, the study of which is called morphology

mRNA messenger RNA

MUTAGEN a mutant-inducing agent, such as radiation, ultraviolet light, or carcinogenic chemicals

MUTAGENESIS the formation of mutants

MUTANT a plant that exhibits some variation (mutation) in characteristics due to rearrangement of other change in genetic make-up

MUTATION a sudden and permanent change in the genetic order of an individual, due to a change in genes, chromosomes, or chromosome number, which will alter some characteristic

MYCORRHIZA a fungus that associates, usually symbiotically, with plant roots

NODE the site on a stem where a bud or leaf is attached

NUCLEIC ACID any of a group of complex organic acids found especially in the nucleus of all living cells as DNA and RNA

NUCLEOTIDE the building blocks of DNA and RNA, each containing a nitrogenous base and its contingent sugar and phosphate

NURSE CULTURE a culture technique whereby a filter paper separates a callus culture (the nurse culture) from a cell culture, which in turn received nutrients and growth-promoting factors diffused through the paper from the callus and from the medium

ORGANIC CHEMICAL any of the compounds made up of molecules containing at least one carbon atom

ORGANOGENESIS the formation of organs, such as leaves, shoots, or roots, from cells or tissues

OSMOSIS the diffusion of a fluid through a membrane from one solution into a solution of higher density or ion concentration. Reverse osmosis, the forcing of a solution through a membrane in the opposite direction, is often used for purifying water.

OSMOTIC PRESSURE the pressure potential existing in the diffusion of a solvent through a semipermeable membrane into a more concentrated solution to equalize the concentrations on both sides of the membrane

OVULE the structure within a flower that develops into a seed after fertilization, consisting of the egg cell, embryo sac, nucellus, and integuments (future seed coats). Ovules are sometimes used as explants and may be fertilized *in vitro*.

PARENCHYMA a large, thin-walled cell that has the capacity to regenerate and differentiate. It is the most common type of plant cell.

PATHOGEN a disease-causing organism

PEDICEL a stem or stalk of a flower

PETIOLE a stem or stalk of a leaf

pH a measure of the degree of acidity or alkalinity of a solution as indicated by the concentration of hydrogen ions. It is measured on a scale of 1 to 14, with levels below 7 indicating acidity and above 7 indicating alkalinity.

PHENOLICS certain organic metabolic by-products. Appearing as purplish black or brown substances, they sometimes exude from new cultures into the surrounding medium and can be toxic to the culture.

PHOTOPERIOD the length of time of exposure to light

PIPET a slender glass tube used to suction small amounts of liquid for the purpose of measuring and transferring such small amounts. Also the verb form, to use such an instrument to transfer small liquid amounts.

PLAGIOTROPIC growth that is horizontal rather than vertical, usually referring to conifers in which the main stem does not grow upright

PLASMOLYSIS the separation of cytoplasm from the cell wall by the removal of water from the protoplast

PLATE to spread a thin film of cells or single cells on an agar medium, usually in a petri dish (petri plate)

POLLEN CULTURE the culture of pollen grains, or the anthers containing pollen grains, t produce haploid plants

POLYPLOID a multiple of the basic number of chromosomes (n) such as triploid ($3n$), tetraploid ($4n$), and so forth

POLYSACCHARIDE a carbohydrate made up of a chain or one or more simple sugars linked together in its chemical structure

PRECIPITATE a solid substance that separates out of media solutions and renders those precipitated elements unavailable as nutrients. Also the verb form, the process of such separation

PREMIX media formulas prepared and sold commercially in (usually) a dry form

PRIMORDIUM, PRIMORDIA (pl.) a plant organ in its earliest stage of differentiation and development

PROPAGULE a small bit of plant that is being propagated; a transfer

PROTEINS a large group of complex organic substances made up of amino acids and other diverse elements. They may serve enzymatic or structural functions.

PROTOPLAST a cell that lacks a cell wall but has a membrane

PROTOPLAST FUSION the uniting of 2 protoplasts

REDUCTION DIVISION *see* meiosis

REGENERATION the formation of new plants or plant parts

RESIN BED with respect to water purification, a tank or cartridge containing charged resin particles to attract ions of opposite charge from impure water

RESISTIVITY a measure of the resistance of water to the flow of an electrical current. Measured in units of ohm-centimeters (ohm-cm), it is the reciprocal of conductivity.

RESOLUTION the ability of a microscope lens system to give individuality to 2 particles that lie in close proximity to one another

ROTATOR a device that rotates culture containers, thus agitating the contents of cultures in liquid media

SCAPE a leafless flower stalk growing from the root crown

SECONDARY PRODUCT a product of plant metabolism that is not primarily related to growth and reproduction, such as medicinals, flavorings, dyes, and pesticides

SENESCENCE aging

SOMATIC vegetative, nonsexual. All plant cells other than reproductive cells are somatic.

SOMATIC EMBRYOGENESIS the formation of embryos from somatic cells

SOMATIC HYBRID a vegetatively derived hybrid, as by protoplast or cell fusion

SPORT a plant or plant part that has undergone a mutation

STAGES OF CULTURE Stage 0, stock plant selection; Stage I, explant establishment; Stage II, multiplication; Stage III, rooting; Stage IV, acclimatization

STERILE TECHNIQUE the protocol for aseptically transferring cultures. Also called aseptic technique.

STOCK PLANT a plant from which other plants are started; a source plant for explants

STOCK SOLUTION a concentrated solution of media chemicals from which portions are used to make media

STOMATE, STOMATA (pl.) a small opening in the epidermis of leaves and stems through which gases and water may pass

SURFACTANT a wetting agent that reduces surface tension, thus allowing solutions to better penetrate and clean plants or laboratory surfaces, for example.

SUSPENSION CULTURE *see* cell suspension culture

SYMBIOSIS the living together of 2 dissimilar organisms, usually to mutual advantage

SYSTEMIC within an organism (system), internal, as a pathogen or systemic weed killer

TARE to allow for the weight of a container, which is deducted from the total weight of the substance and container being weighed

T-DNA transferred DNA

TETRAPLOID 4*n*, double the number of chromosomes in vegetative cells

TISSUE CULTURE literally, the culture of individuals tissues, but usually used more broadly to indicate micropropagation or *in vitro* propagation

TOTIPOTENCE the capability of developing into a whole plant, said of a cell

TRANSFER (1) the process of dividing cultures and placing the propagules in containers of fresh sterile medium; (2) the piece being so transferred

TRANSFER CHAMBER OR HOOD a protected, enclosed area with a sterile atmosphere in which cultures are started, divided, trimmed, and then transferred using sterile technique

TRANSPIRATION the escape of water vapor from a plant, primarily through the leaf stomata

ULTRASONIC cleaner a cleaning device employing ultrasonic vibration, occasionally used for cleaning explants

UNDIFFERENTIATED not differentiated, as in cells or tissues that have not yet been modified for their ultimate structure or function

VARIEGATED plants having multicolored leaves, due to a virus, a nutritional deficiency, or genetic make-up

VECTOR (1) an agent that carries a disease from one organism to another or pollen from one flower to another; (2) in genetic engineering, the plasmid or nucleic acid used to insert genes into receptive cells

VEGETATIVE somatic, nonsexual

VITRIFICATION a phenomenon in which tissues develop a glassy, waterlogged, swollen appearance

WATER OF HYDRATION the amount of water that is chemically attached to a compound. This must be accounted for proportionally if, for example, an anhydrous (lacking the water of hydration) form of a chemical is used instead of the hydrate form.

ZYGOTE a fertilized egg cell; the diploid cell resulting from the union of the male and female sex cells in the process of fertilization

Bibliography

Abdullah, A. A., M. M. Yeoman, and J. Grace. 1985. *In vitro* adventitious shoot formation from embryonic and cotyledonary tissues of *Pinus brutia* Ten. *Plant Cell, Tissue and Organ Culture* 5: 35–44.

Abrie, A. L., and J. van Staden. 2001. Development of regeneration protocols for selected cucurbit cultivars. *Plant Growth Regulation* 35: 263–267.

Adelberg, J., M. Delgado, and J. Tomkins. 2007. *In vitro* sugar and water use in diploid and tetraploid genotypes of daylily (Hemerocallis spp.) in liquid medium as affected by density and plant growth regulators. *HortScience* 42(2): 325–328.

Adelberg, J., M. Kroggel, and J. Toler. 2000. Physical environment *in vitro* affects laboratory and nursery growth of micropropagated hostas. *HortTechnology* 10(4): 754–757.

Adelberg, J. W., M. P. Delgado, and J. T. Tomkins. 2010. Spent medium analysis for liquid culture micropropagation of *Hemerocallis* on Murashige and Skoog medium. *In Vitro Cellular and Developmental Biology—Plant* 46(1): 95-107.

Ahmad, N., and M. Anis. 2005. *In vitro* mass propagation of *Cucumis sativus* L. from nodal segments. *Turk J. Bot.* 29: 237–240.

Akira, I., and H. Itokawa. 1988. Alkaloids of tissue cultures of *Nandina domestica*. *Phytochemistry* 27(7): 2143–2145.

Almehdi, A. A., and D. E. Parfitt. 1986. *In vitro* propagation of peach: 1. Propagation of 'Lovell' and 'Nemaguard' peach rootstocks. *Fruit Varieties Journal* 40(1): 12–17.

Altman, A. 2003. From plant tissue culture to biotechnology: Scientific revolutions, abiotic stress tolerance, and forestry. *In Vitro Cellular and Developmental Biology—Plant* 39(2): 75–84.

Ambrožič-Dolinšek, J., M. Camloh, B. Bohanec, and J. Žel. 2002. Apospory in leaf culture of staghorn fern (*Platycerium bifurcatum*). *Plant Cell Reports* 20(9): 791–796.

Ammirato, P. V. 1987. Organizational events during embryogenesis. In *Plant Tissue and Cell Culture: The Proceedings of the Sixth International Congress of Plant Tissue and Cell Culture, University of Minnesota 1986*, edited by C. E. Green, D. A. Somers, W. P. Hackett, and D. D. Biesboer. New York: Alan Liss.

Anderson, W. C. 1975. Propagation of rhododendrons by tissue culture: Part 1. Development of a culture medium for multiplication of shoots. *Proceedings of the International Plant Propagator's Society* 25: 129–135.

Anderson, W. C. 1977. Rapid propagation of *Lilium* cv. Red Carpet. *In Vitro* 13: 145.

Anderson, W. C. 1978. Rooting of tissue cultured rhododendrons. *Proceedings of the International Plant Propagator's Society* 28: 135–139.

Anderson, W. C. 1980. Tissue culture propagation of red raspberries. In *Proceedings of the Conference on Nursery Production of Fruit Plants Through Tissue Culture: Applications and Feasibility*. Washington, DC: U.S. Department of Agriculture, SEA Agricultural Research Results.

Anderson, W. C. 1984. *Primula* micropropagation multiplies special plants. *Primroses* 42(2): 21–23.

Anderson, W. C., and J. B. Carstens. 1977. Tissue culture propagation of broccoli, *Brassica oleracea* (Italica Group), for use in F1 hybrid seed production. *Journal of the American Society for Horticultural Science* 102: 69–73.

Anderson, W. C., and G. Haner. 1978. *Strawberry Tissue Culture Formula*. Mount Vernon, Washington: Washington State University, Northwestern Washington Research Unit.

Anderson, W. C., and G. W. Meagher. 1977. Cost of propagating broccoli plants through tissue culture. *HortScience* 12: 543–544.

Anderson, W. C., and G. W. Meagher. 1978. Cost of propagating plants through tissue culture using lilies as an example. Corvallis, Oregon: Oregon State University, Ornamentals Short Course.

Anderson, W. C., and K. A. Mielke. 1985. *Outline for Micropropagating Virus-free Bulbous Iris*. Mount Vernon, Washington: Washington State University, Northwestern Washington Research Unit.

Anderson, W. C., P. Miller, K. Mielke, and T. Allen. 1990. *In vitro* bulblet propagation of virus-free Dutch iris. *Acta Horticulturae* 266: 77–81.

Appleton, M. R., and J. van Staden. 1995. Micropropagation of some South African *Hypoxis* species with medicinal and horticultural potential. *Acta Horticulturae* 420: 75–77.

Arditti, J. 1977. Clonal propagation of orchids by means of tissue culture: A manual. In *Orchid Biology: Reviews and Perspectives*, vol. 1, edited by J. Arditti. Ithaca, New York: Cornell University Press.

Arditti, J., E. A. Ball, and D. M. Reisinger. 1977. Culture of flower stalk buds: A method for vegetative propagation of *Phalaenopsis*. *American Orchid Society Bulletin* 46: 236–240.

Arshad, M., J. Silvestre, G. Merlina, C. Dumat, E. Pinelli, and J. Kallerhoff. 2012. Thidiazuron-induced shoot organogenesis from mature leaf explants of scented *Pelargonium capitatum* cultivars. *Plant Cell, Tissue and Organ Culture* 108(2): 315–322.

Ascough, G. D., J. E. Erwin, and J. van Staden. 2009. Micropropagation of Iridaceae: A review. *Plant Cell, Tissue and Organ Culture* 97(1): 1–19.

Avanci, N. C., D. D. Luche, G. H. Goldman, and M. H. Goldman. 2010. Jasmonates are phytohormones with multiple functions, including plant defense and reproduction. *Genetics and Molecular Research* 9(1): 484–505.

Azza M., El-Sheikh, and M. M. Khalafalla. 2010. *In vitro* shoot micropropagation and plant establishment of an ornamental plant dumb cane (*Dieffenbachia compacta*). *International Journal of Current Research* 6: 27–32.

Bachman, G. R. 2008. Care and management of stock plants. In *Plant Propagation, Concepts and Laboratory Exercises*, edited by C. A. Beyl and R. N. Trigiano. Boca Raton, Florida: CRC Press. 225–230.

Ball, E. A. 1978. Cloning *in vitro* of *Sequoia sempervirens*. *Fourth International Congress of Plant Tissue and Cell Culture* (Calgary) 1726: 163.

Barker, P. K., R. A. de Fossard, and R. A. Bourne. 1977. Progress toward clonal propagation of *Eucalyptus* species by tissue culture techniques. *Proceedings of the International Plant Propagator's Society* 27: 546–556.

Barlass, M., and K. G. M. Skene. 1978. *In vitro* propagation of grapevine (*Vitis vinifera* L.) from fragmented shoot apices. *Vitis* 17: 335–340.

Barnett, H. L., and B. B. Hunter. 1998. *Illustrated Genera of Imperfect Fungi*. 4th ed. St. Paul, Minnesota: American Phytopathological Society.

Barnett, J. A., R. W. Payne, and D. Yarrow. 2000. *Yeasts: Characteristics and Identification*. 3rd ed. Cambridge, United Kingdom: Cambridge University Press.

Bates, G. W. 1999. Plant transformation via protoplast electroporation. In *Methods in Molecular Medicine*, vol. 7, *Plant Cell Culture Protocols*, edited by R. D. Hall. 359–366.

Beaty, R. M., E. O. Franco, and O. J. Schwarz. 1985. Hormonally induced shoot development on cotyledonary explants of *Pinus oocarpa*. *In Vitro* 21, Part II: 53A.

Bhojwani, S. S., and P. K. Dantu. 2010. Haploid plants. In *Plant Cell Culture: Essential Methods*, edited by M. R. Davey and P. Anthony. Chichester, United Kingdom: John Wiley and Sons. 61–65.

Bilkey, P. C., B. McCown, and A. C. Hildebrandt. 1978. Micropropagation of African violet petiole cross-sections. *HortScience* 13: 37–38.

Billings, S. G., C.-K. Chin, and G. Jelenkovic. 1988. Regeneration of blueberry plantlets from leaf segments. *HortScience* 23: 763–765.

Borgman, C. A., and K. W. Mudge. 1986. Factors affecting the establishment and maintenance of 'Titan' red raspberry root organ cultures. *Plant Cell, Tissue and Organ Culture* 6: 127.

Bosela, M. J., and C. H. Michler. 2008. Media effects on black walnut (*Juglans nigra* L.) shoot culture growth *in vitro*: Evaluation of multiple nutrient formulations and cytokinin types. *In Vitro Cellular and Developmental Biology—Plant* 44: 316–329.

Boulay, M. 1979. Multiplication et clonage rapide du *Sequoia sempervirens* par la culture *in vitro*. *Etudes et Recherches: Micropropagation d'Arbres Forestiers* (Association Forêt-Cellulose) 12: 49–54.

Bouvier, L., F. R. Fillon, and Y. Lespinasse. 1994. Oryzalin as an efficient agent for chromosome doubling of haploid apple shoots *in vitro*. *Plant Breeding* 113: 343–346.

Bowes, B. G., and E. W. Curtis. 1991. Conservation of the British National *Begonia* Collection by micropropagation. *Phytologist* 119(1): 169–181.

Boxus, P., M. Quoirin, and J. M. Laine. 1977. Large-scale propagation of strawberry plants from tissue culture. In *Applied and Fundamental Aspects of Plant, Cell, Tissue and Organ Culture*, edited by J. Reinert and Y. P. S. Bajaj. New York: Springer-Verlag.

Brand, A. J., and M. P. Bridgen. 1989. 'UConn White', a white-flowered *Torenia fournieri*. *HortScience*. 24(4): 714–715.

Bridgen, M. 1986. Do-it-yourself cloning. *Greenhouse Grower* 1986 (May): 43–47.

Bridgen, M. P. 1994. A review of plant embryo culture. HortScience. 29 (11): 1243–1246.

Bridgen, M. P., and J. W. Bartok. 1987. Designing a plant micropropagation laboratory. *Proceedings of the International Plant Propagator's Society* 37: 462–467.

Bridgen, M. P., R. Craig, and R. Langhans. 1990. Biotechnological breeding of *Alstroemeria*. *Herbertia*. 45(1,2): 93–96.

Bridgen, M. P., J. J. King, C. Pedersen, M. A. Smith, and P. J. Winski. 1991. Micropropagation of *Alstroemeria* hybrids. *Proceedings of the International Plant Propagator's Society* 41: 380–384.

Bridgen, M. P., E. Kollman, and C. Lu. 2009. Interspecific hybridization of *Alstroemeria* for the development of new, ornamental plants. *Acta Horticulturae* 836: 73–78.

Bridgen, M. P. H. 1992. A survey of plant tissue culture laboratories in North America. *Proceedings of the International Plant Propagator's Society* 42: 427–430.

Briggs, B. A., S. M. McCulloch, and L. A. Caton. 1994. In vitro propagation of *Rhododendron*. *Acta Horticulturae* 364: 21–26.

Campbell, R. 1985. *Plant Microbiology*. London: Arnold.

Canli, F. A., and R. M. Skirvin. 2008. *In vitro* separation of a rose chimera. *Plant Cell, Tissue and Organ Culture* 95(3): 353–361.

Caplin, S. M., and F. C. Steward. 1948. Effect of coconut milk on the growth of explants from carrot root. *Science* 108: 655–657.

Carelli, B. P., and S. Echeverrigaray. 2002. An improved system for the *in vitro* propagation of rose cultivars. *Scientia Horticulturae* 92(1): 69–74.

Cassells, A. C. 1991. Problems in tissue culture: Culture contaminants. In *Micropropagation: Technology and Application*, edited by P. C. Debergh and R. H. Zimmerman. Dordrecht, Netherlands: Kluwer Academic. 31–44.

Cassells, A. C. 1992. Screening for pathogens and contaminating microorganisms in micropropagation. In *Techniques for the Rapid Detection of Plant Pathogens*, edited by J. M. Duncan and L. Torrance. Boston: Blackwell Scientific.

Cassells, A. C., and A. Plunkett. 1984. Production and growth analysis of plants from leaf cuttings, and from tissue cultures of disks from mature leaves and young axenic leaves of African Violet (*Saintpaulia ionantha* Wendl.). *Scientia Horticulturae* 23(4): 361–369.

Cassells, A. C., and G. Minas. 1983. Plant and *in vitro* factors influencing the micropropagation of *Pelargonium* cultivars by bud-tip culture. *Scientia Horticulturae* 21: 53–65.

Cassells, A. C., and F. M. Morrish. 1985. Growth measurements of *Begonia rex* Putz. plants regenerated from leaf cuttings and *in vitro* from leaf petioles, axenic leaves, re-cycled axenic leaves and callus. *Scientia Horticulturae* 27(1–2): 113–121.

Cassells, A. C., and F. M. Morrish. 1987. Genetic, ontogenetic and physiological factors and the development of a broad-spectrum medium for *Begonia rex* Putz. cultivars. *Scientia Horticulturae* 31(3–4): 295–302.

Chalfie, M., Y. Tu, G. Euskirchen, W. W. Ward, and D. C. Prasher. 1994. Green fluorescent protein as a marker for gene expression. *Science* 263: 802–805.

Chandra, S. 2012. Natural plant genetic engineer *Agrobacterium rhizogenes*: role of T-DNA in plant secondary metabolism. *Biotechnology Letters* 34(3): 407–415.

Chang, H. N., and S. J. Sim. 1994. Increasing secondary metabolite production in plant cell cultures with fungal elicitors. In *Biotechnological Applications of Plant Cultures*, edited by P. D. Shargool and T. T. Ngo. Cleveland, Ohio: CRC Press.

Chee, R., R. M. Pool, and D. Bucher. 1984. A method for large-scale *in vitro* propagation of *Vitis*. *New York's Food and Life Sciences Bulletin*, No. 109. Geneva, New York: New York State Agricultural Experiment Station.

Cheetham, R. D., C. Mikloiche, M. Glubiak, and P. Weathers. 1992. Micropropagation of a recalcitrant male asparagus clone (MD 22-8). *Plant Cell, Tissue and Organ Culture* 31: 15–19.

Chen, C., X. Hou, H. Zhang, G. Wang, and L. Tian. 2011. Induction of *Anthurium andraeanum* 'Arizona' tetraploid by colchicine *in vitro*. *Euphytica* 181(2): 137–145.

Chen, J., and R. J. Henny. 2007. Ornamental foliage plants: Improvement through biotechnology. In *Recent Advances in Plant Biotechnology and Its Application*, edited by A. Kumar and S. K. Sopory. New Delhi, India: I. K. International Publishing House. 140–156.

Chetty, S., and M. P. Watt. 2011. The development of clone-unspecific micropropagation protocols for 3 commercially important *Eucalyptus* hybrids. Theses. http://hdl.handle.net/10413/3272.

Chiari, A., and M. P. Bridgen. 2000. Rhizome splitting: A new micropropagation technique to increase *in vitro* propagule yield in *Alstroemeria*. *Journal of Plant Cell, Organ and Tissue Culture* 62: 39–46.

Chin, C.-K. 1982. Promotion of shoot and root formation in asparagus *in vitro* by ancymidol. *HortScience* 17: 590–591.

Choi, K.-T., M.-W. Kim, and H.-S. Shin. 1982. Root and shoot formation from callus and leaflet cultures of ginseng (*Panax ginseng*). In *Plant Tissue Culture 1982: Proceedings of the Fifth International Congress of Plant Tissue and Cell Culture*, edited by Akio Fujiwara. Tokyo and Lake Yamanaka, Japan: Japanese Association for Plant Tissue Culture.

Christie, C. B. 1985. Propagation of amaryllids: A brief review. *Proceedings of the International Plant Propagator's Society* 35: 351.

Cooke, R. C. 1977a. Tissue culture propagation of African violets. *HortScience* 12: 549.

Cooke, R. C. 1977b. The use of an agar substitute in the initial growth of Boston ferns *in vitro*. *HortScience* 12: 339.

Cooke, R. C. 1979. Homogenization as an aid in tissue propagation of *Platycerium* and *Davallia*. *HortScience* 14: 21.

Cornu, D., and C. Jay-Allemand. 1989. Micropropagation of hybrid walnut trees (*Juglans nigra* × *Juglans regia*) through culture and multiplication of embryos. *Ann. Sci. For.* 46: 113–116.

Cousineau, J. C., A. K. Anderson, H. A. Daubeny, and D. J. Donnelly. 1993. Characterization of red raspberries and selection using isoenzyme analysis. *HortScience* 28: 1185–1186.

Cristina Pedroso, M., M. Margarida Oliveira, and M. Salome' S. Pais. 1992. Micropropagation and simultaneous rooting of *Actinidia deliciosa* var. *deliciosa* 'Hayward'. *HortScience* 27(5): 443–445.

Cui, J., J. Liu, M. Deng, J. Chen, and R. J. Henny. 2008. Plant regeneration through protocorm-like bodies induced from nodal explants of *Syngonium podophyllum* 'White Butterfly'. *HortScience* 43: 2129–2133.

Curtin, M. E. 1983. Harvesting profitable products from plant tissue culture. *Biotechnology* 1983 (October): 649–657.

Cybularz-Urban, T., E. Hanus-Fajerska, and A. Świderski. 2007. Effect of light wavelength on *in vitro* organogenesis of a *Cattleya* hybrid. *Acta Biologica Cracoviensia Botanica* 49(1): 113–118.

Dagnino, D. S., M. D. Carelli, R. F. Arrabal, and M. A. Esquibel. 1991. Effect of gibberellic acid on *Ipomoea batatas* regeneration from meristem culture. *Pesquisa Agropecuária Brasileira*, Brasília 26(2): 259–262.

Damiano, C. 1980. Strawberry micropropagation. In *Proceedings of the Conference on Nursery Production of Fruit Plants Through Tissue Culture: Applications and Feasibility*. Washington, DC: U.S. Department of Agriculture, SEA Agricultural Research Results.

Davies, D. R. 1980. Rapid propagation of roses *in vitro*. *Scientia Horticulturae* 13: 385–389.

Davis, M. J., R. Baker, and J. J. Hanan. 1977. Clonal multiplication of carnations by micropropagation. *Journal of the American Society for Horticultural Science* 102: 48–53.

De Diego, N., I. A. Montalbán, and P. Moncaleán. 2010. *In vitro* regeneration of adult *Pinus sylvestris* L. trees. *South African Journal of Botany* 76(1): 158–162.

de Fossard, R. A. 1976a. *Tissue Culture for Plant Propagators*. Armidale, Australia: University of New South Wales.

de Fossard, R. A. 1976b. Vegetative propagation of *Eucalyptus ficifolia* F. Muell. by nodal culture *in vitro*. *Proceedings of the International Plant Propagator's Society* 26: 373–378.

de Fossard, R. A. 1977. Progress toward clonal propagation of *Eucalyptus* species by tissue culture techniques. *Proceedings of the International Plant Propagator's Society* 27: 546.

de Fossard, R. A. 1993. Cost of production in commercial micropropagation and research strategies. *Proceedings of the International Plant Propagator's Society* 43: 131–140.

de Fossard, R. A., and H. Fossard. 1988. Micropropagation of some members of the Myrtaceae. *Acta Horticulturae* 227: 346–351.

de Oliveira, E. T., O. J. Crocomo, T. B. Farinha, and L. A. Gallo. 2009. Large-scale micropropagation of *Aloe vera*. *HortScience* 44:1675-1678.

Debergh, P. C., and R. H. Zimmerman, eds. 1991. *Micropropagation: Technology and Application*. Dordrecht, Netherlands: Kluwer Academic.

Debergh, P., J. Aitken-Christie, D. Cohen, B. Grout, S. Arnold, R. Zimmerman, and M. Ziv. 1992. Reconsideration of the term "vitrification" as used in micropropagation. *Plant Cell, Tissue and Organ Culture* 30(2): 135–140.

Devi, W. S., and G. J. Sharma. 2009. *In vitro* propagation of *Arundinaria callosa* Munro—an edible bamboo from nodal explants of mature plants. *The Open Plant Science Journal* 3: 35–39.

Dixon, R. A., and R. A. Gonzales, eds. 1994. *Plant Cell Culture: A Practical Approach*. 2nd ed. Oxford: Oxford University Press.

Dixon, R. A., R. B. Boone, A. Garden, L. S. Daley, and T. L. Righetti. 1990. Analysis of electrophoresis isozyme patterns using a microcomputer-based imaging system. *HortScience* 25: 961–964.

Dobránszki, J., and J. A. Teixeira da Silva. 2010. Micropropagation of apple: A review. *Biotechnology Advances* 28: 462–488.

Doi, M., S. Hamatani, T. Hirata, T. Hisamoto, and H. Imanishi. 1992. Micropropagation system of *Freesia*. *Acta Horticulturae* 319: 249–254.

Dole, J., and H. Wilkins. 2008. Floriculture principles and species. 2nd ed. Upper Saddle River, New Jersey: Prentice Hall.

Donnelly, D. J., and W. E. Vidaver. 1988. *Glossary of Plant Tissue Culture*. Portland, Oregon: Timber Press.

Donnelly, D. J., W. E. Vidaver, and K. Y. Lee. 1985. The anatomy of tissue-cultured red raspberry prior to and after transfer to soil. *Plant Cell, Tissue and Organ Culture* 4: 43–50.

Drew, R. A., J. A. Moisander, and M. K. Smith. 1989. The transmission of banana bunchy-top virus in micropropagated bananas. *Plant Cell, Tissue and Organ Culture* 16: 187–193.

Driver, J. A. 1985. Direct field rooting and acclimatization of tissue culture cuttings. *In Vitro* 21: 57A–80.

Dunstan, D. I. 1981. Transplantation and post-transplantation of micropropagated tree-fruit rootstocks. *Proceedings of the International Plant Propagator's Society* 31: 39–44.

Earle, E. D., and R. W. Langhans. 1975. Carnation propagation from shoot tip culture in liquid medium. *HortScience* 10: 608–610.

Eccher, T., and N. Noè. 1989. Comparison between zip and zeatin in the micropropagation of highbush blueberry (*Vaccinium corymbosum*). *Acta Horticulturae* 241:185–190.

Eeckhaut, T. G. R., S. P. O. Werbrouck, L. W H. Leus, E. J. Bockstaele, and P. C. Deberg. 2004. Chemically induced polyploidisation in *Spathiphyllum wallisii* Regel through somatic embryogenesis. *Plant Cell, Tissue and Organ Culture* 78: 241–246.

Espino, F. J., R. Linacero, J. Rueda, and A. M. Vázquez. 2004. Shoot regeneration in four *Begonia* genotypes. *Biologia Plantarum* 48(1): 101–104.

Feliciano, A. J., and M. de Assis. 1983. *In vitro* rooting of shoots from embryo-cultured peach seedlings. *HortScience* 18: 705–706.

Fernandez, G. E., and J. R. Clark. 1991. *In vitro* propagation of the erect thornless 'Navaho' blackberry. *HortScience* 26(9): 1219.

Fernández, H., and M. A. Revilla. 2003. *In vitro* culture of ornamental ferns. *Plant Cell, Tissue and Organ Culture* 73: 1–13.

Fitter, M., and A. D. Krikorian. 1985. Mature phenotype in *Hemerocallis* plantlets fortuitously generated *in vitro*. *Journal of Plant Physiology* 121: 97–101.

Fraga, M., M. Alonso, and M. Borja. 2004. Shoot regeneration rates of perennial phlox are dependent on cultivar and explant type. *HortScience* 39(6): 1373–1377.

Fraga, M., M. Alonso, P. Ellul, and M. Borja. 2004. Micropropagation of *Dianthus gratianopolitanus*. *HortScience* 39(5): 1083–1087.

Fujita, Y., Y. Hara, C. Suga, and T. Morimoto. 1981. Production of shikonin derivative by cell suspension cultures of *Lithospermum erythrorhizon*. *Plant Cell Reports* 1: 61–63.

Furuya, T., and K. Ushiyama. 1994. Ginseng production in cultures of *Panax ginseng* cells. In *Biotechnological Applications of Plant Cultures*, edited by P. D. Shargool and T. T. Ngo. Cleveland, Ohio: CRC Press.

Gamborg, O. L., R. A. Miller, and K. Ojima. 1968. Nutrient requirements of suspension cultures of soybean root cells. *Experimental Cell Research* 50(1): 151–158.

Gamborg, O. L., T. Murashige, T. A. Thorpe, and I. K. Vasil. 1976. Plant tissue culture media. *In Vitro* 12: 473–478.

Gantait, S., N. Mandal, S. Bhattacharyya, and P. Kanti Das. 2011. Induction and identification of tetraploids using *in vitro* colchicine treatment of *Gerbera jamesonii* Bolus cv. Sciella. *Plant Cell, Tissue and Organ Culture* 106(3): 485–493.

Gao, X., D. Yang, D. Cao, M. Ao, X. Sui, Q. Wang, J. N. Kimatu, and L. Wang. 2010. *In vitro* micropropagation of *Freesia* hybrida and the assessment of genetic and epigenetic stability in regenerated plantlets. *Journal of Plant Growth Regulation* 29(3): 257–267.

Gasser, C., and R. T. Fraley. 1992. Transgenic crops. *Scientific American* 266(6): 62–69.

Gautheret, R. J. 1942. *Manuel Technique de Culture des Tissus Vegetaux*. Paris: Masson.

Gautheret, R. J. 1982. Plant tissue culture: The history. In *Plant Tissue Culture 1982: Proceedings of the Fifth International Congress of Plant Tissue and Cell Culture*, edited by Akio Fujiwara. Tokyo and Lake Yamanaka, Japan: Japanese Association for Plant Tissue Culture.

Gayed, A. M. A., F. M. Saadawy, I. S. Ibrahim, and W. A. Aly. 2006. Studies on the micropropagation of lily *Lilium longiflorum* Thunb. bulbs. II. Multiplication. *Annals Agric. Sci.* 51(2): 471–484.

Gebhardt, K. 1985. Development of a sterile cultivation system for rooting of shoot tip cultures (red raspberries) in Duraplast foam. *Plant Science* 39: 141–148.

George, E. F., and P. D. Sherrington. 1984. *Plant Propagation by Tissue Culture*. Eversley, England: Exegetics.

Gielis, J. 1995. Bamboo and biotechnology. *European Bamboo Society Journal*, May 6: 27–39.

Glick, B. R., and J. E. Thompson. 1993. *Methods in Plant Molecular Biology and Biotechnology*. Boca Raton, Florida: CRC Press.

Glowacka, K., S. Jezowski, and Z. Kaczmarek. 2010a. Impact of colchicine application during callus induction and shoot regeneration on micropropagation and polyploidisation rates of 2 *Miscanthus* species. *In Vitro Cellular and Developmental Biology—Plant* 46: 161–171.

Glowacka, K., S. Jezowski, and Z Kaczmarek. 2010b. The effects of genotype, inflorescence developmental stage, and induction medium on callus induction and plant regeneration in 2 *Miscanthus* species. *Plant Cell, Tissue and Organ Culture* 102: 79–86.

Gosukonda, R. M., A. Porobodessai, E. Blay, C. S. Prakash, and C. M. Peterson. 1995. Thidiazuron-induced adventitious shoot regeneration of sweet potato (*Ipomoea batatas*). *In Vitro Cellular and Developmental Biology—Plant* 31(2): 65–71.

Goldblatt, P. 1980. Polyploidy in angiosperms: monocotyledons. In *Polyploidy*, edited by W. H. Lewis. New York: Plenum Press. 219–239.

Goldfarb, B., and G. E. McGill. 1992. *In vitro* culture and micrografting of white pine meristems. *Proceedings of the International Plant Propagator's Society* 42: 412–414.

Goodwin, P. B., and T. Adisarwanto. 1980. Propagation of potato by shoot tip culture in petri dishes. *Potato Research* 23: 445–448.

Gorecka, K., D. Krzyzanowska, and R. Gorecki. 2003. Micropropagation of asparagus (*Asparagus officinalis* (L) I. Multiplication *in vitro*. *Vegetable Crops Research Bulletin* 59: 27–35.

Gould, J. H., and T. Murashige. 1985. Morphogenic substances released by plant tissue cultures: 1. Identification of berberine in *Nandina* culture medium, morphogenesis, and factors influencing accumulation. *Plant Cell, Tissue and Organ Culture* 4: 29–42.

Grant, B. 1971. *Plant Speciation*. New York: Columbia University Press.

Gray, D. J., E. Hiebert, C. M. Lin, M. E. Compton, D. W. McColley. 1994. Simplified construction and performance of a device for particle bombardment. *Plant Cell, Tissue and Organ Culture* 37: 179–184.

Greenberg, B. M., and B. R. Glick. 1993. The use of recombinant DNA technology to produce genetically modified plants. In *Methods in Plant Molecular Biology and Biotechnology*, edited by B. R. Glick and J. E. Thompson. Boca Raton, Florida: CRC Press.

Griesbsach, R. J. 1989. Selection of a dwarf *Hemerocallis* through tissue culture. *HortScience* 24: 1027–1028.

Griffiths, M. 1994. *Index of Garden Plants*. Portland, Oregon: Timber Press.

Grossnickle, S. C. 2011. Tissue culture of conifer seedlings. In *National Proceedings: Forest and Conservation Nursery Associations 2010*. Proc. RMRS-P-65. Fort Collins, Colorado: USDA Forest Service, Rocky Mountain Research Station. 139–146.

Gubisova, M., J. Gubis, A. Zofajova, D. Mihalik, and J. Kraic. 2012. Enhanced *in vitro* propagation of *Miscanthus* ×*giganteus*. *Industrial Crops and Products* 41: 279–282.

Gugerli, P. 1992. Commercialization of serological tests for plant viruses. In *Techniques for the Rapid Detection of Plant Pathogens*, edited by J. M. Duncan and L. Torrance. Boston: Blackwell Scientific.

Gunver, G., and E. Ertan. 1998. A study on the propagation of figs by the tissue culture techniques. *Acta Horticulturae* 480: 169–172.

Gutiérrez-Miceli, F. A., L. Arias, N. Juarez-Rodríguez, M. Abud-Archila, A. Amaro-Reyes, and L. Dendooven. 2010. Optimization of growth regulators and silver nitrate for micropropagation of *Dianthus caryophyllus* L. with the aid of a response surface experimental design. *In Vitro Cellular and Developmental Biology—Plant* 46(1): 57–63.

Haberlandt, G. 1902. Kulturversuche mit isolierten pflanzenzellen. Sitzungsber. *Akademie der Wissenschaften in Wien, Mathematisch-Naturwissenschaftliche Klasse* 111: 69–92.

Haddadi, F., M. A. Aziz, G. Saleh, A. A. Rashid, and H. Kamaladini. 2010. Micropropagation of strawberry cv. Camarosa: Prolific shoot regeneration from *in vitro* shoot tips using thidiazuron with N6-benzylamino-purine. *HortScience* 45: 453–456.

Hadi, M., and M. Bridgen. 1996. Somaclonal variation as a tool to develop pest-resistant plants of *Torenia fournieri* 'Compacta Blue'. *Plant, Cell, Tissue and Organ Culture* 46: 43–50.

Hadziabdic, D., P. A. Wadl, and S. M. Reed. 2011. Haploid cultures. In *Plant Tissue Culture, Development, and Biotechnology*, edited by R. N. Trigiano and D. J. Gray. Boca Raton, Florida: CRC Press.

Hagiabad, M. S., Y. Hamidoghli, and R. F. Gazvini. 2007. Effects of different concentrations of mineral salt, sucrose and benzyladenine on Boston fern (*Nephrolepis exaltata* Schott cv. Bostoniensis) runner tips initiation. *Journal of Science and Technology of Agriculture and Natural Resources* 11(40A): 137–146.

Han, B. H., B. W. Yae, H. J. Yu, and K. Y. Peak. 2005. Improvement of *in vitro* micropropagation of *Lilium oriental* hybrid 'Casablanca' by the formation of shoots with abnormally swollen basal plates. *Scientia Horticulturae* 103(3): 351–359.

Handley, L. W., and O. L. Chambliss. 1979. *In vitro* propagation of *Cucumis sativus*. *HortScience* 14: 22.

Harris, R. E. 1984. Rapid propagation of saskatoon plants *in vitro*. Agricultural Research Council of Alberta, Canada, Farming for the Future Project 83-0054.

Harris, R. E. 1985. Rooting of *in vitro* shoots of saskatoon (*Amelanchier alnifolia*). Agricultural Research Council of Alberta, Canada, Farming for the Future Project 82-0017.

Harris, R. E., and J. H. Stevenson. 1982. *In vitro* propagation of *Vitis*. *Vitis* 21: 22–32.

Hartmann, H. T., D. E. Kester, F. T. Davies, Jr., and R. L. Geneve. 2010. *Hartmann and Kester's plant propagation: Principles and practices*. 8th ed. Upper Saddle River, New Jersey: Prentice Hall.

Hartney, V. J. 1982. Tissue culture of *Eucalyptus*. *Proceedings of the International Plant Propagator's Society* 32: 98–109.

Hartney, V. J., and E. D. Eabay. 1984. From tissue culture to forest trees. *Proceedings of the International Plant Propagator's Society* 34: 93–99.

Hasagawa, P. M. 1979. *In vitro* propagation of rose. *HortScience* 14: 610–612.

Hasagawa, P. M. 1980. Factors affecting shoot and root initiation from cultured rose shoot tips. *Journal of the American Society for Horticultural Science* 105: 216–220.

Hazell, B. S. 1998. Propagation of apple rootstocks by tissue culture. *Proceedings of the International Plant Propagator's Society* 48: 180–183.

Hebert, C. J., D. H. Touchell, T. G. Ranney, and A. V. LeBude. 2010. *In vitro* shoot regeneration and polyploid induction of *Rhododendron* 'Fragrantissimum Improved'. *HortScience* 45: 801–804.

Heims, D., and T. Sandler. 2010. Tissue culture tips. *Greenhouse Grower* 28: 12.

Heloir, M. C., J. C. Fournioux, L. Oziol, and R. Bessis. 1997. An improved procedure for the propagation *in vitro* of grapevine (*Vitis vinifera* cv. Pinot noir) using axillary-bud microcuttings. *Plant Cell, Tissue and Organ Culture* 49: 223–225.

Hennen, G. R., and T. J. Sheehan. 1978. *In vitro* propagation of *Platycerium stemaria* (Beauvois) Deav. *HortScience* 13: 245.

Heuser, C. W., and D. A. Apps. 1976. *In vitro* plantlet formation from flower petal explants of *Hemerocallis* cv. Chipper Cherry. *Canadian Journal of Botany* 54: 616–618.

Heuser, C. W., and J. Harker. 1976. Tissue culture propagation of daylilies. *Proceedings of the International Plant Propagator's Society* 26: 269–272.

Hill, S. A. 1984. *Methods in Plant Virology*. Boston: Blackwell Scientific.

Holt, J. G., ed. 1994. *Bergey's Manual of Systematic Bacteriology*. 9th ed. Baltimore: Williams and Wilkins.

Horsch, R. B., J. King, and G. E. Jones. 1980. Measurement of cultured plant cell growth on filter paper discs. *Canadian Journal of Botany* 58: 2402–2406.

Hosoki, T., and E. Kajino. 2003. Shoot regeneration from petioles of coral bells (*Heuchera sanguinea* Engelm.) cultured *in vitro*, and subsequent planting and flowering *ex vitro*. *In Vitro Cellular and Developmental Biology—Plant* 39: 135–138.

Hussey, G. 1975. Totipotency in tissue culture explants and callus of some members of the

Liliaceae, Iridaceae, and Amaryllidaceae. *Journal of Experimental Botany* 216: 253–262.

Hussey, G. 1976. Propagation of Dutch iris by tissue culture. *Scientia Horticulturae* 4: 163–165.

Hutchinson, J. F. 1984. Micropropagation of 'Northern Spy' apple rootstocks. *Proceedings of the International Plant Propagator's Society* 34: 38–48.

Hvoslefeide, A. K. 1991. The effect of temperature, daylength and irradiance on the growth of mother plants of *Nephrolepis exaltata* (L.) Schott and on the subsequent growth *in vitro* of runner tip explants. *Scientia Horticulturae* 47(1–2): 137–147.

Ilczuk, A., T. Winkelmann, S. Richartz, M. Witomska, and M. Serek. 2005. *In vitro* propagation of *Hippeastrum* ×*chmielii* Chm.—influence of flurprimidol and the culture in solid or liquid medium and in temporary immersion systems. *Plant Cell, Tissue and Organ Culture* 83: 339–346.

International Union of Pure and Applied Chemistry. 2012. IUPAC Periodic Table of the Elements. http://www.iupac.org/fileadmin/user_upload/news/IUPAC_Periodic_Table-1Jun12.pdf. Accessed 8 August 2012.

Jabbarzadeh, Z., and M. Khosh-Khui. 2005. Factors affecting tissue culture of Damask rose (*Rosa damascena* Mill.). *Scientia Horticulturae* 105(4): 475–482.

Jafari Najaf-Abadi, A., and Y Hamidoghli. 2009. Micropropagation of thornless trailing blackberry (*Rubus* sp.) by axillary bud explants. *Australian Journal of Crop Science* 3(4): 191–194.

Johnson, J. L., and E. R. Emino. 1979. *In vitro* propagation of *Mammillaria elongata*. *HortScience* 14: 605–606.

Jones, J. B. 1986. Determining markets and market potential. In *Tissue Culture as a Plant Production System for Horticultural Crops*, edited by R. H. Zimmerman, R. J. Griesbsach, F. Hammerschlag, and R. H. Lawson. Dordrecht, Netherlands: Martinus-Nijhoff.

Jones, J. B., and C. J. Sluis. 1991. Marketing of micropropagated plants. In *Micropropagation: Technology and Application*, edited by P. C. Debergh and R. H. Zimmerman. Dordrecht, Netherlands: Kluwer Academic. 141–154.

Kale, S. D., and B. M. Tyler. 2011. Assaying effector function in plants using double-barreled particle bombardment. *Methods in Molecular Biology* 712: 153–172.

Kalinina, A., and C. W. Brown. 2007. Micropropagation of ornamental *Prunus* spp. and GF305 peach, a *Prunus* viral indicator. *Plant Cell Reports* 26: 927–935.

Kane, M. E. 2010. Propagation from preexisting meristems. In *Plant Tissue Culture, Development, and Biotechnology*, edited by R. N. Trigiano and D. J. Gray. Boca Raton, Florida: CRC Press.

Kane, M. E., P. Kauth, and S. L. Stewart. 2008. Plant micropropagation. In *Plant Propagation Concepts and Laboratory Exercises*, edited by R. Trigiano and C. Beyl. Boca Raton, Florida: CRC Press. 319–332.

Kaniewski, W. K., and P. E. Thomas. 1988. A 2-step ELISA for rapid, reliable detection of potato viruses. *American Potato Journal* 65: 561–571.

Kanwar, J. K., and S. Kumar. 2008. *In vitro* propagation of *Gerbera*: A review. *Hort. Sci.* (Prague) 35(1): 35–44.

Kawase, K., H. Mizutani, M. Yoshioka, and S. Fukuda. 1995. Shoot formation of floral organs of Japanese iris *in vitro*. *J. Japan. Soc. Hort. Sci.* 64(1): 143–148.

Keever, G. J., and D. A. Findley. 2002. Thidiazuron increases shoot formation in *Nandina*. *Journal of Environmental Horticulture* 20(1): 24–28.

Kevers, C., and T. Gaspar. 1985. Vitrification of carnation *in vitro*: Changes in ethylene production, ACC level and capacity to convert ACC to ethylene. *Plant Cell, Tissue and Organ Culture* 4: 215–223.

Khalil, M. A., M. S. S. El-Shabasi, and M. E. A. Youssef. 2001. *In vitro* studies on *Asparagus officinalis* L. *Annals of Agricultural Science, Moshtohor* 39(1): 465–492.

Khuri, S., and J. Moorby. 1996. Nodal segments or microtubers as explants for *in vitro* microtuber production of potato. *Plant Cell, Tissue and Organ Culture* 45: 215–222.

Kitto, S. L., J. J. Frett, and P. Geiselhart. 1990. Micropropagation and field evaluation of ×*Heucherella* 'Bridget Bloom'. *Journal of Environmental Horticulture* 8(3): 156–159.

Kleyn, J., M. Bicknel, and M. Gilstrap. 1995. *Microbiology Experiments: A Health Science Perspective*. Boston: William C. Brown.

Kleyn, J., W. Johnson, and T. Wetzler. 1981. Microbiological aerosols and Actinomycetes in etiological considerations of mushroom workers' lungs. *Applied and Environmental Microbiology* 41: 1454–1460.

Knauss, J. F. 1976. A tissue culture method for producing *Dieffenbachia picta* cv. Perfection free of fungi and bacteria. *Proceedings of the Florida State Horticultural Society* 89: 293–296.

Knop, W. 1865. Quantitative Untersuchungen uber die Ernahrungsprocesse der Pflanzen. Landwirtsh. Vers. Stn. 70–140.

Knudson, L. 1946. A new nutrient solution for the germination of orchid seeds. *American Orchid Society Bulletin* 15: 214–217.

Konstas, J., and S. Kintzios. 2003. Developing a scale-up system for the micropropagation of cucumber (*Cucumis sativus* L.): the effect of growth retardants, liquid culture, and vessel size. *Plant Cell Reports* 21(6): 538–548.

Koubouris, G., and M. Vasilakakis. 2006. Improvement of *in vitro* propagation of apricot cultivar 'Bebecou'. *Plant Cell, Tissue and Organ Culture* 85: 173–180.

Krul, W. R., and J. Myerson. 1980. *In vitro* propagation of grape. In *Proceedings of the Conference on Nursery Production of Fruit Plants Through Tissue Culture: Applications and Feasibility*. Washington, DC: U.S. Department of Agriculture, SEA Agricultural Research Results.

Kubota, C., and T. Kozai. 1994. Low-temperature storage for quality preservation and growth suppression of broccoli plantlets cultured *in vitro*. *HortScience* 29(10): 1191–1194.

Kumar, P. P., P. Lakshmanan, and T. A. Thorpe. 1998. Regulation of morphogenesis in plant tissue culture by ethylene. *In Vitro Cellular and Developmental Biology—Plant* 34(2): 94–103.

Kumar, V., A. Radha, and S. Kumar Chitta. 1998. *In vitro* plant regeneration of fig (*Ficus carica* L. cv. Gular) using apical buds from mature trees. *Plant Cell Reports* 17(9): 717–720.

Kunachak, A., C.-K. Chin, T. Le, and T. Gianfagna. 1987. Promotion of asparagus shoot and root growth by growth retardants. *Plant Cell, Tissue and Organ Culture* 11: 97–110.

Kunisaki, J. T. 1980. *In vitro* propagation of *Anthurium andraeanum* Lind. *HortScience* 15: 508–509.

Kurosawa, E. 1926. Experimental studies on the nature of the substance secreted by the "bakanae" fungus. *Nat. Hist. Soc. Formosa* 16: 213–227.

Kyte, L., and B. Briggs. 1979. A simplified entry into tissue culture production of rhododendrons. *Proceedings of the International Plant Propagator's Society* 29: 90–95.

Kyte, L., and R. M. Kyte. 1981. Small fruit culture after the test tube. *Proceedings of the International Plant Propagator's Society* 31: 45–47.

Lane, D. W. 1982. *Apple Trees in Test Tubes*. Summerland, British Columbia: Agriculture Canada Research Station.

Lazar, A., P. Cerasela, and P. Sorina. 2010. Studies concerning the "*in vitro*" multiplication of *Nephrolepis exaltata* 'Fluffy Ruffles'. *Journal of Horticulture, Forestry and Biotechnology* 14(3): 191–193.

Lazarte, J. E., M. S. Galser, and O. R. Brown. 1982. *In vitro* propagation of *Epiphyllum chrysocardium*. *HortScience* 17: 84.

Levy, L. W. 1981. A large-scale application of tissue culture: The mass propagation of *Pyrethrum* clones in Ecuador. *Environmental and Experimental Botany* 21: 389–395.

Lewandowski, V. T. 1991. Rooting and acclimatization of micropropagated *Vitis labrusca* 'Delaware'. *HortScience* 26(5): 586–589.

Li, Z. T., S. A. Dhekney, and D. J. Gray. 2011. Genetic engineering technologies. In *Plant Tissue Culture, Development, and Biotechnology*, edited by R. N. Trigiano and D. J. Gray. Boca Raton, Florida: CRC Press.

Lin, Y., C. Lin, C. Chung, M. Su, H. Ho, S. Yeh, F. Jan, and H. Ku. 2011. *In vitro* regeneration and genetic transformation of *Cucumis metuliferus* through cotyledon organogenesis. *HortScience* 46: 616–621.

Linsmaier, E. M., and F. Skoog. 1965. Organic growth factor requirements of tobacco tissue cultures. *Physiologia Plantarum* 18: 100–128.

Litz, R. E., and R. A. Conover. 1977. Tissue culture propagation of some foliage plants. *Proceedings of the Florida State Horticultural Society* 90: 301.

Liu, C., P. Callow, L. J. Rowland, J. F. Hancock, and G. Q. Song. 2010. Adventitious shoot regeneration from leaf explants of southern highbush blueberry cultivars. *Plant Cell, Tissue and Organ Culture* 103(1): 137–144.

Liu, C., X. Xia, W. Yin, L. Huang, and J. Zhou. 2006. Shoot regeneration and somatic embryogenesis from needles of redwood (*Sequoia sempervirens* (D. Don.) Endl.). *Plant Cell Reports* 25(7): 621–628.

Lloyd, G., and B. McCown. 1980. Commercially feasible micropropagation of mountain laurel, *Kalmia latifolia*, by use of shoot-tip culture. *Proceedings of the International Plant Propagator's Society* 30: 421–427.

Loescher, W. H., and C. Albrecht. 1979. Development *in vitro* of *Nephrolepis exaltata* cv. Bostoniensis runner tissues. *Physiologia Plantarum* 47: 250–254.

Lubomski, W. 1989. *In vitro* shoot regeneration from different explants of *Hosta sieboldiana* cv. Gold Standard. *Acta Horticulturae* 251: 229–233.

Ludvova, A., and M. G. Ostrolucka. 1998. Morphogenic processes in callus tissue cultures and de novo regeneration of plants in *Actinidia chinensis* Planch. *Acta Societatis Botanicorum Poloniae* 67: 3–4.

Maene, L. J., and P. C. Debergh. 1983. Rooting of tissue cultured plants under *in vivo* conditions. *Acta Horticulturae* 212: 335–348.

Magyar-Tábori, K., J. Dobránszki, J. A. Teixeira da Silva, S. M. Bulley, and I. Hudák. 2010. The role of cytokinins in shoot organogenesis in apple. *Plant Cell, Tissue and Organ Culture* 101(3): 251–267.

Maki, S. L., M. Delgado, and J. W. Adelberg. 2005. Time course study of ancymidol for micropropagation of *Hosta* in a liquid culture system. *HortScience* 40(3): 764–766.

Makino, R. K., and P. J. Makino. 1978. Propagation of *Syngonium podophyllum* cultivars through tissue culture. *In Vitro* 13: 357.

Martin K. P., D. Joseph, J. Madassery, and V. J. Philip. 2003. Direct shoot regeneration from lamina explants of 2 commercial cut flower cultivars of *Anthurium andreaenum* Hort. *In Vitro Cellular and Developmental Biology—Plant* 39: 500–504.

Martin, K. P., C.-L. Zhang, A. Slater, and J. Madassery. 2007. Control of shoot necrosis and plant death during micropropagation of banana and plantains (*Musa* spp.). *Plant Cell, Tissue and Organ Culture* 88: 51–59

Marutani-Hert, M., T. J. Evens, G. T. McCollum, and R. P. Niedz. 2011. Bud emergence and shoot growth from mature citrus nodal stem segments. *Plant Cell, Tissue and Organ Culture* 106(1): 81–91.

Mata-Rosas, M., R. J. Baltazar-García, and V. M. Chávez-Avila. 2011. *In vitro* regeneration through direct organogenesis from protocorms of *Oncidium tigrinum* Llave & Lex. (Orchidaceae), an endemic and threatened Mexican species. *HortScience* 46: 1132–1135.

Matsuyama, J. 1978. Tissue culture propagation of *Nandina*. *Tissue Culture Association, 29th Annual Meeting* 1978: 357.

Matthews, R. E. F. 1981. *Plant Virology*. New York: Academic Press.

Mauseth, J. 1979. A new method for the propagation of cacti. *Cactus and Succulent Journal* 57: 186–187.

McCabe, D., and P. Christou. 1993. Direct DNA transfer using electrical discharge particle acceleration (ACCELLTM technology). *Plant Cell, Tissue and Organ Culture* 33: 227–236.

McCown, B. H., and G. B. Lloyd. 1983. A survey of the response of *Rhododendron* to *in vitro* culture. *Plant Cell, Tissue and Organ Culture* 2: 77–85.

McCown, B. H., and D. D. McCowan. 1999. A general approach for developing a commercial micropropagation system. *In Vitro Cellular and Developmental Biology—Plant* 35:276–277.

McCown, D. D. 2001. Micropropagation of *Syringa*: Tree vs. shrub lilacs. *Proceedings of the International Plant Propagator's Society* 51: 117–119.

McCullouch, S. M., and B. Briggs. 1982. Preparation of plants for micropropagation. *Proceedings of the International Plant Propagator's Society* 32: 297–303.

McGranahan, G. H., and C. A. Leslie. 1988. *In vitro* propagation of mature Persian walnut cultivars. *HortScience* 23: 220.

McGranahan, G. H., J. A. Driver, and W. Tulecke. 1987. Tissue culture of *Juglans*. In *Cell and Tissue Culture in Forestry*, edited by J. M. Bonga and D. J. Durzsan. Dordrecht, Netherlands: Martinus-Nijhoff.

McNicol, R. J., and J. Graham. 1990. *In vitro* regeneration of *Rubus* from leaf and stem segments. *Plant Cell, Tissue and Organ Culture* 21(1): 45–50.

Meiners, J., M. Schwab, and I. Szankowski. 2007. Efficient *in vitro* regeneration systems for *Vaccinium* species. *Plant Cell, Tissue and Organ Culture* 89: 169–176.

Mele, E., J. Messeguer, and P. Camprubi. 1982. Effect of ethylene on carnation explants grown in sealed vessels. In *Plant Tissue Culture 1982: Proceedings of the Fifth International Congress of Plant Tissue and Cell Culture*, edited by Akio Fujiwara. Tokyo and Lake Yamanaka, Japan: Japanese Association for Plant Tissue Culture.

Meyer, M. M., Jr. 1976. Propagation of daylilies by tissue culture. *HortScience* 11: 485–487.

Meyer, M. M., Jr. 1980. *In vitro* propagation of *Hosta sieboldiana*. *HortScience* 15: 737–738.

Meyer, M. M., Jr. 1982. *In vitro* propagation of *Rhododendron catawbiense* flower buds. *HortScience* 17: 891–892.

Meyer, M. M., Jr., L. H. Fuchigami, and A. N. Roberts. 1975. Propagation of tall bearded iris by tissue culture. *HortScience* 10: 477–480.

Mhatre, M., C. K. Salunkhe, and P. S. Roa. 2000. Micropropagation of *Vitis vinifera* L: towards an improved protocol. *Scientia Horticulturae* 84: 357–363.

Mielke, K. A. 1984. *In vitro* bulbing of bulbous *Iris* cv. Blue Ribbon (*Iris xiphium* × *Iris tingitana*). Master's thesis, Washington State University, Pullman, Washington.

Mielke, K. A., and W. C. Anderson. 1989. *In vitro* bulblet formation in Dutch iris. *HortScience* 24: 1028–1031.

Mii, M., T. Mori, and N. Iwase. 1974. Organ formation from the excised scales of *Hippeastrum hybridum in vitro*. *Journal of the American Society for Horticultural Science* 49: 241–244.

Mikkelsen, E. P., and K. C. Sink. 1978. *In vitro* propagation of Rieger Elatior begonias. *HortScience* 13: 242–244.

Miles, C. 1990. Causes and consequences of vitrification *in vitro* and its impact on micropropagation. *BioScience* 1990 (22 February).

Miller, G. A., D. C. Coston, E. G. Denny, and M. E. Romeo. 1982. *In vitro* propagation of 'Nemaguard' peach rootstock. *HortScience* 17: 194.

Minas, G. J. 2007. Sanitation and micropropagation of geranium (*Pelargonium* spp.) cultivars starting from meristem tip. *Acta Horticulturae* 755: 331–337.

Mohamed, M. A-H. 2011. A protocol for the mass-micropropagation of carnation (*Dianthus caryophyllus* L.). *Journal of Horticultural Science & Biotechnology* 86(2): 135–140.

Mohamed, M. A.-H., and A. A. Alsadon. 2010. Influence of ventilation and sucrose on growth and leaf anatomy of micropropagated potato plantlets. *Scientia Horticulturae* 123: 295–300.

Moncaleán, P., A. Rodriguez, and B. Fernandez. 2001. *In vitro* response of *Actinidia deliciosa* explants to different BA incubation periods. *Plant Cell, Tissue and Organ Culture* 67: 257–266.

Moncaleán, P., M. A. Fal, S. Castanon, B. Fernandez, and A. Rodriguez. 2009. Relative water content, *in vitro* proliferation, and growth of *Actinidia deliciosa* plantlets are affected by benzyladenine. *New Zealand Journal of Crop and Horticultural Science* 37(4): 351–359.

Monette, P. L. 1983. Influence of size of culture vessel on *in vitro* proliferation of grape in a liquid medium. *Plant Cell, Tissue and Organ Culture* 2: 327–332.

Monette, P. L. 1986. Micropropagation of kiwi fruit using non-axenic shoot tips. *Plant Cell, Tissue and Organ Culture* 6: 73–82.

Morel, G. 1960. Producing virus-free cymbidiums. *American Orchid Society Bulletin* 29: 495–497.

Morel, G. 1964. Tissue culture: A new means of clonal propagation of orchids. *American Orchid Society Bulletin* 33: 473–478.

Morel, G. 1974. Clonal multiplication of orchids. In *The Orchids: Scientific Studies*, edited by C. L. Withner. New York: John Wiley & Sons.

Morris, P., A. H. Scragg, N. J. Smart, and A. Stafford. 1985. Secondary product formation by cell suspension cultures. In *Plant Cell Culture: A Practical Approach*, edited by R. A. Dixon. Oxford: Oxford University Press, IRL Press.

Mudge, K. W., J. P. Lardner, and K. L. Eckenrode. 1992. Effects of lighting and CO_2 enrichment on acclimatization of micropropagated woody plants. In *Proceedings of the International Plant Propagator's Society* 42: 421–426.

Mullis, K. B. 1990. The unusual origin of the polymerase chain reaction. *Scientific American* 262(4): 56–65.

Munoz, S. J., V. P. Cravero, F. S. Lopez Anido, M. A. Esposito, and E. L. Cointry. 2006. Influence of selection criteria of elite plants on micropropagation in *Asparagus*. *Asian Journal of Plant Sciences* 5: 910–912.

Murashige, T. 1974. Plant propagation through tissue cultures. *Annual Review of Plant Physiology* 25: 135–166.

Murashige, T., and F. Skoog. 1962. A revised medium for rapid growth and bioassays with tobacco tissue cultures. *Physiologia Plantarum* 15: 473–497.

Murashige, T., M. N. Shaba, P. M. Hasagawa, F. H. Taktori, and J. B. Jones. 1972. Propagation of asparagus through shoot apex culture: 1. Nutrient medium for formation of plantlets. *Journal of the American Society for Horticultural Science* 97: 158–161.

Murashige, T., M. Serpa, and J. B. Jones. 1974. Clonal multiplication of *Gerbera* through tissue culture. *HortScience* 9: 175–180.

Nag, K. K., and H. E. Street. 1973. Carrot embryogenesis from frozen cultured cells. *Nature* 245: 270–272.

Nasircilar, A. G., S. Mirici, O. Eren, O. Karagüzel, and I. Baktir. 2011. Micropropagation of Turkish endemic *Iris stenophylla* Hausskn & Siehe ex Baker subsp. *allisonii* B. Mathew. *Acta Horticulturae* 886: 187–192.

NASS. 2010. Acreage. National Agricultural Statistics Board annual report, June 30. USDA, National Agricultural Statistics Service, Washington, DC.

Ndong, Y. A., A. Wadouachi, B. S. Sangwan-Norreel, and R. S. Sangwan. 2006. Efficient *in vitro* regeneration of fertile plants from corm explants of *Hypoxis hemerocallidea* landrace Gaza—the "African potato." *Plant Cell Reports* 25(4): 265–273.

Nesi, B., D. Trinchello, S. Lazzereschi, A. Grassotti, and B. Ruffoni. 2009. Production of lily symptomless virus-free plants by shoot meristem tip culture and *in vitro* thermotherapy, *HortScience* 44: 217–219.

Nester, E. W., C. E. Roberts, and M. T. Nester. 1995. *Microbiology: A Human Perspective*. Dubuque, Iowa: William C. Brown Publishers.

Nhut, D. T. 1998. Micropropagation of lily (*Lilium longiflorum*) via *in vitro* stem node and pseudo-bulblet culture. *Plant Cell Reports* 17: 913–916.

Nickerson, N. L. 1978. *In vitro* shoot formation in lowbush blueberry seedling explants. *HortScience* 13: 698.

Nishimura, T., and T. Fujimori. 2000. Micropropagation of *Kalmia latifolia* and the use of plant growth retardant for pot-plant production. *Proceedings of the International Plant Propagator's Society* 50: 663–664.

Nitsch, J. P., and C. Nitsch. 1969. Haploid plants from pollen grains. *Science* 163: 85–87.

Nobre, J., and A. Romano. 1998. *In vitro* cloning of *Ficus carica* L. adult trees. *Acta Horticulturae* 480: 161–164.

Norton, M. E., and C. R. Norton. 1985. *In vitro* propagation of Ericaceae: A comparison of the activity of the cytokinins N6-benzyladenine and N6-isopentenyladenine in shoot proliferation. *Scientia Horticulturae* 27(3–4): 335–340.

Nuraini, I., and M. J. Shaib. 1992. Micropropagation of orchids using scape nodes as the explant material. *Acta Horticulturae* 292: 169–172.

Obae, S. G., H. Klandorf, and T. P. West. 2011. Growth characteristics and ginsenosides production of *in vitro* tissues of American ginseng, *Panax quinquefolius* L. *HortScience* 46: 1136–1140.

Ogita, S. 2005. Callus and cell suspension culture of bamboo plant, *Phyllostachys nigra*. *Plant Biotechnology* 22(2): 119–125.

Ogita, S., H. Kashiwagi, and Y. Kato. 2008. *In vitro* node culture of seedlings in bamboo plant, *Phyllostachys meyeri* McClure. *Plant Biotechnology* 25: 381–385.

Oki, L. R. 1981. The modification of research procedures for commercial propagation of Boston ferns. *Environmental and Experimental Botany* 21: 397–400.

Okrslar, V., I. Plaper, M. Kovac, A. Erjavec, T. Obermajer, A. Rebec, M. Ravnikar, and J. Zel. 2007. Saponins in tissue culture of *Primula veris* L. *In Vitro Cellular and Developmental Biology—Plant* 43(6): 644–651.

Okubo, H., C. W. Huang, and F. Kishimoto. 1999. Effects of anti-auxins and basal plate on bulblet formation in scale propagation of amaryllis (*Hippeastrum* ×*hybridum* hort.). *Journal of the Japanese Society for Horticultural Science* 68(3): 513–518.

Olate, E., and M. P. Bridgen. 2004. Techniques for the *in vitro* propagation of *Rhodophiala* and *Leucocoryne* spp. *Acta Horticulturae* 673: 335–342.

Oliphant, J. 1990. The use of paclobutrazol in the rooting medium of micropropagated plants. *Proceedings of the International Plant Propagator's Society* 40: 358–360.

O'Neil, M. J., ed. 2006. *The Merck Index: An Encyclopedia of Chemicals, Drugs, and Biologicals*. 14th ed. New Jersey: Merck.

Oswald, T. H., A. E. Smith, and D. V. Phillips. 1977. Callus and plantlet regeneration from cell cultures of ladino clover (*Trifolium repens* 'Royal Ladino') and soybean (*Glycine max*). *Physiologia Plantarum* 39: 129–134.

Paek, K. Y., and E. C. Yeung. 1991. The effects of 1-naphthaleneacetic acid and N6-benzyladenine on the growth of *Cymbidium forrestii* rhizomes *in vitro*. *Plant Cell, Tissue and Organ Culture* 24: 65–71.

Paek, K. Y., and T. Kozai. 1998. Micropropagation of temperate *Cymbidium* via rhizome culture. HortTechnology. 8(3): 283–288.

Page, Y. M., and J. van Staden. 1984. *In vitro* propagation of *Hypoxis rooperi*. *Plant Cell, Tissue and Organ Culture* 3: 359–362.

Panis, B., and M. Lambardi. 2006. Status of cryopreservation technologies in plants (crops and forest trees). In *The Role of Biotechnology in Exploring and Protecting Agricultural Genetic Resources*, edited by A. Sonnino and J. Ruane. 61–78.

Papachatzi, M., P. A. Hammer, and P. M. Hasagawa. 1980a. *In vitro* propagation of *Hosta plantaginea*. *HortScience* 15: 506–507.

Papachatzi, M., P. A. Hammer, and P. M. Hasagawa. 1980b. Tissue culture propagation of *Hosta decorata* 'Thomas Hogg'. *HortScience* 15: 436.

Papafotiou, M., G. N. Balotis, P. T. Louka, and J. Chronopoulos. 2001. *In vitro* plant regeneration of *Mammillaria elongata* normal and cristate forms. *Plant Cell, Tissue and Organ Culture* 65: 163–167.

Parfitt, D. E., and A. A. Almehdi. 1986. *In vitro* propagation of peach: II. A medium for *in vitro* multiplication of 56 cultivars. *Fruit Varieties Journal* 40(2): 46–47.

Park, S. W., J. H. Jeon, H. S. Kim, Y. M. Park, C. Aswath, and H. Joung. 2004. Effect of sealed and vented gaseous microenvironments on the hyperhydricity of potato shoots *in vitro*. *Scientia Horticulturae* 99(2): 199–205.

Pasqualetto, P. L, R. H. Zimmerman, and I. Fordham. 1986. Gelling agent and growth regulator effects on shoot vitrification of 'Gala' apple *in vitro*. *Journal of the American Society for Horticultural Science* 111: 976–980.

Pérez-Tornero, O., C. I. Tallon, and I. Porras. 2010. An efficient protocol for micropropagation of lemon (*Citrus limon*) from mature nodal segments. *Plant Cell, Tissue and Organ Culture* 100: 263–271.

Phan, C. T., and P. Hagadus. 1986. Possible metabolic basis for the developmental anomaly observed in *in vitro* culture, called "vitreous plants." *Plant Cell, Tissue and Organ Culture* 6: 83–94.

Pierik, R. L. M., and T. A. Segers. 1973. *In vitro* culture of midrib explants of *Gerbera*: Adventitious root formation and callus induction. *Zeitschrift fur Pflanzenphysiologie* 69: 204–212.

Pierik, R. L. M., H. H. M. Steegmans, A. A. Elias, O. T. J. Stiekema, and A. J. van der Velde. 1988. Vegetative propagation of *Syringa vulgaris* L. *in vitro*. *Acta Horticulturae* 226: 195–204.

Pierik, R. L. M., H. H. M. Steegmans, and J. A. J. Vak de Meys. 1974. Plantlet formation in callus tissues of *Anthurium andraeanum* Lind. *Scientia Horticulturae* 2: 193–198.

Pietrosiuk, A., M. Furmanowa, and B. Łata. 2007. *Catharanthus roseus*: micropropagation and *in vitro* techniques. *Phytochemistry Reviews* 6(2–3): 459–473.

Pilon, P. 2011. Bulking *Echinacea* and decreasing plant losses. *Perennial Solutions*. www.perennialsolutions.com/index.cfm/fuseaction/articles.detail/articleID/87/index.htm.

Posada, M., N. Ballesteros, W. Obando, and A. Angarita. 1999. Micropropagation of gerberas from floral buds. *Acta Horticulturae* 482: 329–332.

Prado, M. J., M. T. Herrera, R. A. Vázquez, S. Romo, and M. V. González. 2005. Micropropagation of 2 selected male kiwifruit and analysis of genetic variation with AFLP markers. *HortScience* 40(3): 740–746.

Preece, J. E., and E. G. Sutter. 1991. Acclimatization of micropropagated plants to the greenhouse and field. In *Micropropagation: Technology and Application*, edited by P. C. Debergh and R. H. Zimmerman. Dordrecht, Netherlands: Kluwer Academic. 71–93.

Preil, W., and A. Schum. 1985. Experimental scheme for *in vitro* propagation of *Gerbera*. *European Cooperation in the Field of Science and Technology Research*, Cost Project 87. Washington, DC: U.S. Department of Agriculture.

Pruski, K., J. Nowak, and G. Grainger. 1990. Micropropagation of 4 cultivars of saskatoon berry (*Amelanchier alnifolia* NUTT.). *Plant Cell, Tissue and Organ Culture* 21: 103–109.

Pullman, G. S., S. Johnson, S. Van Tassel, and Y. Zhang. 2005. Somatic embryogenesis in loblolly pine (*Pinus taeda*) and Douglas fir (*Pseudotsuga menziesii*): improving culture initiation and growth with MES pH buffer, biotin, and folic acid. *Plant Cell, Tissue and Organ Culture* 80: 91–103.

Ramirez-Malagon, R., I. Aguilar-Ramirez, A. Borodanenko, L. Perez-Moreno, J. L. Barrera-Guerra, H. G. Nuñez-Palenius, and N. Ochoa-Alejo. 2007. *In vitro* propagation of ten threatened species of *Mammillaria* (Cactaceae). *In Vitro Cellular and Developmental Biology—Plant* 43(6): 660–665.

Rani, K., P. Kharb, and R. Singh. 2011. Rapid micropropagation and callus induction of *Catharanthus roseus in vitro* using different explants. World Journal of Agricultural Sciences 7(6): 699-704.

Rapaka, V. K., J. E. Faust, J. M. Dole, and E. S. Runkle. 2007. Effect of time of harvest on postharvest leaf abscission in lantana (*Lantana camara* L. 'Dallas Red') unrooted cuttings. HortScience 42:304–308.

Rechcigl, M., Jr., ed. 1978. *CRC Handbook*, Series in Nutrition and Food, vol. 3, section G. Cleveland, Ohio: CRC Press.

Redenbaugh, K. 1990. Application of artificial seed to tropical crops. *HortScience* 25: 251–255.

Reed, B. M., and A. Abdelnour-Esquivel. 1991. The use of zeatin to initiate *in vitro* cultures of *Vaccinium* species and cultivars. *HortScience* 26(10): 1320–1322.

Rehwald, L. 1927. *Zeitschrift für Pflanzenkrankheiten un Pflanzenschutz* 37: 65–86.

Revilla, M. A., J. Majada, and R. Rodriguez. 1989. Walnut (*Juglans regia* L.) micropropagation. Ann. Sci. For. 46: 149–151.

Rihan, H., M. Al-Issawi, S. Burchett, and M. P. Fuller. 2011. Encapsulation of cauliflower (*Brassica oleracea* var. *botrytis*) microshoots as artificial seeds and their conversion and growth in commercial substrates. *Plant Cell, Tissue and Organ Culture* 107(2): 243–250.

Rodriguez, R. 1982. Stimulation of multiple shoot-bud formation in walnut seeds. *HortScience* 17: 592.

Rout, G. R., and P. Das. 1997. Recent trends in the biotechnology of *Chrysanthemum*: a critical review. *Scientia Horticulturae* 69(3–4): 239–257.

Sagawa, K., and J. T. Kunisaki. 1982. Clonal propagation of orchids by tissue culture. In *Plant Tissue Culture 1982: Proceedings of the Fifth International Congress of Plant Tissue and Cell Culture*, edited by Akio Fujiwara. Tokyo and Lake Yamanaka, Japan: Japanese Association for Plant Tissue Culture.

Sahijram, L., J. R. Soneji, and K. T. Bollamma. 2003. Invited review: Analyzing somaclonal variation in micropropagated bananas (*Musa* spp.). *In Vitro Cellular and Developmental Biology—Plant* 39: 551–556.

Saito, H., and M. Nakano. 2002. Plant regeneration from suspension cultures of *Hosta sieboldiana*. *Plant Cell, Tissue and Organ Culture* 71: 23–28.

Sakila, S., M. B. Ahmed, U. K. Roy, M. K. Biswas, R. Karim, M. A. Razvy, M. Hossain, R. Islam, and A. Hoque. 2007. Micropropagation of strawberry (*Fragaria* ×*ananassa* Duch.) a newly introduced crop in Bangladesh. *American-Eurasian Journal of Scientific Research* 2(2): 151–154.

Sanikhani, M., S. Frello, and M. Serek. 2006. TDZ induces shoot regeneration in various *Kalanchoë blossfeldiana* Poelln. cultivars in the absence of auxin. *Plant Cell, Tissue and Organ Culture* 85: 75–82.

Santana-Buzzy, N., A. Canto-Flick, L. G. Iglesias-Andreu, M. del C. Montalvo-Peniche, G. López-Puc, and F. Barahona-Pérez. 2006. Improvement of *in vitro* culturing of Habanero pepper by inhibition of ethylene effects. *HortScience* 41(2): 405–409.

Sarowar, S., H. Y. Oh, N. I. Hyung, B. W. Min, C. H. Harn, S. K. Yang, S. H. Ok, and J. S. Shin. 2003. *In vitro* micropropagation of a *Cucurbita* interspecific hybrid cultivar—a root stock plant. *Plant Cell, Tissue and Organ Culture* 75(2): 179–182.

Satyakala, G., M. Muralidhar Rao, and G. Lakshmi Sita. 1995. *In vitro* micropropagation of scented geranium (*Pelargonium graveolens* L'Hér. ex Ait.: syn. *P. roseum* Willd). *Current Science* 68(7): 762–765.

Scaltsoyiannes, A., P. Tsoulpha, K. P. Panetsos, and D. Moulalis. 1997. Effect of genotype on micropropagation of walnut trees (*Juglans regia*). *Silvae Genetica* 46(6): 326–332.

Schenk, R. U., and A. C. Hildebrandt. 1972. Medium and techniques for induction and growth of monocotyledonous and dicotyledonous plant cell cultures. *Canadian Journal of Botany* 50: 199–204.

Schmidt, G. J., and E. E. Schilling. 2000. Phylogeny and biogeography of *Eupatorium* (Asteraceae: Eupatorieae) based on nuclear ITS sequence data. *American Journal of Botany* 87: 716–726.

Schnabelrauch, L. S., and K. C. Sink. 1979. *In vitro* propagation of *Phlox subulata* and *Phlox paniculata*. *HortScience* 14: 607–608.

Schwann, T. 1839. *Mikroskopische Untersuchungen über die Ubereinstimmung in der Struktur und Wachstum der Tiere und Pflanzen*. Reprinted in 1910 as no. 176 of *Klassiker der exakten Wissenschaften*, by W. Ostwald. Leipzig: Engelmann.

Scheen, G., and H. G. Schwenkel. 2002. *In vitro* regeneration in *Primula* ssp. via organogenesis. *Plant Cell Reports* 20(11): 1006–1010.

Scheen, G., and H. G. Schwenkel. 2003. Effect of genotype on callus induction, shoot regeneration, and phenotypic stability of regenerated plants in the greenhouse of *Primula* ssp. *Plant Cell, Tissue and Organ Culture* 72(1): 53–61.

Scully, R. M. 1967. Aspects of meristem culture in the *Cattleya* alliance. *American Orchid Society Bulletin* 36: 103–108.

Seabrook, J. E. A., and B. G. Cumming. 1977. The *in vitro* propagation of amaryllis (*Hippeastrum* spp.) hybrids. *In Vitro* 13: 831–836.

Sharma, D. 2010. Fig mosaic virus (FMaV) elimination and commercial micropropagation in *Ficus carica* 'Sierra'. *Proceedings of the International Plant Propagator's Society* 60: 280.

Shen, X., J. Chen, and M. E. Kane. 2007. Indirect shoot organogenesis from leaves of *Dieffenbachia* cv. Camouflage. *Plant Cell, Tissue and Organ Culture* 89(2–3): 83–90.

Sheng, X., Z. Zhao, H. Yu, J. Wang, Z. Xiaohui, and H. Gu. 2011. Protoplast isolation and plant regeneration of different doubled haploid lines of cauliflower (*Brassica oleracea* var. *botrytis*). *Plant Cell, Tissue and Organ Culture* 107(3): 513–520.

Shimasaki, K., and S. Uemoto. 1990. Micropropagation of a terrestrial *Cymbidium* species using rhizomes developed from seeds and pseudobulbs. *Plant Cell, Tissue and Organ Culture* 22(3): 237–244.

Singh, G., and W. R. Curtis. 1994. Reactor design for plant cell suspension culture. In *Biotechnological Applications of Plant Cultures*, edited by P. D. Shargool and T. T. Ngo. Cleveland, Ohio: CRC Press.

Singh, S., B. K. Ray, S. Bhattacharyya, and P. C. Deka. 1994. *In vitro* propagation of *Citrus reticulata* Blanco and *Citrus limon* Burm.f. *HortScience* 29(3): 214–216.

Skirvin, R. M., and M. A. Norton. 2008. Chimeras. In *Plant Propagation, Concepts and Laboratory Exercises*, edited by C. A. Beyl, C. A. and R. N. Trigiano. Boca Raton, Florida: CRC Press. 163–174.

Skoog, F., and C. O. Miller. 1957. Chemical regulation of growth and organ formation in plant tissues cultured *in vitro*. *Biological Action of Growth Substances Symposium of the Society for Experimental Biology* 11: 118–130.

Smith, E. F., I. Gribaudo, A. V. Roberts, and J. Mottley. 1992. Paclobutrazol and reduced humidity improve resistance to wilting of micropropagated grapevine. *HortScience* 27 (2): 111–113.

Smith, R. H. 1983. *In vitro* propagation of *Nandina*. *HortScience* 18: 304.

Smith, R. H. 1992. *Plant Tissue Culture: Techniques and Experiments*. San Diego: Academic Press.

Smith, R. H., and A. E. Nightingale. 1979. *In vitro* propagation of *Kalanchoe*. *HortScience* 14: 20.

Smith, R. H., J. Burrows, and K. Kurten. 1999. Challenges associated with micropropagation of *Zephyranthes* and *Hippeastrum* spp. (Amaryllidaceae). *In Vitro Cellular and Developmental Biology—Plant* 35(4): 281–282.

Smith, W. A. 1981. The aftermath of the test tube in tissue culture. *Proceedings of the International Plant Propagator's Society* 31: 47–49.

Sommer, H. E., C. L. Brown, and P. P. Kormanik. 1975. Differentiation of plantlets in longleaf pine (*Pinus palustris*) tissue cultured *in vitro*. *Botanical Gazette* 135: 196–200.

Sommer, H. E., and L. S. Caldas. 1981. *In vitro* methods applied to forest trees. In *Plant Tissue Culture: Methods and Applications in Agriculture*, edited by T. A. Thorpe. New York: Academic Press.

Song, J. Y., N. S. Mattson, and B. R. Jeong. 2011. Efficiency of shoot regeneration from leaf, stem, petiole, and petal explants of 6 cultivars of *Chrysanthemum morifolium*. *Plant Cell, Tissue and Organ Culture* 107(2): 295–304.

Sotiropoulos, T. E., K. N. Dimassi, and V. Tsirakoglou. 2006. Effect of boron and methionine on growth and ion content in kiwifruit shoots cultured *in vitro*. *Biologia Plantarum* 50(2): 300–302.

Standaert-de Metsenaere, R. E. A. 1991. Economic considerations. In *Micropropagation: Technology and Application*, edited by P. C. Debergh and R. H. Zimmerman. Dordrecht, Netherlands: Kluwer Academic. 123–140.

Stanley, D. 1995. Tissue-culturing plants on cornstarch. *Agricultural Research* 1995 (July): 11.

Sticklen, M. B. 2008. Plant genetic engineering for biofuel production: towards affordable cellulosic ethanol. *Nature Reviews—Genetics* 9: 433–443.

Stimart, D. P., and P. D. Ascher. 1981a. Developmental responses of *Lilium longiflorum* bulblets to constant or alternating temperatures *in vitro*. *Journal of the American Society for Horticultural Science* 106: 450–454.

Stimart, D. P., and P. D. Ascher. 1981b. Foliar emergence from bulblets of *Lilium longiflorum* Thunb. as related to *in vitro* generation temperatures. *Journal of the American Society for Horticultural Science* 106: 446–450.

Sul, I. W., and S. S. Korban. 1994. Effect of different cytokinins on axillary shoot proliferation and elongation of several genotypes of *Sequoia sempervirens*. *In Vitro Cellular and Developmental Biology—Plant* 30(3): 131–135.

Sul, I. W., and S. S. Korban. 2005. Direct shoot organogenesis form needles of 3 genotypes of *Sequoia sempervirens*. *Plant Cell, Tissue and Organ Culture* 80: 353–358.

Suttle, G. R. L. 1983. Micropropagation of deciduous trees. *Proceedings of the International Plant Propagator's Society* 33: 46–49.

Swanberg, A., and W. Dai. 2008. Plant regeneration of periwinkle (*Catharanthus roseus*) via organogenesis. *HortScience* 43: 832–836.

Taiz, L., and E. Zeiger. 2010. *Plant Physiology*. 5th ed. Sunderland, Massachusetts: Sinauer Associates.

Takayama, S., and M. Misawa. 1980. Differentiation in *Lilium* bulb scales grown *in vitro*. *Physiologia Plantarum* 48: 121–125.

Takayama, S., M. Misawa, Y. Takashige, and H. Tsumori. 1982. Cultivation of *in vitro* propagated *Lilium* bulbs in soil. *Journal of the American Society for Horticultural Science* 107: 830–834.

Tarr, S. A. J. 1972. *The Principles of Plant Pathology*. London: Macmillan.

Taylor, N. L., M. K. Anderson, K. H. Wuesenberry, and L. Watson. 1976. Doubling the chromosome number of *Trifolium* species using nitrous oxide. *Crop Science* 16: 516–518.

Thomas, D. des S., and T. Murashige. 1979. Volatile emissions of plant tissue cultures: I. Identification of the major components. *In Vitro* 15: 654–662.

Thompson, D. C., and J. B. Zaerr. 1981. Induction of adventitious buds on cultured shoot tips of Douglas fir (*Pseudotsuga menziesii* [Mirb.] Franco). In *IUFRO Colloque International sur la Culture "in Vitro" des Essence Forestières*. Nangis, France: Association Forêt-Cellulose.

Thorpe, T. A. 2007. History of plant tissue culture. *Molecular Biotechnology* 37: 169–180.

Thorpe, T. A., ed. 1981. *Plant Tissue Culture: Methods and Applications in Agriculture*. New York: Academic Press.

Toki, S. 1997. Rapid and efficient *Agrobacterium*-mediated transformation in rice. *Plant Molecular Biology Reporter* 15(1): 16–21.

Tombolato, A. F. C., C. Azevedoh, and V. Nagal. 1994. Effects of auxin treatments on *in vivo* propagation of *Hippeastrum hybridum* hort. by twin scaling. *HortScience* 29: 922.

Tortora, G. J., B. R. Funke, and C. L. Case, eds. 1982. *Microbiology: An Introduction*. Menlo Park, California: Benjamin-Cummings.

U.S. Department of Agriculture, Agricultural Research Service. 1994. *The USDA-ARS Plant Genome Research Program*. Athens, Georgia: University of Georgia.

Uchendu, E. E., G. Paliyath, D. C. W. Brown, and P. K. Saxena. 2011. *In vitro* propagation of North American ginseng (*Panax quinquefolius* L.). *In Vitro Cellular and Developmental Biology—Plant* 47: 710–718.

Vacin, E. F, and F. W. Went. 1949. Some pH changes in nutrient solutions. *Botanical Gazette* 110: 605–613.

van Aartrijk, J., and P. C. G. van der Linde. 1986. *In vitro* propagation of flower-bulb crops. In *Tissue Culture as a Plant Production System for Horticultural Crops*, edited by R. H. Zimmerman, R. J. Griesbsach, F. Hammerschlag, and R. H. Lawson. Dordrecht, Netherlands: Martinus-Nijhoff.

van der Linde, P. C. G., G. M. G. M. Hol, G. J. Blom-Barnhoorn, J. van Aartrijk, and G. J. de Klerk. 1988. *In vitro* propagation of *Iris* ×*hollandica* Tub. cv. Prof. Blaaw. I. Regeneration on bulb-scale explants. *Acta Horticulturae* (ISHS) 226: 121–128.

van Overbeek, J., M. E. Conklin, and A. F. Blakeslee. 1941. Factors in coconut milk essential for growth and development of very young *Datura* embryo. *Science* 94: 350–351.

Van Tuyl, J. M., and K.-B. Lim. 2003. Hybridization and polyploidization as tools in ornamental plant breeding. *Acta Horticulturae* 612:13–22.

Vasudevan, A., N. Selvaraj, A. Ganapathi, S. Kasthurirengan, V. Ramesh Anbazhagan, and M. Manickavasagam. 2004. Glutamine: a suitable nitrogen source for enhance shoot multiplication in *Cucumis sativus* L. *Biologia Plantarum* 48(1): 125–128.

Verma, P., and A. K. Mathur. 2011. Direct shoot bud organogenesis and plant regeneration from pre-plasmolysed leaf explants in *Catharanthus roseus*. *Plant Cell, Tissue and Organ Culture* 106(3): 401–408.

von Arnold, S., and T. Eriksson. 1981. *In vitro* studies of adventitious shoot formation in *Pinus contorta*. *Canadian Journal of Botany* 59: 870–872.

Wadl, P. A., S. M. Reed, and D. Hadziabdic. 2011. Haploid cultures. In *Plant Tissue Culture, Development, and Biotechnology*, edited by R. N. Trigiano and D. J. Gray. Boca Raton, Florida: CRC Press.

Wang, A. 1990. Callus induction and plant regeneration of American ginseng. *HortScience* 25: 5.

Watad, A. A., K. G. Raghothama, M. Kochba, A. Nissim, and V. Gaba. 1997. Micropropagation of *Spathiphyllum* and *Syngonium* is facilitated by use of interfacial membrane rafts. *HortScience* 32(2): 307–308.

Wehner, T., and R. D. Locy. 1981. *In vitro* adventitious shoot and root formation of cultivars and lines of *Cucumis sativus* L. *HortScience* 16: 759–760.

White, P. R. 1943. *A Handbook of Plant Tissue Culture*. Tempe, Arizona: Jacques Cattell.

White, P. R. 1963. *The Cultivation of Animal and Plant Cells*. 2nd ed. New York: Ronald.

Wimber, D. E. 1963. Clonal multiplication of cymbidiums through tissue culture of the shoot meristem. *American Orchid Society Bulletin* 32: 105–107.

Wong, S. 1981. Direct rooting of tissue cultured rhododendrons into an artificial soil mix. *Proceedings of the International Plant Propagator's Society* 31: 36–37.

Wood, M. 1989. Blast those genes! *Agricultural Research* (June). 2 pp.

Wu, J., S. A. Miller, H. K. Hall, and P. A. Mooney. 2009. Factors affecting the efficiency of micropropagation from lateral buds and shoot tips of *Rubus*. *Plant Cell, Tissue and Organ Culture* 99(1): 17–25.

Yamamoto, Y., Y. Magaya, and Y. Maruyama. 1998. Mass propagation of *Primula sieboldii* through leaf segment culture. *Proceedings of the International Plant Propagator's Society* 48: 269.

Yang, H. J., and W. J. Clore. 1973. Rapid vegetative propagation of asparagus through lateral bud culture. *HortScience* 8: 141–143.

Yang, H. J., and W. J. Clore. 1974. Development of complete plantlets from vigorous shoots of stock plants of asparagus *in vitro*. *HortScience* 9: 138–140.

Yanny, D. 1988. A study to determine the more economically viable alternatives for greenhouse operations between micropropagation and lateral shoot production of plants. Bachelor's thesis, Cardinal Stritch College, Wisconsin.

Zenk, M. H., H. El Shagi, H. Arens, J. Stockigt, E. W. Weiler, and B. Deus. 1977. Formation of the indole alkaloids serpentine and ajmalicine in cell suspension cultures of *Catharanthus roseus*. In *Plant Tissue Culture and Its Bio-Technological Application*, edited by W. Barz, E. Rienhard, and M. H. Zenk. Berlin: Springer-Verlag.

Zhang, Q., J. Chen, and R. J. Henny. 2006. Regeneration of *Syngonium podophyllum* 'Variegatum' through direct somatic embryogenesis. *Plant Cell, Tissue and Organ Culture* 84: 181–188.

Zhang, Z., H. Dai, M. Xiao, and X. Liu. 2008. *In vitro* induction of tetraploids in *Phlox subulata* L. *Euphytica* 159(1–2): 59–65.

Zillis, M., and D. Zwagerman. 1979. Clonal propagation of hostas by scape section culture *in vitro*. *HortScience* 14: 80.

Zillis, M., D. Zwagerman, D. Lamberts, and L. Kurtz. 1979. Commercial propagation of herbaceous perennials by tissue culture. *Proceedings of the International Plant Propagator's Society* 29: 404–413.

Zimmerman, R. H. 1991. Micropropagation of temperate zone fruit and nut crops. In *Micropropagation: Technology and Application*, edited by P. C. Debergh and R. H. Zimmerman. Dordrecht, Netherlands: Kluwer Academic.

Zimmerman, R. H., and O. C. Broome. 1980a. Apple cultivar micropropagation. In *Proceedings of the Conference on Nursery Production of Fruit Plants Through Tissue Culture: Applications and Feasibility*. Washington, DC: U.S. Department of Agriculture, SEA Agricultural Research Results.

Zimmerman, R. H., and O. C. Broome. 1980b. Blueberry micropropagation. In *Proceedings of the Conference on Nursery Production of Fruit Plants Through Tissue Culture: Applications and Feasibility*. Washington, DC: U.S. Department of Agriculture, SEA Agricultural Research Results.

Zimmerman, R. H., and O. C. Broome. 1980c. Micropropagation of thornless blackberry. In *Proceedings of the Conference on Nursery Production of Fruit Plants Through Tissue Culture: Applications and Feasibility*. Washington, DC: U.S. Department of Agriculture, SEA Agricultural Research Results.

Ziv, M. 1986. *In vitro* hardening and acclimation of tissue culture plants. In *Plant Tissue Culture and Its Agricultural Applications*, edited by L. A. Withers and P. G. Alderson. London: Butterworth.

Photo and Illustration Credits

CanStockPhoto.com/freesoulproduction, page 62
George Chan of LumiGrow, page 42
Enio Tiago de Oliveira and Luiz A. Gallo, page 119
Rob Flynn, U.S. Department of Agriculture, Natural Resources Conservation Service, page 36
iStockphoto.com/Collpicto, page 158 left
iStockphoto.com/grebcha, page 128 left
iStockphoto.com/Vasiliy Koval, page 158 right
Ingmar Karge/Wikimedia Commons/CC-BY-SA-3.0/GFDL, page 25 right
John Kleyn, pages 129 left and right, 132 left and right
Dennis Kunkel Microscopy, page 134
Lydiane Kyte, page 135
Marjorie Leggitt, pages 2–3, 11, 12, 15, 18, 22, 23, 25 left, 90 left and right, 101, 105, 128 right (based on drawing from the Department of Microbiology, University of Washington, Seattle), 169–223
123RF.com/iimages, page 86
Brent and Deborah McCown, page 144
Proaudio55, page 47

Edwin Remsburg, courtesy of U.S. Department of Agriculture, Agricultural Research Service, page 68
Belynda Rinck, pages 19, 108, 116, 117
Holly Scoggins, pages 17, 21, 27, 35, 50, 51 left and right, 52 bottom, 54 bottom, 56 left, 60, 82, 87, 91, 99, 106, 109, 118, 120, 121, 143, 160
Shutterstock.com/Dimarion, page 152
Terra Nova Nurseries, pages 104, 148
Thermo Scientific, page 52 top
Tom Webster, pages 29 left and right, 39, 43, 44, 46 (based on original drawing from Thermo Scientific *Waterbook*, courtesy Thermo Fisher Scientific, Waltham, Massachusetts), 48, 49, 54 top, 55, 56 right, 64, 126, 127, 136, 153, 228; pages 151 and 156 (based on drawings from *Microbiology: A Human Perspective*, by E. W. Nester, C. E. Roberts, and M. T. Nester, 1995, Wm. C. Brown Communications, Inc., Times Mirror Higher Education Group, Dubuque, Iowa)

Index

(+)-α-Tocopherol, 70
abscisic acid (ABA), 71
abscission, leaf, 72
ACCEL system, 158
acclimatization. *See* transplanting and acclimatization
Actinidia, 172–173
Actinidiaceae, 172
activated charcoal (AC), 74, 110
adenine sulfate, 69, 71
adventitious buds, 25, 70
adventitious growth, 24, 25
adventitious shoots, 27, 91, 161, 164
aeration, 110–111
African potato, 193–194
African violet, 18, 78, 91, 140, 216–217
agar, 73–74
  addressing media problems, 110, 111
  disposal of, 106
  preparation requirements, 40, 48
  rooting and transplanting, 104, 105, 115, 118
agargel, 74
agitators, 51–52, 73, 87, 167
Agrobacterium, 32, 73, 155, 157
  rhizogenes, 155
  tumefaciens, 32, 36, 122–123, 154, 155
air layering, 23
airlift fermentor, 167
air quality, 39, 44, 131–132
air requirements, 40–41, 53, 110–111
air testing, 131–132
alcohol disinfection, 50–51, 94, 97
alkaloid, 69
alleles, 152
allelopathic plants, 166
Allium, 152
Alstroemeria, 73, 173–174
Alstroemeria family, 173
alumroot, 191
Amaryllidaceae, 191
amaryllis, 191–192

amaryllis family, 191
Amelanchier, 174
amino acids, 72, 153–154
ammonium reduction, 79
ampicillin, 73
ancymidol (A-Rest), 71
Anderson's medium, 75
androgenesis, 163
Anthemis, 181
anther culture, 35, 163
Anthurium, 174–175
antibiotics, 72–73, 108
aperture, 229
apical meristem, 15, 91, 145
Apium, 67, 162
Apocynaceae, 178
apple, 20, 27, 74, 112, 117, 200–201
apricot, 211–212
Arabidopsis, 158, 159
Araceae, 174, 185, 218
Araliaceae, 205
aralia family, 205
Argyranthemum, 181
arrowhead vine, 218–219
artificial seeds, 161–162
arum family, 174, 185, 218
aseptic culture, 96–106
  cleaning and treating explants, 93–94
  clean up, 106
  cryopreservation of, 166
  defined, 13, 122
  establishment, 12, 104
  manipulation, 99–100, 103–106, 146–147
  methods overview, 101–103, 160–164
  sterile technique for, 96–100
asexual reproduction, 21
Asparagaceae, 175, 192
Asparagus, 175–176
asparagus family, 175, 192
aspirator pump, 49
Asteraceae, 158, 181, 189
aster family, 181, 189
Atmos-Bags, 54–55
atom, 62

atomic weights, 63–64
autoclave, 106, 109
  deciding to use, 30, 141
  in laboratory design, 41, 44, 49–50
  supplies suited for, 55, 56
automatic pipetter, 49
autotrophic, 31
auxins, 22, 70, 109–110, 115, 118, 227
auxin stock, 82
axenic plants, 139
axillary bud/shoot, 22, 25, 70–71, 164
azalea, 47, 71, 212–213

BA/BAP, 71, 115
backbulbs, 179
bacteria, 13, 108
  dealing with, 93–94, 122–123, 128–130, 133
  DNA insertion, 154–157
  virus-infected, 134
bacteriophage, 134
Bacti-Cinerator, 51
balance, 47–48
bamboo, fish pole, 208–209
bamboo, golden, 208–209
banana, 140, 162, 203–204
banana family, 203
barberry family, 204
Barnstead classic still, 46
basal salt, 75, 86
beakers, 55, 62
bearded iris, 195
beets, 67
Begoniaceae, 176
begonia family, 176
Begonia rex-cultorum, 24, 118
  chimeras, 26
  culture guide, 176–177
  culture overview, 18, 79
  explant selection, 89, 91
  media mixing, 83, 85–86, 177
  stock solution, 80
benzyladenine (BA), 71, 91, 115
Berberidaceae, 204

Beta, 67
biolistics, 32, 37, 157–158
biomass fuel, 168
bioreactors, 167–168
biotechnology. *See* plant biotechnology
blackberry, 104–105, 117, 214–215
"black cloth" technique, 90
bleach solution, 51, 93, 94, 95, 97–98
bleeding, 87, 109
blueberry, 47, 64, 220
boron (B), 67
botanical basis. *See* tissue culture, botanical basis
BPA (synthetic cytokinin), 71
Brassicaceae, 158, 177
Brassica, 67, 177–178
brassinosteroids (BRs), 71
bridges, 87. *See also* rafts
Briggs Nursery laboratory, 44
broccoli, 177–178
brokerage services, 139
Brownian motion, 130
bulbs, 23–24, 93, 140
Bunsen burner, 50, 51
business, 138–148
    costs, 141–144
    crop planning, 146–147
    history of micropropagation, 13–14, 34
    marketing, 138–139
    opportunities for, 18, 19
    overview of, 16–20
    patents and trademarks, 145–146
    product mix decisions, 144–145
    record keeping, 147–148
    secondary products, 36, 166–168
    shipping, 148
    worldwide production, 38, 139–141

Cactaceae, 186, 201
cactus, golden star, 201
cactus, orchid, 186–187
cactus family, 186 201
calcium (Ca), 66
calcium hypochlorite disinfectant, 94

Calendula, 181
callus, 25–26, 160
cambium, 23
Canna, 19
caps, vented, 110
caps, with filters, 55, 115
Capsicum, 111
carbenicillin, 73
carbohydrates, 68–69
carbon dioxide, 67, 111, 117
carnation, 18, 90, 93, 107, 115, 140, 143, 184–185
carnation family, 184
carrot, 33–34, 75, 160–162
Caryophyllaceae, 184
casein hydrolysate, 74
Catharanthus, 168, 178–179
Cattleya, 179–180
cauliflower, 67, 177–178
cefotaxime, 73, 108
celery, 67, 162
CEL-GRO tissue culture rotator, 52
cell culture, 161–162, 167, 168
cell suspension cultures, 161
cell trays, 118–119
charcoal. *See* activated charcoal
chemical inventory, 147–148
chemical units, 62–63
chemistry basics, 61–65
cherry, 211–212
chimera, 26–27
Chinese gooseberry family, 172
chlorine (Cl), 67
chlorophyll synthesis, 67, 68, 107
Chlorophytum, 21
chlorotic leaves, 66, 67, 113
choline, 69
chromatography, 111
chromosomal crossover, 152
chromosomes, 26, 35, 150–153, 163
Chrysanthemum, 90, 143, 181
chrysanthemum (mum), 90, 133, 143, 181
citric acid cycle, 69
citrus, 68, 181
Citrus ×*limon*, 181–182
cleaning practices, 97, 106
cloning, 14–15
clothing requirements, 97, 143

clove pink, 184–185
clover, ladino, 70
cobalt (Co), 68
coconut milk, 32, 33, 74–75, 180
codons, 153–154
coenzyme activity, 69
colchicine, 153, 163
cold acclimation, 166
cold sterilization, 50
cold storage, 112
colony, 124
commercial applications. *See* business
compound, defined, 62
compound list, 77
compound microscope, 53
conductivity measurements, 45–46
conductivity meter, 46
coneflower, 167
conifer cultures, 91
container costs, 142–143
container covers, 110, 115
contaminant detection, 122–137
    indexing, 133
    and laboratory sanitation, 123–124
    microbiological stain, 128, 130
    overview of, 122–123
    streak plate, 124–128
    testing laboratory air, 131–132
    viable air particle, 133
    virus detection methods, 134–137
    wet mount slides, 128, 129
contamination prevention, 119, 143
    cleaning and treating explants, 93–95
    in clean up phase, 106
    growing room, 107–109
    laboratory design and, 39, 40
    microscopes used in, 53
    sanitation techniques, 123–124
    sterile technique for, 96–100
    sterilization equipment, 49–51
Convolvulaceae, 194
copper (Cu), 67
coral bells, 191
cormels, 24
corms, 24, 93, 140

corn milk, 75
cornstarch, 74
Corymbia, 182–183
costs, 16, 107, 139, 141–144
cotton swab-streak plate procedure, 124–128
CPA (para-chlorophenoxyacetic acid), 72
Crassulaceae, 197
creeping phlox, 208
crop planning, 146–147
crop records, 147
crown gall disease, 122, 154, 221
cryopreservation, 166
cucumber, 183
Cucumis, 183
Cucurbitaceae, 183
culture growing room, 40–43, 118
culture techniques. *See* aseptic culture
Cupressaceae, 217
cupric/cuprous state, 67
cuticle, 116
cuttings, 15–16, 22
C-Value paradox, 152
cyanocobalamin, 69
Cynara, 115
Cymbidium, 183–184
cypress family, 217
cytokinins, 34, 70–71, 109–110, 111, 115, 227
cytokinin stock, 82

dahlia, 34
daisy, Transvaal, 189–190
Darwin, Charles, 31, 33
Datura, 32, 33, 75
Daucus, 33, 35, 160, 162
day length, 90
daylily, 190
D-Biotin, 69
decision-making ladder, 144–145
dedifferentiation, 34
deionizer, 45, 46–47
denaturation process, 37
Dendrobium, 184
desiccation, 166
determinant growth, 145
Dianthus, 18, 90, 93, 107, 115, 140, 143, 184–185

Dieffenbachia, 185–186
dilution method, 47–48
dioecious plants, 172
Dioscorides, 28
diploid plants, 152
diseases, 15, 16, 19, 36, 91–93, 141
dishwasher sterilization, 52, 106, 141
dishwashing liquid surfactants, 94
disinfecting explants, 93–94
disinfection, laboratory, 123
distilled water, 45–46
division, 103–105, 146–147
D-Mannitol, 69
DNA (deoxyribonucleic acid), 32, 37, 150–152. *See also* genetic engineering
  bacterial, 134, 154–157
dogbane family, 178
double antibody sandwich (DAS) technique, 135–136
D-Pantothenic acid, 69
D-Sorbitol, 69
dumb cane, 185–186
dumping, 139
Dutch iris, 196
dwarf brush cherry, 117

Echinacea, 121, 167–168
electrical equipment, 41
electrophoresis, 17, 158
electroporation, 155–157
element, 62
elicitors, 166, 167, 168
ELISA (enzyme-linked immunosorbent assay) procedure, 135–137
elkhorn fern, 210
embryogenesis, 33, 161
embryoids, 25–26, 35
embryo rescue/culture, 32, 33, 35, 161, 164
employees, 97, 124, 141–142, 143
endogenous contaminants, 93
endospore, 122
English primrose, 211
enzymes, 37
epicuticular wax, 116
Epiphyllum chrysocardium, 186–187
epiphytes, 119, 201–202
episodic growth, 145

equipment. *See* laboratory equipment
Ericaceae, 198, 212, 220
Erlenmeyer flasks, 55, 62
establishment (Stage I), 12, 93–95, 104, 105–106
ethanol, 168
ethylene, 72, 111
etiolation, 93
Eucalyptus, 91, 140, 182–183
Eugenia, 117
exogenous contaminants, 93
exons, 153
explant selection (Stage 0), 12, 89–91, 92
explants, 14, 89–95
  cleaning and treating, 49, 93–95
  detecting pathogens, 91–93
  problems and solutions, 113
  selling, 139, 141, 148
  sterile technique for, 93–95, 98–100
  stock plant selection and preparation, 89–91

Fabaceae, 67
false agave, 116
fern, Boston, 18, 78, 103, 204–205
ferns, 17, 140
ferric/ferrous state, 67
Ficus, 187–188
field diaphragm, 229–230
fig, 187–188
filter, furnace, 40, 54
financial accounting, 141, 148
fingerprinting, 17, 158
flagella, 130
flamingo flower, 174–175
flower parts diagram, 90
fluorescent lighting, 42–43, 44, 118
fog systems, 118, 120
folic acid, 69
food crops, 140, 150, 152–153, 158–159
food production (plant cell), 31, 32
foot-candle, 41
foreign trade, 19, 38, 92, 139–141
forests, 164–165
formula comparison table, 224–225

Fragaria, 133, 188
   cold storage of, 143
   explant cleaning treatment, 94, 95
   meristem culture procedure, 101–102
   multiplication, 103, 104, 105
   pH requirements, 47
   polyploidy in, 152–153
   rooting, 106
   sample crop production plan, 146
Freesia, 188–189
friable callus, 35
fruit trees, 18, 19–20, 23
full-sib seedlings, 165
fungi, 122, 128, 133
fungicides, 118
Furcraea, 116

Gamborg's B5 medium, 161, 162
gametoclonal variation, 164
gametophyte stage, 205
gellan gum, 74
gelling agents, 73–74
gene guns, 37, 157–158
genetically modified organisms (GMOs), 150
genetic engineering, 149–159
   concerns around, 14–15, 150
   defined, 150
   history of, 32, 37
   inserting desired DNA, 154–159
   mitosis and meiosis, 151–152
   overview of, 149–150
   and plant DNA, 150–151
   ploidy and plant improvement, 152–153
   protein formation, 153–154
   R&D employees, 142
genome sequencing, 158–159
gentamicin sulfate, 73, 109
Geraniaceae, 206
geranium, 206
Gerbera, 107, 140, 189–190
germplasm storage, 93, 165–166
Gesneriaceae, 216
gesneria family, 216
G418 (Geneticin), 73
gibberellins, 71, 91

ginseng, 168, 205–206
glass bead sterilizer, 51
glass jars, 55–56, 110
glass stills, 46
Glycine, 70, 72
gourd family, 183
government regulations, 38, 92, 141
grafting, 23, 103
grape family, 221
grapes, 103, 117, 221
grass, Chinese silver, 202–203
grass family, 160, 202, 208
grass tree family, 190
growing media, 115–116, 118–119
growing room, 107–114
   air requirements, 110–111
   contaminants, 107–109
   design requirements, 40–43, 107
   general guidelines/table, 112–114
   lighting requirements, 41–43
   maintaining humidity, 119–120
   media problems, 109–110
   refrigeration, 111–112
   temperature requirements, 43
   vitrification, 111
growing room shelves, 55
growth regulators. *See* plant growth regulators
gum agar. *See* agar
guttation, 117
gynogenesis, 163

habituation, 70
halide stock, 81
haploid clones, 35, 163
heart of gold, 186–187
heath family, 198, 212, 220
heat treatment, 73, 93, 101, 163
heavenly bamboo, 204
Helleborus, 145
Hemerocallis, 190
HEPA filters, 53–54, 96–97
herbaceous plant cultures, 144, 145
heterotrophic, 31, 117
Heuchera, 191
Hippeastrum, 191–192
history of tissue culture. *See* tissue culture history

hobbyists, 17–18, 39
homozygous diploids, 163
hood. *See* transfer chamber
Hooke, Robert, 29, 229
hormones. *See* auxins; cytokinins; gibberellins; plant growth regulators
Hosta, 27, 71, 121, 192–193
hostas, 107, 192–193
hot plate, 48
humidity domes, 119
humidity requirements, 115, 118, 119–120
hybridizing, 18
hydathodes, 117
hydrochloric acid (HCl) solutions, 65
hydrogen peroxide disinfectant, 94
hygromycin B, 73
hygroscopic materials, 86
hyperhydricity, 72, 109, 111, 113
hyperplasia, 36
Hypoxidaceae, 193
Hypoxis, 193–194

IAA (indole-3-acetic acid), 33, 68, 70
IBA (indole-3-butyric acid), 70
Inca lily, 73, 173–174
indeterminant growth, 145
indexing systemic contaminants, 133
indicator plants, 92, 134, 135, 137
infrared sterilizer, 51
inorganic chemicals, 65–68, 76
inositol stock, 82
instrument holder, 57
instrument sterilization, 97
International Plant Propagators' Society (IPPS), 142, 144, 232
international trade, 19, 38, 92, 140–141
introns, 153
in vitro growing. *See* tissue culture overview
in vitro plants. *See* explants
in vivo (whole plant), 93
iodine (I), 68
ion, 62–63
Ipomoea, 194–195

INDEX   263

Iridaceae, 188, 195, 196
Iris, 115, 195, 196
iris family, 188, 195, 196
iron (Fe), 67
iron stock, 81
irradiance, 117
isopropyl alcohol disinfection, 94, 95
ITS (internal transcribed spacer), 158

jar covers, 110, 115
jasmonates, 72
jimson weed, 33, 75
journals, 139, 144
Juglandaceae, 196
Juglans, 166, 196–197
juvenility, 91, 165

Kalanchoe, 18, 197–198
Kalmia, 47, 71, 117, 198–199
kanamycin, 73
kinetin (KIN), 34, 71, 115
kiwifruit, 172–173
Knudson's C medium, 184, 207
Krebs cycle, 69

labeling, 53, 98
laboratory contamination prevention, 39, 40, 97, 123–124, 131–132
laboratory design, 39–43
    examples of, 43–45
    location and, 38–39
laboratory equipment, 45–57. *See also* autoclave; microscope
    balance, 47–48
    dishwasher, 52, 106
    explant cleaning, 49
    growing room shelves, 55
    hot plate/stirrer, 48
    labeler, 53
    media dispenser, 48–49
    media sterilization, 49–50
    overview of, 57–59, 141
    pH meter, 47
    refrigerator, 52
    rotator/shaker, 51–52, 73, 87, 167
    transfer chamber/hood, 53–54

transfer tool sterilization, 50–51, 97
water purification, 45–47
laboratory supplies, 55–59
labor costs, 141–142
laminar airflow transfer chamber, 53, 54, 97
L-Ascorbic acid, 69–70
laurel, 198–199
layering, 22–23
leaf cuttings, 24, 91
leaf mesophyll, 35
Leeuwenhoek, Antonie van, 28–29, 228–229
legumes, 67
lemon, 181–182
Leucanthemum, 181
L-form amino acids, 72
light meters, 41, 43
light requirements, 41–44, 107, 120–121, 143–144
lilac, 117, 145, 219–220
Liliaceae, 199
Lilium, 103, 107, 199–200
lily, 103, 107, 199–200
lily family, 199
lily-of-the-Incas, 173–174
limiting factors theory, 31, 65
liquid cell suspension cultures, 73
liquid media, 34–35, 51, 87, 180
Liriope, 115
Lithodora, 167
living gene banks, 165–166
lux, 41
Lycopersicon, 33

macerated tissue culture, 103
macronutrients, 65–66
Madagascar periwinkle, 168, 178–179
magnesium (Mg), 66
Malus, 27, 74, 112, 117, 200–201
Mammillaria, 201
manganese (Mn), 67
manipulation, 99–100, 103–106, 146–147
Maranta, 115
marketing, 138–139
Matricaria, 181

measurements, 61–62, 86, 226–227
media, streak plate testing, 124
media dispenser, 48–49
media formulas. *See* individual names
media ingredients, 60–78
    chemistry review, 61–65
    inorganic chemicals, 65–68, 76
    organic chemicals, 68–76
    purchasing chemicals, 75
media preparation, 79–88
    instructions and checklist, 74
    liquid media, 87
    making stock solutions, 79–82
    media mixing, 83–86
    from prepared mixes, 86
    simple recipe, 78
media preparation areas, 40
media sterilization, 30
medium, media. *See also* MS medium
    bleeding, 87, 109
    clean up, 106
    development history, 30–33
    handling problems in, 108, 109–110
    liquid media, 34–35, 51, 87, 180
    precipitate removal, 80
    premixed, 18, 60–61, 141,
    sterilization equipment, 49–50
    undefined, 74–75
meiosis, 26, 151–152
meiotic division, 26
membrane filter apparatus sampling, 131–132
meniscus, 86
meristematic tissue, 25
meristem culture, 15, 53, 93, 101–102, 104
metabolites, 36, 166–168
methylolurea, 73
metric system, 61, 226–227
microbiological stain, 128, 130
microcuttings, 19, 22, 148
micrografting, 103
micronutrients, 66–68
Micro Plantae laboratory, 44–45
micropropagation. *See* tissue culture overview

micropropagation stages, 12, 14, 34, 105–106, 170. *See also* individual stages
microscope, 28–29, 30
   compound, 53, 128
   electron, 116, 134
   stereo, 53, 103, 104
   use and care, 53, 128, 228–230
microtubers, 218
Milli-Q water station, 47
Miltonia, 201–202
mineral deficiency, 66–68
Miscanthus, 202–203
mist systems, 119, 120
mites, 108
mitosis, 151–152
mobility, 65
modified basal salt, 75
mold spores, 128, 129
molecular weight (MW), 63–64
molecule, 62
molybdenum (Mo), 67–68
Moraceae, 187
morning glory family, 194
morphogenesis, 35
mountain laurel, 47, 71, 198–199
mouse ear cress, 158
mRNA, 153
MS (Murashige and Skoog) medium, 171
   Begonia rex media, 79, 80, 85–86
   culture guide notes, 170
   development of, 34
   reducing ammonium and salts in, 79, 88
   salt stocks, 80–81
   as standard medium, 60, 75
mulberry family, 187
multiplication (Stage II), 12, 105–106
   division and transfer, 103–105, 146–147
   exponential rate in, 16–17
   problems and solutions, 109, 110, 113, 114
   sterile technique for, 99–100
Musa, 140, 162, 203–204
Musaceae, 203
mustard family, 158, 177

mutagens, 36
mutations, 26, 36, 152
myo-inositol, 69
Myrtaceae, 182
myrtle family, 182

Nandina, 204
neomycin, 73
Nephrolepidaceae, 204
Nephrolepis, 18, 78, 103, 204–205
Nicotiana, 25, 33, 34, 160
nicotinic acid, 69, 82
nightshade family, 217
nitrate reduction, 115
nitrate stock, 80–81
nitrogen fixation, 68
nitrogen (N), 65–66, 67
nonzygotic embryogenesis, 161–162
nurse culture, 103

Odontoglossum, 201–202
ohm-centimeters (ohm-cm), 45
Olea, 67
Oleaceae, 219
olive family, 219
olives, 67
Oncidium, 201–202
onion, 152
orbital shakers, 52
orchid, 75, 107, 140
   coconut milk cultures, 75
   culture guide, 179–180, 183–184, 201–202, 207–208
   micropropagation history, 14, 31–34
   substrate, 119
orchid, boat, 183–184
orchid, moth, 207–208
Orchidaceae, 179, 183, 201, 207
orchid cactus, 186–187
orchid family, 179, 183, 201, 207
organic chemicals, 68–76
   amino acids, 72
   antibiotics, 72–73
   carbohydrates, 68–69
   gelling agents, 73–74
   plant growth regulators, 70–72
   undefined culture media constituents, 74–75

   vitamins, 69–70
organogenesis, 161, 165
ornamental plants, 26, 140, 145–146
orpine family, 197
osmosis, 47
osmotic media, 166
osmoticum, 69
ovule, 152

packing methods, 148
paclobutrazol (PBZ), 72
Panax, 168, 205–206
Panicum, 118, 168
para-aminobenzoic acid (PABA), 69
parenchyma cells, 25
Particle Inflow Gun (PIG), 157–158
patented plants, 141, 145–146
pathogens, 122
PBA, 71
peach, 211–212
pear, 74, 112
pectin formation, 66
Pelargonium, 93, 206
penicillin-streptomycin solution, 109
pepper, habanero, 111
periwinkle, 168, 178–179
pests, 16, 36
petri dish sampling, 131
Petunia, 73
pH adjustment, 64–65, 110
Phalaenopsis, 75, 184, 207–208
pharmaceutical products, 166–168
phenolic compounds, 70, 109
Philodendron, 115
phloroglucinol, 72
phlox family, 208
Phlox subulata, 208
pH meter, 47
phosphate stock, 81
phosphorus (P), 66
photoperiod, 90, 120
photosynthesis, 67, 117
photosynthetically active radiation (PAR), 41
Phyllostachys, 208–209
phytohormones, 70–72

phytosanitary certification, 38, 141
*Phyto*Technology Laboratories, 75, 77
Pinaceae, 201
pine, 103, 208
pine family, 208
pink, moss, 208
pink family, 184
Pinus, 103, 165, 209–210
plantain lily, 192–193
plant biotechnology, 149–168. *See also* genetic engineering
   culture methods, 160–164
   germplasm storage, 165–166
   overview of, 34–37, 149
   reforestation, 164–165
   secondary products, 36, 166–168
   tissue culture and, 149
plant genome sequencing, 158–159
plant growth regulators (PGRs), 70–72, 91, 109–110, 227
plantlet establishment. *See* transplanting and acclimatization
plantlet sales, 19, 148
plantlet sorting, 118, 119
plant patents, 145–146
Plant Preservative Mixture (PPM), 93–94
plant stem cuttings, 15–16, 22
plant tissue culture, defined, 13
plasmid, 154
Platycerium, 210
plum, 211–212
Poaceae, 160, 202, 208
Polemoniaceae, 208
pollen, 152
polyethylene glycol (PEG), 157
polymerase chain reaction (PCR), 32, 37, 151, 153
polymyxin B, 73, 109
polyploidy plants, 152–153
polypod family, 210
Polypodiaceae, 210
polysaccharide, 68
pool disinfectant, 94
potassium hydroxide (KOH) solution, 64–65
potassium (K), 66

potato, 34, 36, 110, 111, 133, 140, 163, 217–218
potato extract, 75
precipitates, 45
precocious germination, 31
premixes. *See* medium/media, premixed
pressure cooker, 40, 49, 106, 141
primrose, 211
primrose family, 211
Primula, 211
Primulaceae, 211
production plans, 146–147
product mix decisions, 144–145
prokaryotes, 122
propagule, 99
protein formation, 153–154
protoplast cultures, 35–36, 162–163
protoplasts, DNA insertion, 157
Prunus, 211–212
pyridoxine, 69, 82
Pyrus, 74, 112
Pythium, 168

quality control, 112, 123, 128, 142

rafts, 87. *See also* bridges
raspberry, American, 215–216
recombinant plasmid DNA, 155–157
record keeping, 112, 141, 147–148
red dye production, 167
red-flowering gum, 182–183
red stems, 109, 113
reduction division, 26
redwood, coastal, 217
Reemay, 120
reforestation seedlings, 164–165
refrigeration requirements, 52, 111–112, 143
research and development (R&D), 142
resin bed exchangers, 46
resistivity measurements, 45
resolution, 53, 228
restriction endonuclease enzymes, 155
reverse osmosis, 46–47

rex begonias. *See Begonia rex-cultorum*
rhizomes, 24
rhizosphere, 122
Rhododendron, 47, 64, 68, 71, 103–105, 143, 145, 212–213
rhododendrons, 18, 47, 71, 110, 212–213
ribavirin, 73
riboflavin, 69
rifampicin, 73, 108
RNA (ribonucleic acid), 153, 155
root cuttings, 24
rooted liners, 148
rooting hormone, 70
rooting (Stage III), 12, 105–106
   crop planning, 146–147
   inducing, 115–116
   omitting, 117–118, 170
   problems and solution, 109, 113
root primordia, 26
roots, aerial, 25
roots, tips, 25, 33, 65, 66
rootstock cloning, 20
rootstock grafting, 23
Rosa, 115, 214
Rosaceae, 174, 188, 200, 211, 214, 215
rose, 214
rose family, 174, 188, 200, 211, 214, 215
rotators, 51–52, 73, 87, 167
row covers, 120, 121
Rubus, 22, 104–105, 117, 158, 214–215
rue family, 181
runners, 24
Rutaceae, 181

Sabouraud's dextrose agar (SDA), 124–127
safety considerations, 41, 97
Saintpaulia, 18, 78, 91, 140, 216–217
salicylic acid, 72
salt reduction, 79, 115
sanitation techniques, 123–124
Sansevieria trifasciata, 91
saskatoon serviceberry, 174

Saxifragaceae, 191
saxifrage family, 191
scaling, 24
Schefflera, 115
scions, 20
scooping, 24
scoring, 24
secondary products, 36, 166–168
seed-grown plants, 15–16, 21
seedlings, 14, 16, 90, 153, 166
seed potato certification, 133
seeds, 14–15
seed/seedling diagram, 90
Selenicerens chrysocardium. See Epiphyllum chrysocardium
senescence, 70
sequencing table, 158–159
Sequoia, 91, 217
serological test, 134–135
serviceberry, 174
sexual reproduction, 21
shakers, 51–52, 73, 87, 167
shikonin, 167
shipping methods, 148
shoot primordia, 26
siemens (S), 45
snake plants, 91
sodium hydroxide (NaOH) solution, 64–65
soil-less substrates, 117, 118, 119
Solanaceae, 217
Solanum, 34, 110, 111, 133, 140, 217–218
somaclonal variation, 26, 32, 163–164
somatic cells, 26
somatic embryogenesis, 35, 161–162, 165
soybeans, 36, 70, 72, 157
Speedling planter tray, 56, 57
spontaneous generation theory, 29–30
spontaneous mutation, 27
sports, 27
squills, 24
Stage 0. *See* explant selection
Stage I. *See* establishment
Stage II. *See* multiplication stage
Stage III. *See* rooting stage

Stage IV. *See* transplanting and acclimatization
staghorn fern, 210
stain, microbiological, 128–130
staining, 128, 130
star lily family, 193
sterile technique, 96–100
sterilization, 49–51, 55, 123
still, 45, 46, 47
still-air box, 39, 40, 45, 53–54, 96–99
stirrer, 48
stir tanks, 167–168
stock plant refrigeration, 111
stock plant selection and preparation, 12, 89–91
stock plants pathogens, 91–93
stock solutions, 60, 79–82, 110
stolons, 24
stomata, 116
strawberry. *See Fragaria*
streak plate procedure, 124–128
streptomycin, 73, 109
sucrose, 68–69, 116
sulfate stock, 81
sulfur (S), 66
supplies, 55–59, 232–235
surfactant, 94
suspension cultures, 34–35
sweet potato vine, 194–195
switchgrass, 118, 168
sympodial, 179
Syngonium, 18, 140, 218–219
synthetic seeds, 35, 161–162
Syringa, 117, 219–220
systemic acquired resistance (SAR), 72
systemic contaminants, 109, 133

Tanacetum, 181
taxol, 168
Taxus, 168
T-DNA, 154
temperature conversion, 227
temperature requirements, 43, 107, 121
tents, humidity, 118, 119–120
test tubes, 55–57, 98, 99, 106, 143
tetracycline, 73, 108

tetraploids, 27
thermotherapy, 93
thiamine, 69, 82
Thidiazuron (TDZ), 71
TIBA (2,3,5-triiodobenzoic acid), 72
tissue culture, 12–20
   benefits, 15–17
   cell division and mutation, 25–27
   culture methods, 160–164
   genetic stability concerns, 14–15
   hobbyists, 17–18
   overview, 21, 24–25
   protocol patents, 145
   purposes not suited for, 15–16
   stages of, 12, 14, 34
   suitable plant parts, 21–24
tissue culture history, 28–37
   commercial, 13–14, 34
   highlights of, 32
   leading figures in, 28–34
   and new technologies, 34–37
tissue culture sales, 139, 141
tobacco, 25, 32, 33, 34, 160, 163
tobacco mosaic virus (TMV), 122, 134
Todd planter tray, 56, 57
tomato, 33, 36
totipotency, 30, 32
Touch-O-Matic burner, 50–51
trace elements, 66–68
trademarks, 145–146
trade organizations, 37, 232
Tradescantia, 78
transfer, 103–105, 146–147
transfer chamber/hood, 39, 40, 45, 50, 53–54, 96–99
transferring cultures, 100, 103–106
transfer room, 40, 98–100
transfer tool sterilization, 50–51, 97–100
translocation, 69
transpiration, 116
transplanting and acclimatization (Stage IV), 12, 115–121
   omitting Stage III, 117–118
   problems and solution, 114

transplanting (continued)
  reducing water loss, 116–117
  shipping plants, 148
  succeeding with, 116–117
Transvaal daisy, 189–190
trays, planting, 57, 118–119
Trifolium, 70
triploidy, 152, 154
TSA (trypticase soy agar) medium, 124–127
tubers, 24
Tulipa, 134
tulips, 134
Tween 20, 94, 170
2iP (IPA), 71, 115
tyndallization, 30

ultrasonic cleaner, 49
ultraviolet lights, 44
U.S. Patent and Trademark Office (USPTO), 145

Vaccinium, 47, 64, 220
Vacin and Went's medium, 75, 184, 207
vacuum pumps, 49
variegation, 26
vectors, 122
vegetative reproduction, 14, 21, 22, 24, 25

vernalization, 121
viable, 133
vine family, 221
viral contaminants, 122, 134–137
virus DNA insertion, 155–157
viruses, 101, 122, 134, 154–157
virus-free plants, 101
Vitaceae, 221
vitamins, 69–70
vitamin stock, 82
Vitis, 91, 103, 117, 221
vitrification (cold cooling), 166
vitrification (hyperhydricity), 72, 109, 111, 113
volume percent concentration (% v/v), 98

walnut, 166, 196–197
walnut family, 196
wandering Jew, 78
water disinfection, 94
water holding capacity (WHC), 119
water loss reduction, 116–117
water of hydration, 63
water purification equipment, 45–47
water quality, 45–47
weighing, 47–48, 62
wet mount slides, 128, 129
wetting agents, 94

wheat, 152
whiptail, 67
White, Philip R., 33, 167
woody plant cultures, 14, 140
  CO_2 introduction, 117
  eliminating bacteria, 73, 109
  hardwood forest trees, 165
  multiplication, 16, 104
  success potential with, 144, 145
  thidiazuron growth regulation, 71
woody plant medium (WPM), 60, 75, 88
woody plant reproduction, 22, 23
working distance, 229
worldwide microplant trade, 19, 38, 92, 139–141
wounding, 22, 72

Xanthorrhoeaceae, 190

yeast contaminants, 128–130
yeast culture media, 31, 33, 75
yellowing leaves, 66, 67
yellow star plant, 193–194
yew, 168

zeatin, 71, 75
zinc (Zn), 68
zygotic embryos, 103, 161, 164

About the Authors

Lydiane Kyte (1919–2008) became involved in the budding science of tissue culture while managing the expansion from one million to twenty million containerized conifer seedlings for a timber replanting program. She then built a plant tissue culture program for a prominent woody ornamental lab and later developed a micropropagation business with her husband. She earned a B.S. in botany from the University of Washington.

Holly Scoggins is an associate professor of horticulture and director of the Hahn Horticulture Garden at Virginia Tech in Blacksburg. She has published extensively in the academic horticultural literature and is also one of the four "Garden Professors" at the blog of that name (https://sharepoint.cahnrs.wsu.edu/blogs/urbanhort/default.aspx). Holly holds a Ph.D. in horticultural science from North Carolina State University.

John Kleyn (1923–2005) was an associate professor of biology at the University of Puget Sound in Tacoma, Washington, and author of a preeminent microbiology lab manual, now in its seventh edition. He received a Ph.D. in bacteriology from Cornell University

Mark Bridgen is a professor of horticulture at Cornell University and director of Long Island Horticultural Research and Extension Center. He is also a prolific writer. Using biotechnological techniques, he has developed several *Alstroemeria* hybrids now available at select nurseries. Mark earned a Ph.D. in plant physiology from Virginia Polytechnic Institute and State University.